Low-Carbon Consumption in China: Residential Behavior, Corporate Practices and Policy Implication

Zhaohua Wang · Bin Zhang

Low-Carbon Consumption in China: Residential Behavior, Corporate Practices and Policy Implication

Science Press
Beijing

Springer

Zhaohua Wang
School of Management and Economics
Beijing Institute of Technology
Beijing, China

Bin Zhang
School of Management and Economics
Beijing Institute of Technology
Beijing, China

ISBN 978-981-15-2791-3 ISBN 978-981-15-2792-0 (eBook)
https://doi.org/10.1007/978-981-15-2792-0

Jointly published with Science Press
The print edition is not for sale in China. Customers from China please order the print book from: Science Press.

This Springer imprint is published by the registered company Springer Nature Singapore Pte Ltd.
The registered company address is: 152 Beach Road, #21-01/04 Gateway East, Singapore 189721, Singapore

Preface

With the rapid development of China's economy, the consumption capacity of China is increasing. According to data from the World Bank, China's GDP (in Purchasing Power Parity) ranked first in 2013, surpassing the United States for the first time, and accounted for 15.86% of the world's total. In 2017, the proportion accounted by China increased to 18.14%, while it decreased to 15.13% for the United States. Also, as China has surpassed the United States in terms of CO_2 emissions to become the top emitter in the world, and the CO_2 emissions from residential consumption ranked only second to that from industrial production in China, it is critical to develope low-carbon consumption policies for China.

This book focuses on the practices of low-carbon consumption in China from the following three perspectives: residential behavior, corporate practices, and policy implications. Some hot issues concerning residents' electricity consumption, transportation, low-carbon product purchasing, and E-waste recycling are analyzed to provide a reference for low-carbon consumption. Corporate practices in improving energy efficiency, CO_2 reduction, as well as interfirm collaboration are further explored to achieve low-carbon consumption in the industrial sector. To sum up, the issues studied in this book include the following aspects:

(1) CO_2 emissions from residential consumption in China. Facing increasing residential consumption along with the rapid development of our economic level, the related CO_2 emissions from residential consumption show an increasing trend in recent decades. The emissions from residential consumption, however, rank second only to emissions from industrial sectors: how to take effective measures to mitigate emissions from this sector merits our attention. Thus, we first explain the situations of energy consumption and related CO_2 emissions from residential consumption from 2000, and then the indirect impacts of residents' lifestyles on the ecological footprint of indirect energy consumption are examined, along with the factors driving indirect CO_2 emissions from residents. Based on the environmental extended input–output model, the indirect CO_2 emissions from residential consumption have been conducted at the national level. Combining the structural decomposition analysis (SDA) and logarithmic mean divisia index (LMDI) approaches, the influencing factors and their respective contributions to emission

changes have also been examined. Analysis of CO_2 emissions from residential consumption can provide a basis for related mitigation and reduction measures, leading to a low-carbon society in China and urging residents to engage with low carbon consumptions.

(2) Household electricity consumption and saving behavior in China. With the rapid economic growth of China, the disparity between the energy consumption and production is increasing: this is especially salient in electricity consumption and supply. Taking Beijing as an example, the total consumption of electricity in 2007 was about 67.5 billion kW·h, and however, only 22.4 billion kW·h was generated in Beijing, with more than 65% of electricity imported from neighboring provinces (autonomous regions, municipalities). It is, therefore, necessary for China to improve the efficiency of energy use and encourage energy conservation. Thus, we conduct questionnaire surveys and utilize the logit regression model to analyze empirically household electricity-saving behavior. Furthermore, we build a cointegration equation and a panel error correction model to analyze the direct rebound effect of urban residential electricity consumption, with China's 30 provincial government panel data from 1996 to 2010 applied. Considering the consistency and availability of statistical data, Xizang, Taiwan, Hong Kong and Macao are not included in this study. Correspondingly, some inferences and suggestions are offered for the promotion of electricity-saving behaviors and formulation of energy policies.

(3) Low-carbon transportation for the residential sector in China. Nowadays, there has been much interest in sustainable transport. Evaluating CO_2 emission efficiency and its marginal abatement cost in transportation sectors has been a hot topic: however, few studies have examined the driving forces behind household transportation emissions from the perspective of individual travel characteristics. This chapter examines the features and driving factors of CO_2 emissions from household daily travel in Beijing from 2000 to 2012. It first investigates the changes in personal travel characteristics and CO_2 emissions, and then discusses the effects of population, economic activity, transport capacity, vehicle emission intensity, and individual travel characteristics which include the effects of transport intensity, transport mode share, and vehicle-use intensity on CO_2 emissions based on decomposition analysis. By applying an LMDI method, CO_2 emissions due to passenger daily travel are decomposed into population scale, economic activity, transport intensity, transport mode share, transport capacity, vehicle-use intensity, and vehicle emission intensity effects.

(4) Residents' willingness in purchasing low-carbon products. We conduct nationwide questionnaire surveys in China and empirically analyze the features and determinants of residential willingness in purchasing new energy vehicles and energy-efficient household appliances with a multinomial logit regression model and structure–function model applied. Furthermore, some policy suggestions are offered to promote energy-efficient household appliances and new energy vehicles.

(5) E-waste recycling behavior in China. This chapter discusses the recycling behavior of electronic waste in China. First of all, this chapter introduces the characteristics of electronic waste, the generation and flow of electronic waste, it

also analyzes the reasons for the dilemma posed by E-waste in China from the aspects of laws, policies, and industrial systems. Then, we research the willingness and behavior toward E-waste recycling in China. The result shows that a significant proportion of residents in Beijing are still not very willing to participate in E-waste recycling. At last, we discuss the determinants of residential E-waste recycling behavioral intention in China.

(6) The motivation of energy-intensive industries on CO_2 reduction: learning from the iron and steel industry (ISI). With international communities having recognized the severity of climate change, the pressure for reduction of CO_2 emissions has become more prominent. China is one of the world's biggest emitters, accounting for 25% of global CO_2 emissions. The pressure on China to reduce CO_2 emissions has increased significantly. The iron and steel industry (ISI) ranks as the third biggest CO_2 emitter in China (after the power and construction material sectors), due to its coal-dominated energy structure and high consumption of limestone. It is imperative to mitigate the incremental pressures on CO_2 emissions by improving energy efficiency in ISI; however, it is unclear which pressures predominate as drivers or barriers for Chinese iron and steel enterprises in implementing CO_2 reduction. We explore a conceptual model of the relationship between CO_2 reduction practice and the determinants thereof. The model is then tested using the data collected from 85 iron and steel enterprises of China utilizing a questionnaire survey, designed to address the following three questions: What CO_2 reduction practices are being employed by Chinese ISI? Which determinants drive or hinder the implementation of these practices? Do these practices have a positive effect on the environmental and economic performance of ISI?

(7) Energy efficiency and CO_2 emission abatement technology. The technology for energy efficiency and CO_2 emission abatement is important in achieving low carbon consumption. Our study explores the relationship between energy technology patents and CO_2 emissions in China. Considering the potential different role of fossil-fuelled, and carbon-free, technological innovations in reducing CO_2 emissions, we distinguish patents for fossil-fuelled technologies from patents for carbon-free energy technologies. The former mainly refers to the patents relevant to fossil-fuelled (coal, oil, natural gas, etc.) technologies in energy sectors and energy use sectors; the latter mainly refers to patents relevant to nuclear and renewable energy technologies. Specifically, this chapter tries to answer the following research questions: Is there a significant relationship between patents for fossil-fuelled technologies and CO_2 emissions? Is there a significant relationship between patents for carbon-free energy technologies and CO_2 emissions? Are the relationships the same in eastern, central, and western China? To avoid omitted variable bias, GDP is also included in the current study due to its important role in affecting CO_2 emissions and energy technology patents. A dynamic panel data approach is applied to examine such relationships.

(8) Interfirm collaboration on CO_2 emission abatement. Interfirm cooperation is a new option for industrial firms to reduce their CO_2 emissions, and particularly for firms within the same industrial chain, they have similar technical backgrounds: some even have mutual supply–demand business. Cooperation in CO_2 reduction

within these firms is easily conducted and less costly. More firms now realize the importance of interfirm collaboration in CO_2 reduction. Different modes and types of collaboration have been emerging. As a single firm's effect on emission reduction is limited, what if we combine multiple firms' power and collaborate on CO_2 reduction? Industrial firms have taken measures to collaborate with other firms to deal with environmental problems, such as eco-design cooperation and industrial symbiosis. Some of these measures are also helpful for CO_2 reduction. This chapter is designed to enrich the current research to elucidate the status of interfirm cooperation on CO_2 reduction within industrial chains. We focus on solving the following questions: How do the industrial firms cooperate on carbon reduction within industrial chains? What kinds of determinants drive or hinder the implementation of this carbon reduction cooperation within industrial chains? Do these carbon reduction cooperations have any effects on their environmental and economic performance?

(9) Low-carbon policies in China. There have been many policies aimed at low-carbon development in China. In this chapter, we review the main low-carbon policies in China. Then, we take environmental regulation policy as an example to explore its policy effect in China. Lastly, we employ a questionnaire survey study to assess public acceptance of certain low-carbon policies in China. Here, we take the tiered electricity price policy for our empirical analysis. We anticipate that this chapter will provide support for decision-makers, and promote the exchange of our findings with low-carbon policy research peers.

We gratefully acknowledge the financial support from the National Science Fund for Distinguished Young Scholars (Grant No. 71625003), National Natural Science Foundation of China (Grant Nos. 91746208, 71403021, 71774014, 71521002, 71573016), National Key Research and Development Program of China (Grant Nos. 2016YFA0602500), and Beijing Institute of Technology special fund for "Double First Class University Plan". We especially thank the following researchers for their spirited and patient participation in many rounds of presenting and reworking research: Mengtian Xue, Wei Liu, Hualin Zeng, Fanchao Yin, Chao Feng, Tongfan Liu, Milin Lu, Lin Yang, Weijun He, Xiaoyang Dong, Yuandong Zhao, Senyu He, Qingyu Sun, Yiming Li, and Yuantao Yang.

Beijing, China

Zhaohua Wang
Bin Zhang

Contents

Chapter 1
CO_2 Emissions from Residential Consumption in China

China's economy has developed rapidly, leading to the rapid growth of population and consumption rates. The GDP of China ranked first in terms of purchasing power parity (PPP) in 2013, surpassing the US for the first time, and accounting for 15.90% of the world's total (US: 15.86%). In 2017, the proportion of GDP of China increased to 18.14%, while it decreased to 15.13% for the US. The GDP per capita by PPP of China has progressively increased from $2,932 in 2000 to $16,842 in 2016. Consequently, China has surpassed the US in terms of CO_2 emissions and become the top emitter in the world,[1] the CO_2 emissions from residential energy consumption in China rank second only after that from industrial production. This draws attention to the amount of CO_2 emissions from residential areas in China.

This chapter first describes the characteristics from the perspective of residential consumption, and then investigates the characteristics of residential energy consumption and its related CO_2 emissions. Furthermore, this chapter analyzes the driving forces of indirect CO_2 emissions from residents at the national level.

[1]The results of Carbon Dioxide Information Analysis Center (CDIAC) shows the total CO_2 emissions of China were about 10.3 billion tons in 2014, about twice of that of the second ranked the US.

This chapter quotes from the following literature:

Wang Z, Yang L. 2014. Indirect CO_2 emissions in household consumption: evidence from the urban and rural areas in China. Journal of Cleaner Production, 78: 94–103.

Wang Z, Liu W, Yin J. 2015. Driving forces of indirect CO_2 emissions from household consumption in China: an input-output decomposition analysis. Natural Hazards, 75(2): 257–272.

Wang Z, Yang Y. 2016. Features and influencing factors of CO_2 emissions indicators in the perspective of residential consumption: Evidence from Beijing, China. Ecological Indicators, 61: 634–645.

Z. Wang and B. Zhang, *Low-Carbon Consumption in China: Residential Behavior, Corporate Practices and Policy Implication*,
https://doi.org/10.1007/978-981-15-2792-0_1

1.1 Residential Energy Consumption and Direct CO_2 Emissions

1.1.1 Current Residential Consumption in China

Since the foundation of China, especially after the reform and opening-up, the economy has developed at high speed, and often maintained a GDP growth rate of over 10%. In recent years, China's economy has entered the "New Normal". Although the growth rate had slowed down, it still remains to be more than 6%. The living standards and consumption capacity of Chinese residents have not fallen therewith, as Fig. 1.1 shows. After we enter the twenty-first century, both the consumption level of Chinese urban and rural residents increases with each passing year. The consumption expenditure per capita of Chinese residents progressively increased year-by-year from 3,721 yuan in 2000 to 12,730 yuan in 2015 (at 2000 constant prices), with an annual growth rate of 8.55%, in which, the annual growth rate for urban residents is 6.66%, while 7.69% for the rural. The difference in growth rates is related to the increasing urbanization: as living standards improves, a growing number of rural residents move into urban areas, and the urbanization rate of China has risen from 36.22% in 2000 to 56.10% in 2015.[2]

The ratio of residential consumption expenditure of urban to rural remains between 3.1 and 3.8, with the biggest difference occurring in 2007. The ratio has continued to decline since 2010 and reached its lowest level at 3.16 in 2015: this implies that the difference in consumption level per capita between urban and rural residents is gradually shrinking.

1.1.2 Characteristics of Energy Consumption in China

Along with the rapid development of China's economy, total energy consumption is also rising rapidly. The GDP grew from 10.03 trillion yuan in 2000 to 39.90 trillion yuan in 2015 (at 2000 constant price), with an annual growth rate of 9.64% (Table 1.1). Meanwhile, the total energy consumption grew from 1,410 Mtce in 2000 to 4,022 Mtce in 2015, with an annual growth rate of 7.24%. China's production technology has been improved continuously to reduce CO_2 emissions. As a result, China's energy consumption per unit GDP has been declining continuously since the early twenty-first century: it has decreased from 1.57 tce/10^4 yuan in 2005 to 1.01 tce/10^4 yuan in 2015 (a cumulative decline of 35.67%).

The residential consumption level and energy consumption in China keeps rising (Fig. 1.1 and Table 1.1), moreover, the total energy consumption of both urban and rural residents continues to increase as well, from 68.0 and 58.2 Mtce in 2000 to

[2]The latest data from the National Bureau of Statistics of China shows the urbanization rate is 59.58% in 2018.

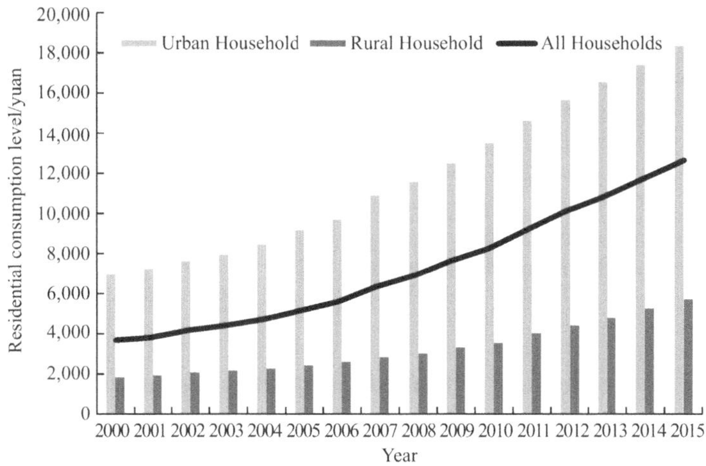

Fig. 1.1 Residential consumption level in China during 2000–2015

Table 1.1 Economic outputs and energy consumption in China during 2000–2015

Year	Total energy consumption/Mtce	GDP/trillion yuan	Energy consumption per unit of GDP/ ($tce/10^4$ yuan)
2000	1410	10.03	1.41
2001	1483	10.86	1.37
2002	1619	11.85	1.37
2003	1893	13.03	1.45
2004	2207	14.35	1.54
2005	2508	15.99	1.57
2006	2751	18.02	1.53
2007	2993	20.57	1.45
2008	3065	22.57	1.36
2009	3213	24.69	1.30
2010	3436	27.31	1.26
2011	3702	29.90	1.24
2012	3815	32.27	1.18
2013	3948	34.78	1.14
2014	4003	37.32	1.07
2015	4022	39.90	1.01

Sources China Statistical Yearbook 2016; China Energy Statistical Yearbook 2016
Notes The energy consumption is calculated in line with the calorific value calculation. The GDP in each year is at the constant price of 2000

211.7 and 151.0 Mtce in 2015, with annual growth rates of 7.86% and 6.56%, respectively. Under their joint actions, the annual growth rate of total energy consumption in China is 7.29% from 2000 to 2015. From the perspective of energy structure, and take 2015 as an example, the total energy consumption of China's residents is 363 Mtce, 7.16% higher than that in 2014, the consumption of large amounts of energy includes: electricity (93.0 Mtce), coal (69.7 Mtce), natural gas (46.8 Mtce), LPG (43.7 Mtce), gasoline (38.3 Mtce), and heat (32.0 Mtce). The increased energy consumption is mainly contributed by gasoline (7.1 Mtce), LPG (6.5 Mtce), and electricity (4.8 Mtce) (Fig. 1.2).

1.1.3 Energy-Related CO_2 Emissions from Residential Consumption in China

From the past decade of the twentieth century, China's total CO_2 emissions have been continuously rising, and have surpassed those of the US since 2006 to become the world's top CO_2 emitter. The emissions from China have even exceeded the sum from the US and EU in recent years. Research shows that the CO_2 emission factor of coal in China is 40% lower than the IPCC default value, and the total emission of China in 2013 is 14% lower than those calculated by other research institutions around the world (Liu et al. 2015). However, these results still cannot change the fact that China is the world's largest CO_2 emitting nation. We calculated the direct CO_2 emissions from residential energy consumption in China over the years 2000–2015 using the updated CO_2 emission factors of various fossil fuels obtained from the latest research (Liu et al. 2015) (Fig. 1.3).

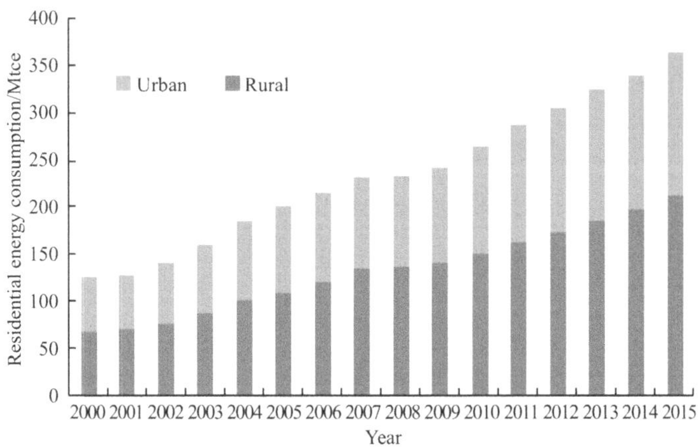

Fig. 1.2 Residential energy consumption in China during 2000–2015. *Sources* China Energy Statistical Yearbooks 2014–2016. *Note* The energy consumption is calculated in line with the calorific value calculation

From Fig. 1.3, we can see that the direct energy-related CO_2 emissions from residential consumption kept increasing except for a slight decline in 2008. Direct emissions grew from 193.4 Mt in 2000 to 418.1 Mt in 2015, with an annual growth rate of 5.27%. Emissions from urban and rural residents grew from 96.7 to 235.3 Mt and 96.7 to 182.8 Mt, respectively, from 2000 to 2015. This calculation did not include the consumption of electricity, heat, or other kinds of energy that emit CO_2 albeit indirectly. Thus, the total emissions in 2008 showed a slight decline in emissions from both urban and rural residents. However, the overall trend in emissions from residents still increased steadily.

1.2 Residential Indirect Energy Consumption and Ecological Footprint

Energy is the most basic driving force behind global economic development. It is estimated that global energy consumption in the process of industrialization has increased about 30 folds in the twentieth century (Lu 2002). China is in an intermediate stage of industrialization and urbanization: the effect of energy supply becomes more and more apparent to the society and the environment from both input and output perspectives along with the rapid development of the economy and the improvement of living standards. The ultimate goal of industrial production is for consumption and as a result, a reasonable ratio of consumption to GDP (i.e.,

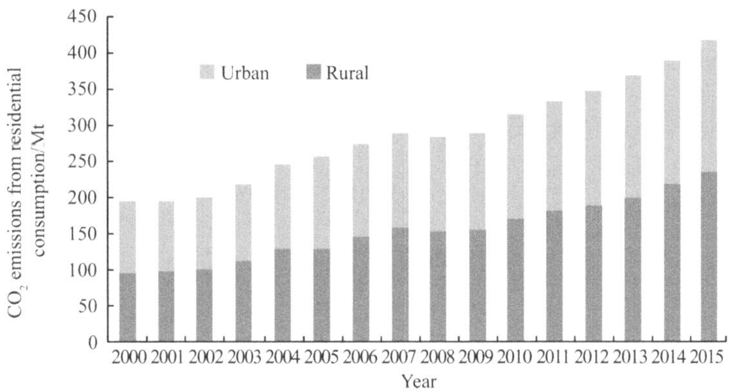

Fig. 1.3 Direct CO_2 emissions from residential consumption from 2000 to 2015

consumption rate) is necessary for sustainable development. The ecological footprint (EF) has received considerable attention and been broadly used to mirror the situation (Holmberg et al. 1999; Erb 2004). Here, the perspective of EF is chosen to study residential energy consumption here.

1.2.1 Methods of Analysis

1.2.1.1 Consumer Lifestyle Approach (CLA)

The main idea of CLA was described by Bin and Dowlatabadi (2005). This approach aims to decompose all components of a household's lifestyle. The details of daily residential energy consumption (Table 1.2) show that energy consumption can be classified into direct and indirect energy use behaviors. Direct energy consumption refers to the demand for energy consumed directly by resident's energy demands such as energy used for cooking, lighting, and heating. Indirect energy consumption refers to the energy used during the production process of various goods and services such as clothing, food, housing, and travel. In this section, the average annual residential consumption expenditure is used to calculate indirect energy consumption by urban and rural resident's behaviors. Consumer behavior is divided into eight categories based on China Statistical Yearbook 2011 and categorized into separate sectors (Table 1.3). Those categories are used to estimate the indirect energy use and CO_2 emissions associated with residential behavior. This may cause a mismatch in energy use by sectors in China Statistical Yearbook 2011 and commodities terms. Furthermore, the carbon intensity for each of these consumer expenditures can be calculated by the energy use, CO_2 emissions, and industrial added value of each sector.

The CO_2 emissions associated with indirect energy consumption in urban and rural areas for each category are calculated as follows:

$$CO_{2\_\text{urban}} = \sum_i (\text{CI}_i \cdot X_i) \cdot N_{\text{urban}} \tag{1.1}$$

Table 1.2 Classification of resident behavior for energy consumption

Resident behavior	Resident behavior categorization
Direct energy use behaviors	Lighting; cooking; entertainment; heating and cooling; cleaning; transportation
Indirect energy use behaviors	Food; clothing; household facilities and services; education, cultural and recreation services; medicine and medical services; residence; transport and communication services; miscellaneous commodities and services

Table 1.3 Sectors related to household consumer behaviors

Consumer expenditure	Related sectors
Food	Food processing; food production; beverage production
Clothing	Textile industry; garments and other fiber products; leather, furs, down and related products
Household facilities, articles, and services	Timber processing, bamboo, cane, palm fiber, and straw products; furniture manufacturing; electronic equipment and machinery
Education, cultural, and recreation services	Papermaking and paper products; printing and record medium reproduction; cultural education and sports articles
Medicine and medical services	Medical and pharmaceutical products
Residence	Production and supply of electric power, steam and hot water; production and supply of gas; production, and supply of tap water; construction; nonmetal mineral product and metal products
Transport and communication services	Electronic and telecommunications equipment; transportation equipment
Miscellaneous commodities and services	Tobacco processing; wholesale, retail trade and catering

$$CO_{2\_rural} = \sum_i (CI_i \cdot X_i) \cdot N_{rural} \quad (1.2)$$

CO_{2_urban} refers to the indirect CO_2 emissions from urban residents and CO_{2_rural} refers to the indirect CO_2 emissions from rural residents. X_i refers to the per capita expenditure. N_{urban} refers to the number of urban residents, while N_{rural} refers to the number of rural residents. CI_i refers to the carbon intensity of the sector i ($CI_i = C_i/G_i$, C_i refers to the sum of the CO_2 emissions from the industries in sector i, while G_i refers to the sum of value added of the industries in sector i).

1.2.1.2 Energy Ecological Footprint

The Energy Ecological Footprint (EEF) refers to the land needed to absorb the pollution caused by the combustion of fossil energy. A conceptual breakthrough in improving EEF analysis to express the carbon absorption ability of an ecosystem called Net Primary Productivity (NPP)[3] is introduced here to reflect the eco-environmental impacts of energy consumption combined with EEF. The NPP-EEF model suggests three primary changes: (a) incorporating the entire surface of a region's biocapacity; (b) reserving a fraction of NPP for other species; and

[3]NPP marks the first visible step in carbon accumulation and quantifies the conversion of atmospheric CO_2 into plant biomass. It is a rate process that tracks the net flux of carbon from the atmosphere into green plants per day, week, or year.

Table 1.4 Global average NPP of different ecosystems

Ecosystem class	Arable land	Forest land	Grassland	Water area	Sea	Construction land	Wet land
NPP/[tC/(hm^2 year)]	4.243	6.583	4.835	5.344	0.959	0.997	11.800

Source Venetoulis and Talberth (2008)

(c) changing assumptions about carbon sequestration rates (Venetoulis and Talberth 2008). On the basis of regional comprehensive carbon ability, results from NPP-EEF model are more suitable to evaluate actual conditions pertaining to regional nature and society. The method for assessing EEF is developed based on land-use type, land area, and global NPP. The equation was constructed by Fang et al. (2012) as follows:

$$\mathrm{EEF} = \frac{\mathrm{CO_2}}{\overline{\mathrm{NPP}}} = \frac{\mathrm{CO_2}}{\sum_{j=1}^{m} A_j \cdot \mathrm{NPP}_j \Big/ \sum_{j=1}^{m} A_j} \tag{1.3}$$

where EEF is ecological footprint of indirect energy consumption, CO_2 denotes CO_2 emissions from indirect energy consumption, $\overline{\mathrm{NPP}}$ is regional average NPP, NPP_j is NPP of the j-class ecosystem, and A_j is the land area occupied by the j-class ecosystem in China. The corresponding NPP parameters of various ecosystems are listed in Table 1.4.

1.2.1.3 IPAT Model for Influencing Factor Analysis

The IPAT model was developed by Ehrlich and Holdren (1971). The equation is widely recognized for analyzing the effects of human activities on the environment: $I = PAT$. I represents the environmental impact, P represents population, A denotes affluence, and T shows technological level. On the basis of this model, Dietz and Rosa (1994) developed the STIRPAT (Stochastic Impacts by Regression on Population, Affluence and Technology) model, which rejects the unit elasticity assumption and adds randomness for the convenience of empirical analysis. Moreover, the STIRPAT model can be used to decompose the population and technology terms to examine the multiple influencing factors impacting on the environment, such as urbanization, industrial structure, and energy structure. The STIRPAT model can be summarized as

$$I = aP^bA^cT^de \tag{1.4}$$

where I, P, A, and T have the same meaning as in the IPAT model, a, b, c, and d represent indicial terms to be estimated, and e represents the random error. In empirical studies, Eq. (1.4) is often applied in logarithmic form:

$$\ln I = a + b(\ln P) + c(\ln A) + d(\ln T) + e \tag{1.5}$$

It appears that the IPAT model can be regarded as a special form of STIRPAT, when $a = b = c = d = 1$. In this form, b, c, and d can be described as the percentage change in environmental impact caused by 1% change in one factor when the others remain unchanged. The form is equivalent to the elastic coefficient in economics.

Table 1.5 lists the explanation of variables in the STIRPAT model applied in this section. As a result, the extended STIRPAT model can be expressed as

$$\ln I = a + b(\ln P) + c(\ln A) + d(\ln E) + f(\ln T) + g(\ln S) + h(\ln B) + e \tag{1.6}$$

Energy consumption data are collected from China Energy Statistical Yearbooks (NBSC, 2000–2011); CO_2 emission coefficients of various types of fuels are estimated by the method proposed by IPCC (2006). Data on the gross output of different sectors are taken from the China Industry Economy Statistical Yearbooks (NBSC, 2000–2011) and data on consumption expenditure of different sectors are collected by urban and rural households. Data acquired from the China Statistical Yearbooks (this yearbook has been integrated into China Industry Statistical Yearbooks since 2013) (NBSC, 2000–2011) are demographic data, per capita income, GDP, urbanization level, tertiary industry proportion, and Engel coefficient. Data on land-use are collected from the China Land and Resources Statistical Yearbooks. Moreover, data concerned with GDP such as consumption expenditure, per capita income, and industrial added value are converted to a 1999 constant price.

Table 1.5 The definitions and measurement methods of variables used in Eq. (1.6)

Variable	Symbol	Measurement
Indirect EEF	I	EF of household indirect energy consumption
Urbanization level	P	The percentage of the urban population in the total population
Consumption level	A	Per capita disposable income of residents of urban and rural areas
Consumption structure	E	Engel coefficient of residents of urban and rural areas
Energy intensity	T	Energy use per unit GDP
Tertiary industry proportion	S	The share of the tertiary industry output value over the total GDP
Secondary industry proportion	B	The share of the secondary industry output value over the total GDP

1.2.2 EEF of Residential Indirect Energy Consumption

The carbon intensity can be obtained after calculating the CO_2 emissions from various sectors for each of these consumption expenditures (Table 1.6). It shows that the falling trend in energy intensity slows down and the influence of technological progress on improving energy efficiency decreases.

The average NPP of China is given in Table 1.7, and Fig. 1.4 shows the indirect energy consumption of residents that are converted to EEF according to Eqs. (1.1)–(1.3). The NPP of China rises slowly because of three main factors: expansion of investment in infrastructure construction, destruction of habitat, and overexploitation of resources. It may not help too much by planting trees to absorb CO_2 from the atmosphere to limit climate change. Although the average NPP of China is increasing, the total EEF continues to increase. The energy intensity of related sectors is falling, while EEF is rising. The total indirect energy use shows an upward trend in urban areas. This is because the growth of expenditure on residence, education, cultural, and recreation services, medicine and medical services are more rapid than in 1999 with a decrease in their respective energy intensities. In rural areas, although the expenditures on these consumption items are rising, energy use caused by these behaviors is falling due to the reduction in energy intensity and the lower growth rate of these expenditures. This is consistent with the results obtained by other studies (Wei et al. 2007; Feng et al. 2011; Zhang et al. 2011).

The comparison between the NPP-EEF model and the traditional EEF model is shown in Fig. 1.5: EEF calculated by the traditional method is significantly higher than the results from the NPP-EEF model. The new model also reflects the powerful carbon cycle function of the ecosystem. Thus, it could be concluded that the analysis of influencing factors based on different approaches would be distinctly different.

Owing to the differences in consumption expenditures and carbon intensity of the eight categories, the share of rural and urban households' EEF varies significantly (Figs. 1.6 and 1.7). From 1999 to 2010, the most energy-intensive consumer behaviors for both urban and rural households are residence, food, and education, cultural, and recreation services. In addition, household facilities, articles and services, medicine and medical services, transport and communication services, and miscellaneous commodities and services account for diminishing shares overall. The share of clothing shows a steady trend. The EEF caused by urban household's residence, food, and household facilities, articles and services is falling while others are rising. The remainder consumption items remain broadly flat over the study period (Fig. 1.6). However, EEF caused by rural households' residence remains broadly flat over the previous year, EEF obtained from education, cultural and recreation services, medicine and medical services, and clothing and household facilities, articles and services is rising. Only that EEF obtained from the food industry was falling (Fig. 1.7).

Table 1.6 Carbon intensity of different sectors during 1999–2010 (Unit: $tC/10^4$ yuan)

Items	1999	2000	2001	2002	2003	2004	2005	2006	2007	2008	2009	2010
Food	1.21	1.18	1.12	0.95	0.84	0.73	0.62	0.51	0.45	0.41	0.36	0.30
Clothing	0.88	0.86	0.80	0.77	0.74	0.71	0.72	0.70	0.66	0.64	0.60	0.57
Household facilities, articles and services	0.41	0.39	0.37	0.35	0.34	0.32	0.31	0.27	0.25	0.23	0.19	0.16
Education, cultural and recreation services	2.35	2.24	2.06	1.85	1.70	1.66	1.65	1.54	1.35	1.24	1.14	1.03
Medicine and medical services	0.82	0.76	0.70	0.65	0.59	0.55	0.50	0.46	0.40	0.37	0.30	0.26
Residence	5.81	5.41	5.10	4.88	5.04	4.81	4.49	4.16	3.72	3.46	3.12	2.78
Transport and communication services	0.46	0.40	0.38	0.32	0.27	0.28	0.25	0.24	0.26	0.22	0.19	0.17
Miscellaneous commodities and services	0.19	0.18	0.17	0.17	0.18	0.17	0.16	0.17	0.16	0.16	0.15	0.15

Table 1.7 Average Chinese NPP during the year of 1999–2010 [Unit: tC/(hm^2 year)]

Item	1999	2000	2001	2002	2003	2004	2005	2006	2007	2008	2009	2010
$\overline{NPP}$	5.2659	5.2715	5.2765	5.2853	5.2921	5.3052	5.3104	5.3250	5.3393	5.3457	5.3548	5.3639

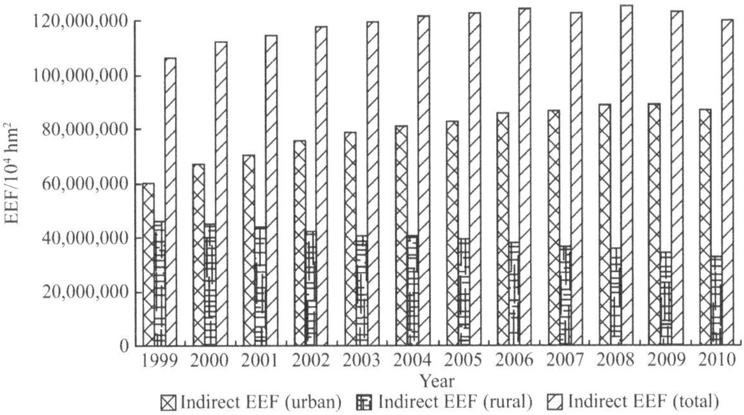

Fig. 1.4 Indirect EEF of urban, rural, and total household (1999–2010)

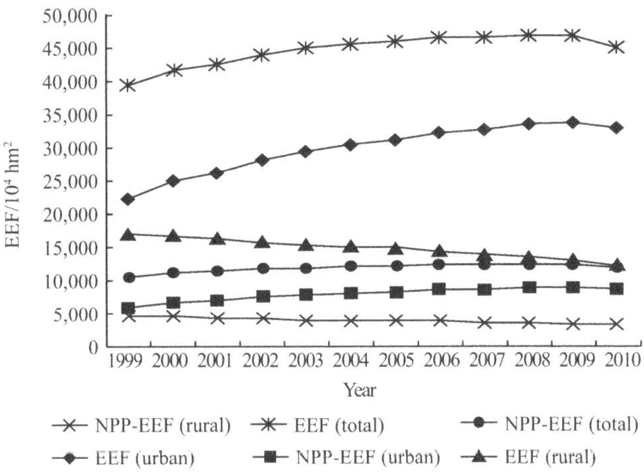

Fig. 1.5 Comparison between the NPP-EEF model and the traditional EEF model (1999–2010)

1.2.3 Influencing Factors of Indirect Energy Consumption by Improved STIRPAT Model

The correlations between the variables for both rural and urban residents are shown in Tables 1.8 and 1.9. The results indicate that ln I had a significant correlation with ln P, ln A, ln E, ln T, ln S, and ln B at the 0.01 significance level. Moreover, the absolute values of correlation coefficients are above 0.8. This indicates that the independent variables have a problem of multicollinearity and information overlaps.

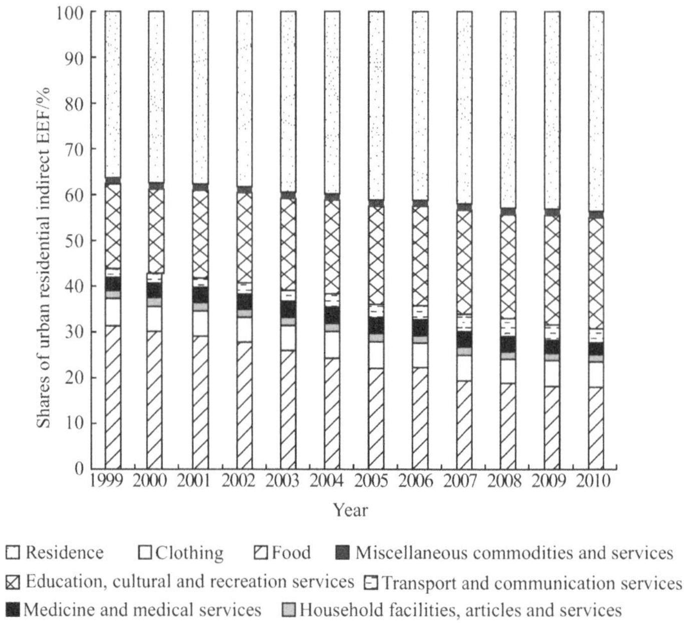

Fig. 1.6 Shares of urban residential indirect EEF during 1999–2010 by industry

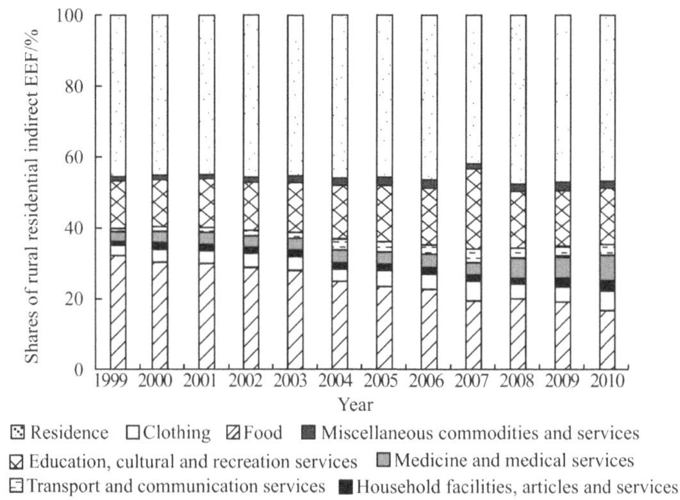

Fig. 1.7 Shares of rural residential indirect EEF during 1999–2010 by industry

Table 1.8 Correlation between variables of urban residents

Variables	ln *I*	ln *P*	ln *A*	ln *E*	ln *T*	ln *S*	ln *B*
ln *I*	1	0.969[a]	0.961[a]	0.970[a]	−0.948[a]	0.960[a]	0.933[a]
ln *P*		1	0.993[a]	−0.920[a]	−0.972[a]	0.930[a]	0.956[a]
ln *A*			1	−0.918[a]	0.971[a]	0.905[a]	0.975[a]
ln *E*				1	0.921[a]	−0.899[a]	−0.900[a]
ln *T*					1	−0.920[a]	−0.934[a]
ln *S*						1	0.873[a]
ln *B*							1

[a]Coefficient is significant at the 1% level

Table 1.9 Correlation between variables of rural residents

Variables	ln *I*	ln *P*	ln *A*	ln *E*	ln *T*	ln *S*	ln *B*
ln *I*	1	−0.982[a]	−0.996[a]	0.992[a]	0.984[a]	−0.886[a]	−0.812[a]
ln *P*		1	0.989[a]	−0.969[a]	−0.972[a]	0.930[a]	0.859[a]
ln *A*			1	−0.987[a]	−0.977[a]	0.884[a]	−0.824[a]
ln *E*				1	0.985[a]	−0.884[a]	−0.812[a]
ln *T*					1	−0.920[a]	−0.812[a]
ln *S*						1	0.808[a]
ln *B*							1

[a]Coefficient is significant at the 1% level

Partial least squares (PLS) estimation is employed to overcome the above problems. The concepts of extracting components and information are used to optimize the limitations of the PLS method (Wold 1975). The regression coefficients of the PLS method are listed in Table 1.10. The results obtained here differ from those of other studies in that consumption structure (Engel coefficients) becomes the main driving factor rather than population, energy intensity, or income. China is at the key stage of rapid industrialization and urbanization. Its consumption structure and industrial structure are being upgraded at a faster pace and its market holds considerable potential for growth. In addition, for urban residents, urbanization level, consumption level, tertiary industry proportion, and secondary industry proportion positively influence the EEF, while the influences of consumption structure and energy intensity are negative. For rural residents, consumption structure, energy intensity, tertiary industry proportion, and secondary industry proportion positively influence EEF, while urbanization level and consumption level do so negatively.

The positive coefficient of urbanization indicates that urban expansion increases environmental pressures on the environment. The impact of urbanization development can be seen in the following aspects: first, it attracts an increasing number of young people moving away from rural areas with traditional lifestyles. Second, factors such as the increased area, rate of expansion, and expansion intensity of built-up areas are greater in urban areas than in rural areas. Soil erosion, soil fertility

Table 1.10 PLS regression results of urban and rural residents

Variables	Unstandardized coefficients		Standardized coefficients	
	Urban residents	Rural residents	Urban residents	Rural residents
a (Constant)	13.096	9.618	63.255	165.599
ln *P*	0.212	−0.102	0.131	−0.184
ln *A*	0.035	−0.351	0.106	−0.362
ln *E*	−1.218	0.392	−0.440	0.389
ln *T*	−0.040	0.259	−0.034	0.335
ln *S*	1.167	0.021	0.363	0.061
ln *B*	0.116	0.125	0.022	0.228

degradation, soil degradation and desertification, environmental deterioration, and other issues caused by irrational use of land resources become more serious. Third, the improvement of people's living standards and the expansion of the urban population will cause a steep increase in consumption among tertiary industries and energy. As a result, it will influence CO_2 emissions. Overall, the effects of urbanization have been reflected in population structure, consumption level and structure, economic scale, and industrial structure (Zhao et al. 2008).

The consumption level has a minimal positive contribution to EEF for urban residents, while the Engel coefficient exerts a negative influence on EEF. The results indicate that economic factors no longer hinder household consumption in urban areas as much as the rapid development of urbanization. Although most other categories have been rising, food's relative share of consumption has been declining. Nowadays, people tend to pursue more abstract satisfaction than material joy that is mostly serviced by the tertiary industry. Per capita income exerts notable negative impacts on EEF for rural residents, while the Engel coefficient exerts a positive influence on EEF for rural residents. This is because the economic factor becomes the dominant factor in curbing consumption. Food and residence constitute significant components in their expenditure. The share of total household consumption on food has been falling significantly since the 1970s, while there has also been an increase in spending on other consumer products, although, the rate of increase is less rapid than that among urban residents as relating to the supply of electricity and low consumption of metal and nonmetallic minerals. Factors such as consumption habits, income gap, and living environment influence rural residents' lifestyles. In conclusion, the gap between urban and rural sector causes the consumption structure to be significantly different in this new era of economic development.

Energy intensity is negatively correlated with the EEF for urban residents. The elasticity coefficient is −0.04, which means that EEF will increase by 0.04% when energy intensity decreases by 1%. The faint "rebound effect" shows that although energy intensity has decreased in line with industrial structural adjustment and technological progress, the increase in CO_2 emissions cannot be offset by rapid consumer expenditure. This perhaps results from the economic and technological development stage of China. In the future, the positive effects of energy intensity will emerge together with industrial upgrading and labor promotion. Tertiary industry occupies a proportion that has a notable positive influence on the EEF of urban residents, while that of secondary industries is smaller. A new trend of progressive increase in consumption of nonmaterial products occurs when urban residents' leisure time and incomes rise. Tertiary industry is mostly dependent on secondary industries in China, which also leads to an indirect negative consequence for the EEF. The EEF has been significantly inhibited by energy intensity for rural residents, while being less positively affected by tertiary and secondary industries, because the nonmaterial consumption of rural residents is lower than that of urban residents, although it has increased gradually in the past few years. Lifestyles, in rural areas, are relatively simple; consumer expenditure on entertainment grows much more slowly and vital services and facilities are not well equipped, which causes the impact of tertiary and secondary industries on the EEF to be lower.

1.3 Driving Forces of Indirect CO_2 Emissions from Residential Consumption in China

Indirect CO_2 emissions are usually much larger than direct emissions from residential consumption. Many factors cause these actions, such as industrial upgrading, technological progress, residential income, urbanization level, population (Zha et al. 2010; Dong and Yuan 2011; Yao et al. 2012; Rosas et al. 2010; Liu et al. 2012). In addition, some researchers studied the impact of family factors, for example, family characteristics, personal characteristics, and consumption behaviors on CO_2 emissions from the microscopic point of view (Wei et al. 2007; Bartusch et al. 2012; Yu et al. 2012). The growth of indirect CO_2 emission is affected significantly by factors from production, consumption, and population system perspectives. However, those three aspects are usually treated separately by scholars. In this section, an environmental-extended input–output model is constructed to calculate the indirect CO_2 emissions of residential consumption from 1992–2007 in China. Then, the driving forces behind it will be examined by combining the SDA and LMDI approach.

1.3.1 Methodology: Environmental-Extended Input–Output Analysis

The environmental-extended input–output (EEIO) model, an extension of the economic input–output model incorporating the environmental satellite account of each sector, is used to calculate the indirect CO_2 emissions from residential consumption. The model is given in Eq. (1.7):

$$E = F(I - A)^{-1}Y \tag{1.7}$$

where E is the indirect CO_2 emissions from various sectors of residential consumption, F is the CO_2 emission per unit of total output for each sector in the input–output table, and $(I - A)^{-1}$ is the Leontief inverse matrix, which is key to input–output analysis, and we take B as $(I - A)^{-1}$ for convenience, Y is the final consumption of consumption goods and services from residents in each sector. Furthermore, we decompose Y into four factors to examine the effects of changes in consumption patterns and population on the indirect CO_2 emissions as follows:

$$Y = SLUP \tag{1.8}$$

where S is the proportion of the consumption in a specific industrial sector to the total consumption, L is the per capita household consumption, U is the proportion of the rural or urban populations in the total population, and P is the total population. As mentioned above, the model above is expanded to examine the effects of changes in consumption patterns and population as shown in Eq. (1.9):

$$E = FBSLUP \tag{1.9}$$

Then, the classical LMDI approach (Ang 2005) is applied to measure the impact of changes of various variables on indirect CO_2 emissions from residential consumption. The driving forces behind emission changes from residential consumption can be investigated and decomposed as follows:

$$\Delta E = E^T - E^0 = \sum_i F^T B^T S^T L^T U^T P^T - \sum_i F^0 B^0 S^0 L^0 U^0 P^0 \tag{1.10}$$

where i denotes the sector number.

The total emission change can be divided into changes of the decomposed drivers stated as follows:

$$\Delta E = E(\Delta F) + E(\Delta B) + E(\Delta S) + E(\Delta L) + E(\Delta U) + E(\Delta P) \tag{1.11}$$

where the emission intensity effect:

$$E(\Delta F) = \sum \frac{E^T - E^0}{\ln E^T - \ln E^0} \ln\left(\frac{F^T}{F^0}\right) \tag{1.12}$$

the production structure effect:

$$E(\Delta B) = \sum \frac{E^T - E^0}{\ln E^T - \ln E^0} \ln\left(\frac{B^T}{B^0}\right) \tag{1.13}$$

the consumption structure effect:

$$E(\Delta S) = \sum \frac{E^T - E^0}{\ln E^T - \ln E^0} \ln\left(\frac{S^T}{S^0}\right) \tag{1.14}$$

the consumption level effect:

$$E(\Delta L) = \sum \frac{E^T - E^0}{\ln E^T - \ln E^0} \ln\left(\frac{L^T}{L^0}\right) \tag{1.15}$$

the urbanization effect:

$$E(\Delta U) = \sum \frac{E^T - E^0}{\ln E^T - \ln E^0} \ln\left(\frac{U^T}{U^0}\right) \tag{1.16}$$

the population effect:

$$E(\Delta P) = \sum \frac{E^T - E^0}{\ln E^T - \ln E^0} \ln\left(\frac{P^T}{P^0}\right) \tag{1.17}$$

From Eqs. (1.12) to (1.17), we can measure the contributions of each factor to the total change of indirect CO_2 emissions. Meanwhile, $E(\Delta F)$ and $E(\Delta B)$ are considered as production system indices because they reflect the effects of changes in production patterns. Similarly, $E(\Delta S)$ and $E(\Delta L)$ are considered as the consumption system indices, and $E(\Delta U)$ and $E(\Delta P)$ are considered as the population system indices.

The energy consumption data of each sector and residents are gathered from China Energy Statistical Yearbooks (NBSC, 1993, 1998, 2003, 2006, 2008). The input–output tables (IOTs) for the years 1992, 1997, 2002, 2005, and 2007 are obtained from China Statistical Yearbooks (NBSC, 1996, 1999, 2006, 2008, 2010). NBSC regularly publishes benchmark IOT every five years, and extends them based on economic upscaling and balancing methods in the five-year period. Here, the IOTs for the years 1992, 1997, 2002 and 2007 are the surveyed ones, and the IOT for the year 2005 is the extended version of benchmark table of the year 2002. The 2007 IOT is the latest benchmark table when we conduct this study. The original IOTs are converted to 2000 constant price using the method established by Liu and Peng (2010). Finally, all the IOTs are aggregated into 22 sectors.

1.3.2 Indirect CO_2 Emissions from Residential Consumption in China

1.3.2.1 Total Indirect CO_2 Emissions from Residential Consumption

Indirect CO_2 emissions from Chinese residential consumption increased continuously during the study period, growing from 390.57 million tons (MtC) in 1992 to

837.43 MtC in 2007 (increasing by 1.14 times). In particular, a rapid growth occurred after 2002, the first year after China joined the WTO. Emissions from the tertiary industry increased from 106.5 MtC in 1992 to 342.41 MtC in 2007, an increase of 2.22 times in fifteen years. At the same time, CO_2 emissions from the secondary industry increased by 98%, while those from the primary industry decrease by 11%.

It is observed that significant differences exist between rural and urban households. The urban household indirect CO_2 emissions showed continuing growth to 641.93 MtC in 2007, which was 2.98 times of that in 1992. However, the indirect CO_2 emissions from rural households exhibited a slight fluctuant trend, and the average annual growth rate was only 0.7% during the study period. Besides, the proportion of indirect CO_2 emissions from urban households increases steadily, accounting for 76.65% in 2007, so that the proportion arising from rural households showed a continued decreasing trend. The comparison between urban and rural residents of per capita indirect CO_2 emission indicates an increasing trend therein in both urban and rural residents, however, there was an exception of a slight decrease in 2005 while the growth in urban residents was twice that among rural residents.

1.3.3 Driving Forces of Indirect CO_2 Emissions from Residential Consumption

Indirect CO_2 emissions from residential consumption are decomposed (Tables 1.11 and 1.12), and contributions of each driving factor to the change of indirect CO_2 emissions are listed in Table 1.11. There are four factors exhibiting positive effects on the growth of indirect emissions during all the study intervals: the consumption structure effect, the consumption level effect, the urbanization effect, and the population size effect. On the contrary, the CO_2 emission intensity effect contributes to a decrease in indirect CO_2 emissions. Moreover, the impact of the production structure is complex for its negative contributions during 2002–2005 and positive contributions during other study periods.

Table 1.12 shows the effect of the studied drivers on indirect emissions from different sectors during the study period (a more detailed analysis follows).

1.3.3.1 Effects from Production System

The change of carbon intensity exerts a significant negative effect on the emissions. This effect offset emissions by 428.25 MtC, and its contribution rate reached 95.84% during 1992–2007. The result suggests that reducing carbon intensity through continuous technological progress is an effective measure to restrain the growth of indirect CO_2 emissions from residential consumption in China. Besides, a

Table 1.11 Decomposition results of indirect CO_2 emissions from residential consumption (Unit: MtC)

Period	Production system		Consumption system		Population system		Changes of indirect CO_2 emissions
	$E\ (\Delta F)$	$E\ (\Delta B)$	$E(\Delta S)$	$E\ (\Delta L)$	$E(\Delta U)$	$E(\Delta P)$	
1992–1997	−107.05	14.14	13.70	134.46	27.19	23.15	105.58
1997–2002	−114.04	142.76	22.86	80.00	58.63	22.23	212.44
2002–2005	−88.23	−79.66	14.86	140.63	37.88	12.86	38.34
2005–2007	−175.34	31.98	11.67	194.59	19.48	8.11	90.49
1992–2007	−428.25	110.14	58.08	506.87	134.19	65.83	446.86

sector with the largest contribution to the decrease in emissions was the Financial Intermediate, Business, Real Estate, and Other Services sectors, with a value of 98.49 MtC.

Figure 1.8 shows the carbon intensity and the absolute value of the intensity changing rate for each sector (1992–2007). The carbon intensity of most industrial sectors shows a downward trend, and the average reduction therein is 54.92%. Among these industrial sectors, the decrease in carbon intensity in 16 sectors exceeds 50%. Furthermore, the four biggest declines come from the Manufacture of Transport Equipment sector (dropped 87.75%), the Manufacture of Communication Equipment, Computers and Other Electronic Equipment sector (dropped 87.19%), the Manufacture of Measuring Instruments and Machinery for Cultural Activity and Office Work sector (dropped 85.31%), and the *Manufacture of electrical machinery and equipment* sector (dropped 82.47%).

From the perspective of different industries, the most obvious declining of CO_2 emission intensity is among the secondary industry (decreased by 63.19%). The intensity of primary, and tertiary industries declined by 33.41% and 38.20% from 1992 to 2007, respectivley. The dramatic decrease in CO_2 emission intensity is inseparable from China's insisting on the new industrialization pathway. In the process of new industrialization, China has pursued high-tech, low-pollution, and low-resource production and gained remarkable results.

1.3.3.2 Effects from Consumption System

Consumption structure exerts a positive effect on the indirect CO_2 emissions, although the extent of its impact is somewhat less than other effects. The effects from consumption structure reach 58.08 MtC, which accounts for 13% of the total increment over the period 1992–2007.

Table 1.12 Decomposition results of each sector during the whole period (1992–2007) (Unit: MtC)

No.	Sector	Production system		Consumption system		Population system	
		$E(\Delta F)$	$E(\Delta B)$	$E(\Delta S)$	$E(\Delta L)$	$E(\Delta U)$	$E(\Delta P)$
1	Agriculture, Forestry, Animal Husbandry and Fishery	−40.08	17.08	−49.03	52.36	5.53	7.03
2	Mining	−2.93	1.36	−4.28	2.81	0.26	0.39
3	Manufacture of Foods, Beverage and Tobacco	−67.25	18.11	−11.58	78.74	23.30	10.16
4	Cotton Textile, Apparel and Leather Products	−23.09	14.79	−15.41	38.51	10.08	5.00
5	Manufacture of Wood and Furniture	−4.34	−0.07	−1.47	4.33	1.27	0.56
6	Manufacture of Paper, and Articles for Culture, Education and Sport Activities	−4.96	0.26	−1.12	4.41	1.42	0.57
7	Coking, Gas and Processing of Petroleum	−2.02	1.80	4.14	3.30	1.24	0.42
8	Chemical Industry	−36.03	13.20	−10.24	37.18	7.49	4.89
9	Construction, Manufacture of Nonmetallic Mineral Products	−8.49	−0.05	5.36	8.01	2.59	1.03
10	Mineral Products	−7.84	1.28	−6.17	7.86	1.87	1.03
11	Manufacture of General and Special Purpose Machinery	−0.90	−0.05	−0.60	0.82	0.25	0.11
12	Manufacture of Transport Equipment	−15.13	−5.27	21.17	12.71	3.61	1.64
13	Manufacture of Electrical Machinery and Equipment	−23.48	3.06	7.36	21.24	6.81	2.73
14	Manufacture of Communication Equipment, Computers and Other Electronic Equipment	−16.88	−3.69	14.97	14.50	4.02	1.88
15	Manufacture of Measuring Instruments and Machinery for Cultural Activity and Office Work	−1.59	−0.23	1.06	1.37	0.25	0.18
16	Manufacture of Artwork and Other Manufacturing	−9.97	2.06	0.47	11.45	4.25	1.45
17	Electricity and Steam Production and Supply	−18.85	10.28	18.72	19.54	5.97	2.52
18	Gas Production and Supply	−2.69	0.49	0.17	2.83	1.13	0.36
19	Hot Water Production and Supply	−3.00	0.87	1.94	3.27	1.31	0.41
20	Transport, Storage and Post	−9.95	−0.13	−6.56	13.31	3.12	1.74

(continued)

Table 1.12 (continued)

No.	Sector	Production system		Consumption system		Population system	
		$E(\Delta F)$	$E(\Delta B)$	$E(\Delta S)$	$E(\Delta L)$	$E(\Delta U)$	$E(\Delta P)$
21	Wholesale and Retail Trades, Hotels and Catering Services	−30.30	8.66	13.25	43.16	11.73	5.60
22	Financial Intermediate, Business, Real Estate and Other Services	−98.49	26.33	75.93	125.18	36.67	16.15
	Total	−428.26	110.14	58.08	506.89	134.17	65.85

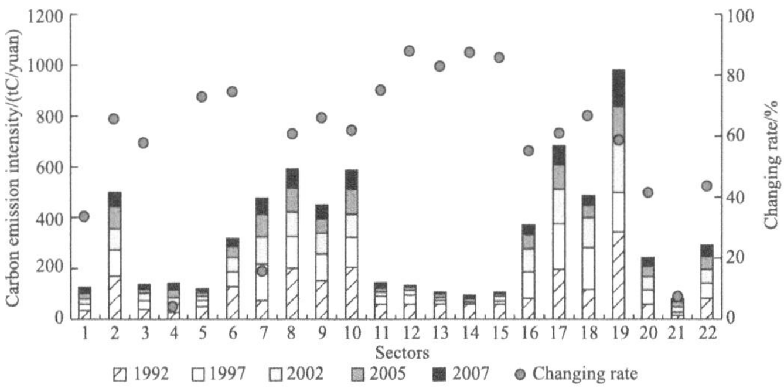

Fig. 1.8 The CO_2 emission intensity and the absolute value of the intensity changing rate for each sector from 1992 to 2007

As shown in Table 1.12, the consumption structure effect causes an increase in indirect CO_2 emissions for 10 industrial sectors. Given the sum of the contributions of the positive effects is greater than that of the negative effects, the total contribution appears to exert a positive effect during the study period. In terms of different industries, the share of consumption in the primary industry declined from 28.01% in 1992 to 10.33% in 2007, while the shares of consumption in secondary and tertiary industries increased rapidly from 42.64% and 29.35% in 1992 to 44.76% and 44.91% in 2007, respectively. However, the emission intensity of secondary and tertiary industries is much higher than that of the primary industry. Hence, the changes in consumption structure stimulate the increase of indirect CO_2 emissions.

The decomposition results show that consumption level exerts the most significant positive effect on the increase of emissions during the study period, with a contribution of 506.87 MtC and contribution ratio of 113.43%. Moreover, the per capita household consumption rises rapidly (increased by 116%), this indicates that the rapid increase in household consumption level has become the most important driving force behind the growth of household indirect CO_2 emissions in China. The increasing trend exists in both urban and rural household consumption, and the growth in per capita consumption level in urban residents is always larger than that of rural residents. It could be the main reason that indirect CO_2 emissions from urban households are much higher than those of their rural counterparts.

In fact, the effect of consumption level reflects the change in residents' incomes; incomes and living conditions have greatly improved resulting from the rapid growth of China's economy, which leads to acceleration of energy consumption to accommodate more comfortable lives. For example, household appliances have become more widespread, and the number of private car owners has risen, therefore, both household energy consumption and CO_2 emissions grow rapidly during the sample period.

1.3.3.3 Effects from Population System

As shown in Tables 1.11 and 1.12, the urbanization effect contributes to the increase of indirect CO_2 emissions in 1992–2007 with a contribution rate of 30.03%. It shows that the per capita indirect CO_2 emissions from urban residential consumption are much higher than that from rural residents, and the gap between urban and rural per capita emissions becomes wider year-by-year. In the case of the urban and rural populations, the proportion of urban population increases annually, the contribution of urbanization effect can be assumed as the influence of population migration on indirect CO_2 emissions during the urbanization process. Based on these results, changes in urban and rural population structure stimulate the rise of indirect CO_2 emissions from residents even though other factors remain unchanged.

The population size exerts a positive effect on indirect CO_2 emissions from residential consumption. In the whole study period, this factor caused a 65.83 MtC emissions increment with a contribution rate of 14.73%. China has a large population base although the per capita CO_2 emission base is small. Therefore, the increase in per capita emission would be amplified by the size of the population and promote increasing CO_2 emissions. According to Table 1.12, the contribution of this effect is slightly higher than that of consumption structure, but far lower than both consumption level effect and carbon intensity effect.

1.4 Summary

In this chapter, we firstly analyzed the energy consumption and its related CO_2 emissions from residents in China. In addition, some features related to CO_2 emissions from urban and rural residents in China as well as useful indicators of emissions performance are obtained. Then, the indirect impact of residents' lifestyle on the ecological footprint of indirect energy consumption is examined based on CLA and NPP. The result finds that the EEF in urban China has been increasing while that in rural areas it has been continuously declining during 1999–2010. The most energy-intensive choices are residence, food and education, cultural and recreation services. Food accounts for the highest proportion of indirect energy consumption in rural households, apart from the residence itself. In addition, influencing factors are analyzed by the PLS method and compared with the STIRPAT model. The analysis finds that the PLS methods are more reasonable and acceptable. Multicollinearity among these driving indicators may have a negative impact on the final results of the STIRPAT model, while multicollinearity had little impact on the PLS method. The major driving forces are level of urbanization, per capita disposable income, Engel coefficient, energy intensity, tertiary industry proportion, and secondary industry proportion. It is found that the Engel coefficient makes the greatest contribution to EEF in both urban and rural households. The other important driving forces for urban residents are tertiary industry proportion, while those for rural residents are per capita income and energy intensity.

In the second part of this chapter, we analyzed the CO_2 emissions from residents from the perspective of indirect emissions and influencing factors in China during 1992–2007 based on EEIO analysis together with SDA and LMDI. The indirect CO_2 emissions from residential consumption increase rapidly, and the emission structure has undergone a fundamental change along with the changes in the structure of residential consumption. The largest share of CO_2 emissions moves from the agriculture sector to the construction and service sectors. Decomposition results show that the factors of consumption structure, consumption level, urbanization, production structure, and population size cause the rise of indirect CO_2 emissions from residential consumption, while the CO_2 emission intensity of the industrial sector exhibits a continuous negative effect.

References

Ang, B.W. 2005. The LMDI approach to decomposition analysis: A practical guide. *Energy Policy* 33 (7): 867–871.

Bartusch, C., M. Odlare, F. Wallin, et al. 2012. Exploring variance in residential electricity consumption: Household features and building properties. *Applied Energy* 92: 637–643.

Bin, S., and H. Dowlatabadi. 2005. Consumer lifestyle approach to US energy use and the related CO_2 emissions. *Energy Policy* 33 (2): 197–208.

Dietz, T., and E.A. Rosa. 1994. Rethinking the environmental impacts of population, affluence and technology. *Human Ecology Review* 1 (1): 277–300.

Dong, X.Y., and G.Q. Yuan. 2011. China's greenhouse gas emissions' dynamic effects in the process of its urbanization: A perspective from shocks decomposition under long-term constraints. *Energy Procedia* 5: 1660–1665.

Ehrlich, P.R., and J.P. Holdren. 1971. Impact of population growth. *Science* 171: 1212–1217.

Erb, K.H. 2004. Actual land demand of Austria (1926–2000): A variation on ecological footprint assessments. *Land Use Policy* 21 (3): 247–259.

Fang, K., D.M. Dong, Z. Lin, et al. 2012. Calculation method of energy ecological footprint based on global net primary productivity. *Acta Ecological Sinica* 32 (9): 2900–2909.

Feng, Z.H., L.L. Zou, and Y.M. Wei. 2011. The impact of household consumption on energy use and CO_2 emissions in China. *Energy* 36 (1): 656–670.

Holmberg, J., U. Lundqvist, K.H. Robèrt, et al. 1999. The ecological footprint from a systems perspective of sustainability. *International Journal of Sustainable Development & World Ecology* 6 (1): 17–33.

IPCC. 2006. IPCC guidelines for national greenhouse gas inventories. Hayama, Japan: The Institute for Global Environmental Strategies.

Liu, Q.Y., and Z.L. Peng. 2010. *Comparable price input–output series table and analysis of China from 1992 to 2005*. Beijing: China Statistics Press.

Liu, W.L., C. Wang, and A. Mol. 2012. Rural residential CO_2 emissions in China: Where is the major mitigation potential? *Energy Policy* 51: 223–232.

Liu, Z., D. Guan, W. Wei, et al. 2015. Reduced carbon emission estimates from fossil fuel combustion and cement production in China. *Nature* 524 (7565): 335–338.

Lu, T.H. 2002. *Researches on sustainable development indicators system*. Xiamen: Xiamen University.

NBSC. 1993. China Energy Statistical Yearbook 1993. Beijing: China Statistics Press.

NBSC. 1996. China Statistical Yearbook 1996. Beijing: China Statistics Press.

NBSC. 1998. China Energy Statistical Yearbook 1998. Beijing: China Statistics Press.

NBSC. 1999. China Statistical Yearbook 1999. Beijing: China Statistics Press.

NBSC. 2003. China Energy Statistical Yearbook 2003. Beijing: China Statistics Press.

NBSC. 2006. China Energy Statistical Yearbook 2006. Beijing: China Statistics Press.

NBSC. 2006. China Statistical Yearbook 2006. Beijing: China Statistics Press.

NBSC. 2008. China Energy Statistical Yearbook 2008. Beijing: China Statistics Press.

NBSC. 2008. China Statistical Yearbook 2008. Beijing: China Statistics Press.

NBSC. 2010. China Statistical Yearbook 2010. Beijing: China Statistics Press.

Rosas, J., C. Sheinbaum, and D. Morillón. 2010. The structure of household energy consumption and related CO_2 emissions by income group in Mexico. *Energy for Sustainable Development* 14 (2): 127–133.

Venetoulis, J., and J. Talberth. 2008. Refining the ecological footprint. *Environment Development and Sustainability* 10 (4): 441–469.

Wei, Y.M., L.C. Liu, Y. Fan, et al. 2007. The impact of lifestyle on energy use and CO_2 emissions: An empirical analysis of China's residents. *Energy Policy* 35: 247–257.

Wold, H. 1975. Soft modeling by latent variables: The non-linear iterative partial least squares (NIPALS) approach. *Journal of Applied Probability* 12: 117–142.

Yao, C.S., C. Chen, and M. Li. 2012. Analysis of rural residential energy consumption and corresponding carbon emissions in China. *Energy Policy* 41: 445–450.

Yu, B.Y., J.Y. Zhang, and A. Fujiwara. 2012. Analysis of the residential location choice and household energy consumption behavior by incorporating multiple self-selection effects. *Energy Policy* 46: 319–334.

Zha, D.L., D.Q. Zhou, and P. Zhou. 2010. Driving forces of residential CO_2 emissions in urban and rural China: An index decomposition analysis. *Energy Policy* 38 (7): 3377–3383.

Zhang, X., S.W. Niu, C.S. Zhao, et al. 2011. The study on household energy consumption and carbon emissions in China's urbanization. *China Soft Science Magazine* 9: 65–75.

Zhao, W., J.S. Liu, and F.E. Kong. 2008. Effects of urbanization on supply and demand of regional ecological footprint. *Journal of Applied Ecology* 19 (1): 120–126.

Chapter 2
Household Electricity Consumption and Saving Behavior in China

China is one of the major energy-consuming countries. In 2010, China surpassed the U.S. as the world's biggest energy consumption country, whose total energy consumption was 3249.39 million tons of standard coal equivalent. However, with its rapid economic growth, the disparity between energy consumption and production is becoming increasingly greater in recent years. Primary energy production was only 2969.16 million in 2010, much lower than energy consumption (NBSC 2011). Such glaring disparity is especially salient in electricity consumption and supply. Taking Beijing as an example, the total consumption of electricity in 2007 was about 67.5 billion kW h. However, only 22.4 billion kW h was generated in Beijing, with more than 65% of electricity imported from neighboring provinces (autonomous regions, municipalities) (Beijing Bureau of Statistics of China 2009). It is, therefore, urgent for China to improve the efficiency of energy use and encourage energy conservation.

In this chapter, we conduct questionnaire surveys and utilize the logit regression model to empirically analyze household electricity-saving behavior. Furthermore, we build a cointegration equation and a panel error correction model to analyze the direct rebound effect of urban residential electricity consumption, with China's 30 provincial government panel data from 1996 to 2010 applied. Correspondingly, some inferences and suggestions are offered for the promotion of electricity-saving behaviors and formulation of energy policies.

This chapter takes the following literature for reference:

Wang Z, Zhang B, Yin J, et al. 2011. Determinants and policy implications for household electricity-saving behavior: evidence from Beijing, China. Energy Policy, 39(6): 3550–3557.

Wang Z H, Lu M, Wang J. 2014. Direct rebound effect on urban residential electricity use: an empirical study in China. Renewable and Sustainable Energy Reviews, 30(2): 124–132.

Z. Wang and B. Zhang, *Low-Carbon Consumption in China: Residential Behavior, Corporate Practices and Policy Implication*,
https://doi.org/10.1007/978-981-15-2792-0_2

2.1 Features and Determinants of Household Electricity-Saving Behavior

Households have been identified by researchers as an important target group for energy conservation. To illustrate, households account for 25% of the total energy consumption in the US, 26% in Japan, and 50% in Saudi Arabia (Saidur et al. 2007). Given this, reducing waste or inefficiency of household energy use can be an effective means of decreasing global energy demand and related adverse environmental impacts. As Murata et al. (2008) argued, "28% reduction in China could be achieved by the year 2020 by means of improving citizens' energy efficiency in household appliances use." Again, to take Beijing as an example, household electricity consumption accounts for 15.8% of the total consumption in 2007, which is equivalent to 10.7 billion kW h. There is a great potential to relieve the pressure of electricity shortage by further encouraging residents' electricity-saving behavior. Since households can make a great contribution to energy conservation, in order to effectively encourage household energy-saving behaviors, it is necessary to identify what the key behavioral antecedents are.

The literature suggests that different types of energy-saving behavior are related to different behavioral antecedents (e.g., Stern and Oskamp 1987; McKenzie-Mohr et al. 1995; Hansla et al. 2008). As to household energy-saving behavior, scholars have indicated various determinants including psychological factors (Becker et al. 1981), and socio-demographic issues (Gatersleben et al. 2002; Moll et al. 2005). Although these antecedents have received varying levels of attention in the literature, few, if any, researches of household electricity-saving behavior have addressed them in an integrated way.

In addition, most previous studies set out to explore the possibility to improve the efficiency of energy use in the Western world. A paucity of studies have addressed household electricity-saving behavior in a Chinese context (Min and Mills 1997; Price and Sinton 2002; Lu 2006; Andrews-Speed 2009). As explaining human behavior is a difficult task, the adaptability of Western management theory to non-Western cultural contexts is often questioned (Bryman and Bell 2007). To address energy challenges in China in a timely manner, it is necessary to conduct research to identify the key determinants of energy conservation behavior. China's unique political and cultural background must be taken into account; for example, the impact of policy and statutes issued by the Chinese government needs to be considered.

Given the lack of a comprehensive study on antecedents of electricity-saving behavior, and that almost no study has been executed to explore this topic in the Chinese context, the current research attempts to address three issues: (1) What is the present situation of household electricity conservation in Beijing? (2) What are the major determinants that influence household electricity-saving behavior in China? (3) Does policy direction have any impact on encouraging household electricity-saving behavior?

2.1.1 Hypothesis Framework for Determinants of Household Electricity Saving Behavior

Most electricity-conservation studies have focused on analyzing residents' willingness to increase their efforts to save electricity. Their electricity-saving behavior is usually measured by examining the usage of energy-saving appliances (e.g., Lu 2006; Saidur et al. 2007; Widen et al. 2009) or consumption of electricity generated by renewable energy (e.g., Longo et al. 2008; Zografakis et al. 2010). Some studies also put emphasis on energy-saving measures for assessment of household electricity-saving behavior (e.g., Banfi et al. 2008). Given that China is one of the largest consumer markets of appliances, this research is dedicated to exploring electricity-saving behavior in household appliance usage. With reference to previous literature, a literature review on the antecedents of household appliance usage was conducted in the current study. Determinants influencing electricity-saving behavior are summarized below to capture the homogenous nature of most studies in this domain.

The energy-saving behavior may largely depend on psychological variables and socio-demographic variables (Abrahamse and Steg 2009). The theory of planned behavior (TPB) (Ajzen 1991) is often used to examine pro-environmental behavior from the perspective of psychology. The use of energy-saving light bulbs (Meijnders et al. 2001), the use of bleach research (Harland and Staats 1999), car use (Bamberg and Schmidt 2003), and bus use for commuting (Heath and Gifford 2002) seem to be effectively explained by variables from TPB.

As shown in Fig. 2.1, attitude, perceived behavioral control, subjective norms, and residue effect can influence people's intentions regarding electricity saving behavior. As shown in Fig. 2.1, the TPB postulates four conceptually independent antecedents of intention. The first is the attitudes, which can be measured by the degree to which an individual has a favorable or unfavorable evaluation of the particular behavior. In the current study, attitude refers to the degree of people's awareness of performing electricity-saving behavior, which largely depends on the evaluation of preference to electricity saving and the information the individual holds about such behavior. Pro-environment awareness plays a significant role in energy use and conservation (Samuelson 1990). As Ek and Söderholm (2010) confirmed, residents' attitude toward the environment is an important factor to predict their electricity-saving activities. This perspective has been further supported by Zografakis et al. (2010) research; they argue that people with more energy saving information and stronger awareness of climate change are more likely to purchase renewable energy and participate in energy-saving activities (Zografakis et al. 2010).

The second antecedent is the degree of perceived behavioral control, which refers to ease or difficulty of performing particular behavior. It largely depends on the weighing of the costs and benefits in the process of performing specific behavior, such as financial cost, effort, and time (Lindenberg and Steg 2007). Feng et al. (2010) conducted a research related to the relationship between electricity

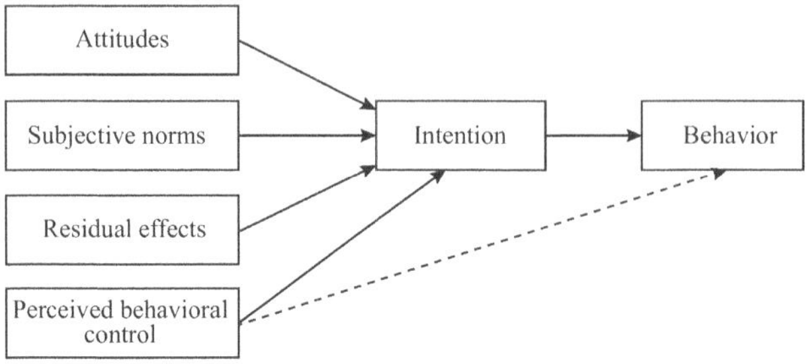

Fig. 2.1 The framework of TPB including residual effects. *Note* Ajzen I. 1991. The theory of planned behavior. Organizational Behavior and Human Decision Processes, 50: 182

prices and the levels of consumption in China; the results indicate that the economical benefit has a great influence on electricity-saving behavior. This conclusion has been confirmed by several researches related to the effect of financial cost on household energy consumption (Banfi et al. 2008; Scarpa and Willis 2010). Besides, the comfort or discomfort (e.g., thermal comfort, air quality and noise protection) that residents felt when engaging in certain kinds of electricity-saving behaviors have a significant influence on their further energy-saving activities (Banfi et al. 2008).

Subjective norm refers to the perceived social pressure to perform or refrain from behavior (Abrahamse and Steg 2009). From this point of view, the individuals' perception of the external environment affects their behavior a lot. Ek and Söderholm (2010) have investigated social interactions that attach great importance to electricity-saving behavior. Other people's attitudes and behavior in electricity saving may influence individuals' willingness for electricity-saving activities. Besides, media promotion for environment protection and climate change may reduce residents' unnecessary electricity consumption.

By summarizing all the points made above, one can see that household electricity-saving behavior can be predicted by a number of determinants, which mainly consist of psychological and socio-demographic factors. As culture has great importance in predicting people's behavior, residents' electricity-saving behavior in different countries may present more or less different characteristics. Since policy instruments and social propaganda are more frequently implemented to regulate the economy by the Chinese government compared with many Western countries, they may play a significant role in influencing residents' daily lives, including their electricity-saving behaviors. Nevertheless, pro-environmental awareness seems to have a limited effect on predicting Chinese residents' electricity-saving behavior. Compared with Western residents, Chinese residents

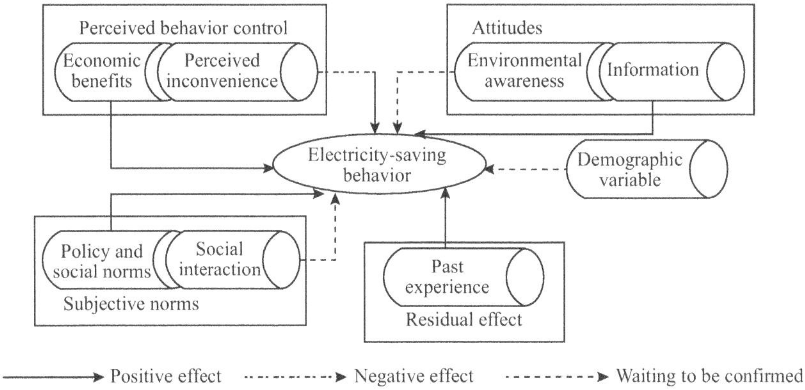

Fig. 2.2 The theoretical frame of the research

pay more attention to the economic benefits or convenience. Besides, social interaction, as another predictor of household electricity-saving behavior also requires further discussion (Zeng 2005).

Considering these differences, a hypothesis framework combined with the determinants that previous studies have explored is displayed in Fig. 2.2. We suppose that economic benefits, policy and social norms, as well as past experience may have a positive influence on household electricity saving, while the discomfort caused by electricity-saving activities, may have a negative effect. Nevertheless, the effects of environmental awareness and social interactions regarding electricity-saving experience in the household may be not obvious in Beijing and need to be further confirmed. In addition, some socio-demographic variables are also taken into account to predict household electricity-saving behavior.

2.1.2 Methodology for Determinants of Household Electricity Saving Behavior

2.1.2.1 Sample and Data Collection

Restated, the present study aims to identify the determinants of household electricity-saving behavior in Beijing. A questionnaire was duly devised, and using a random sampling method, a sample of 1500 residents living in the residential quarter of Beijing's urban area was selected. Some neighborhood committees, which are the community management organizations of residential quarters in China, were enlisted to help distribute questionnaires to the residents on the spot. We explain our survey style as follows:

Unlike their foreign counterparts, Chinese urban residents mainly live in a compact community called a residential quarter. Taking residents living in these residential quarters as our research subjects would not only be convenient for data collection, but the sample would also reasonably represent the researched population—Beijing households.

Usually, the response rate of the postal survey in China is relatively low. Filling in questionnaires is time-consuming; it is also inconvenient for the respondents to mail them back. As a result, most respondents show little interest. The neighborhood committees are often very familiar with residents living in their residential quarters. Enlisting their help can improve the credibility and raise the response rate of the survey.

The design of the questionnaire was based on the hypotheses framework as shown in Fig. 2.2, with the content divided into three main sections: awareness of household electricity saving, factors related to household electricity use and saving behavior, and basic individual information. The construct to measure behavioral determinants consists of 16 items. For each item, a five-point Likert scale was used to enable respondents to indicate the extent to which they agree with these items (1 represents not at all important, 2 represents not important, 3 represents not thinking about it, 4 represents important, and 5 represents extremely important). And the question to evaluate household willingness toward electricity saving is "Do you often take household electricity saving actions (e.g., pull off the plug after using appliances or choose to buy energy efficient household appliances)?" To answer this question, respondents were asked to tick one of two boxes: 1—YES; 0—NO.

Probably because the researchers enlisted the help of neighborhood committees, the response rate was quite high (62.26%), which could be considered satisfactory for such a comprehensive survey. 118 out of 943 returned questionnaires were deleted, having failed to answer more than 40% of the total questions. Finally, a total of 816 usable responses were received. 311 responses of these 816 questionnaires were followed up with phone calls to collect answers for unfilled questions in first wave responses. To assess the potential nonresponse bias, the differences in the mean values of the determinants between the 311 responses and the rest 505 responses were checked. The results showed no significant differences between usable respondents and the others at the 0.05 level. This indicates that nonresponse bias is not a major problem in our study and the results from the sampled residents could be generalized to represent other residents in Beijing.

2.1.2.2 Modeling Household Willingness to Participate in Electricity Saving

An econometric model was developed to identify the antecedents of household electricity-saving behavior in Beijing. When the dependent variable is in 0–1 style, researchers have a choice between logistic regression and probit regression. According to Borsch-Supan (1990), the Logit model is the better choice if the response decision is made based on the maximization of utility. Given that residents' willingness in electricity saving mainly depends on the expected utility from their saving behavior, the Logit model was selected in current research. The following specification was used:

$$\text{Logit}(R) = \frac{1}{1+e^{-z}} \tag{2.1}$$

$$z = \beta_0 + \sum_{i=1}^{n} \beta_i x_i + \varepsilon_i \tag{2.2}$$

where z is the latent and continuous measure of residents' willingness in electricity saving, x_i the vector of observations of explanatory variables, β_i the vector of parameters to be estimated, ε_i the random error term (assumed to follow a standard normal distribution), R the observed and coded discrete willingness variable.

Table 2.1 presents the descriptive statistics for both dependent and independent variables used in the econometric model. To form a single indicator factor, each independent variable is measured by the mean of their evaluating items. This can reduce the model complexity and allow a more accurate assessment of the determinants of electricity-saving behavior. Some demographic variables (e.g., age, dwelling area, education) are also introduced in our regression model as control variables. This is to account for the possibility that the distinctions of the respondents may influence the extent of their electricity conservation behavior, which has also been discussed in our hypothesis framework.

2.1.3 *Empirical Analysis for Features of Household Electricity-Saving Behavior*

According to our questionnaire survey, there is a relatively high awareness among Beijing residents about the environmental burden and resource scarcity. About 65.32% of respondents have clearly noticed the negative environmental effects due to energy-intensive consumption, especially the climate change; nearly half of respondents paid special attention to the national policies and regulations regarding energy conservation. However, such awareness does not be translated into electricity-saving behavior. The results indicate that a large number of residents in

Table 2.1 Variables included in the analysis: descriptive statistics

Items	*N*	Mean	S.D.
Dependent variables			
Household willingness in electricity saving	816	0.61	0.489
Independent variables			
Economic benefits		2.68	0.2
Less expenditure from electricity saving	816	2.73	0.729
Subsidy for energy conservation appliance use	814	2.64	0.824
Policy and social propaganda		2.59	0.658
Government support	815	2.2	0.766
Social norms	815	2.21	0.998
Media propaganda for electricity saving	815	3.34	1.203
Past experience		2.17	0.122
Experience for electricity shortfall	816	2.25	0.938
Habits of electricity saving in daily lives	816	2.08	1.022
Perceived inconvenience		2.04	0.202
Discomfort from electricity saving	816	2.18	0.781
Perceived time-wasted from electricity saving	816	1.9	0.833
Social interaction		2.84	0.205
Influence from friends' electricity saving behavior	816	2.7	0.806
Community activities of electricity saving	816	2.99	0.971
Environmental awareness		2.75	0.245
Awareness of energy crisis	813	2.81	0.886
Global climate change awareness	813	2.79	0.922
Awareness of environmental protection	813	2.66	0.815
Information		2.48	0.138
Knowledge of methods of electricity saving	814	2.48	0.851
Comprehension of energy efficiency labels	814	2.62	1.28
Awareness of policy or regulation in energy saving	814	2.34	0.797

Beijing paid little attention to electricity use in their daily lives. About 69% of respondents said "I do not know my electricity bill of last month," and 45% of respondents do not even know current electricity price. Besides, only 39.46% of respondents indicated that they have selected the electricity-saving appliances as their main options for home appliances.

The estimates from the questionnaire survey also show that the perceived inconvenience for participating in electricity-saving activities, together with the high cost of electricity-saving appliances, might slump residents' willingness for electricity saving. According to the research results, inconvenience in purchase and usage (accounting for 27.59%) is second to high price (28.16%) as the main reason for respondents' reluctance to use them. And 22.41% of respondents feel uncertain about the durability of electrically efficient appliances and point to the inconvenience that might come with such appliances.

Moreover, we especially estimated the impact of economic factors on household electricity-saving behavior. In our questionnaire, respondents were asked to answer the question "which one would you like to buy: an electricity-saving appliance that requires an extra payment, or an appliance that is the same with the above-mentioned one in all aspects except that it is not electricity-saving and does not need the extra payment?" 20.27% of respondents said they were unwilling to choose the electricity-saving appliances. And most respondents (about 57.37%) expressed that the extra payment they deem acceptable would be under 10% of the original price. Only 4.67% of respondents would like to pay more than 20%. Then, we made a further crosstab analysis to find out whether it is related to household income level and self-perception of individual electricity consumption level. Figure 2.3 shows that there are no significant differences among the respondents of different income levels in their acceptance to pay extra for an electricity-saving appliance (Pearson Chi-square = 10.690, Sig. = 0.556). However, the respondents who have recognized themselves as an excessive consumer of electricity seem more willing to switch to electricity-saving appliances (Pearson Chi-square = 23.238, Sig. = 0.026), as shown in Fig. 2.4. It is indicated that economic benefits for encouraging electricity saving should be combined with an effort to strengthen residents' awareness of their excessive electricity use.

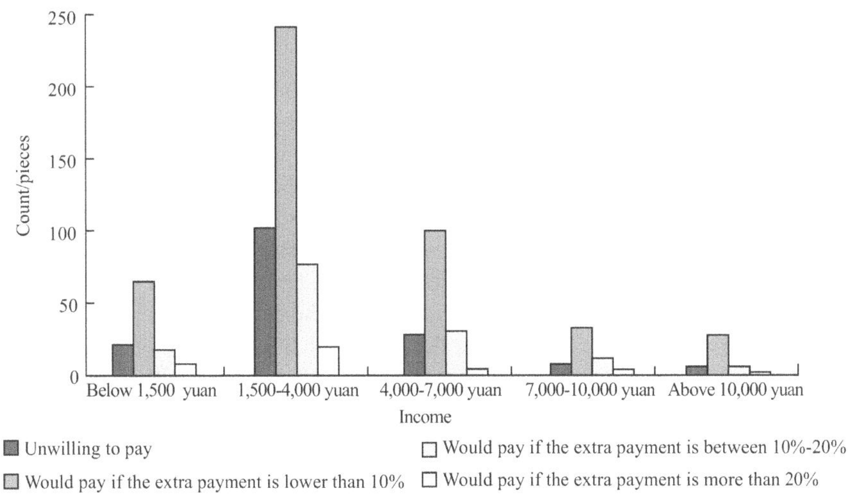

Fig. 2.3 Crosstab between household willingness to pay extra money for electricity saving appliances and income level

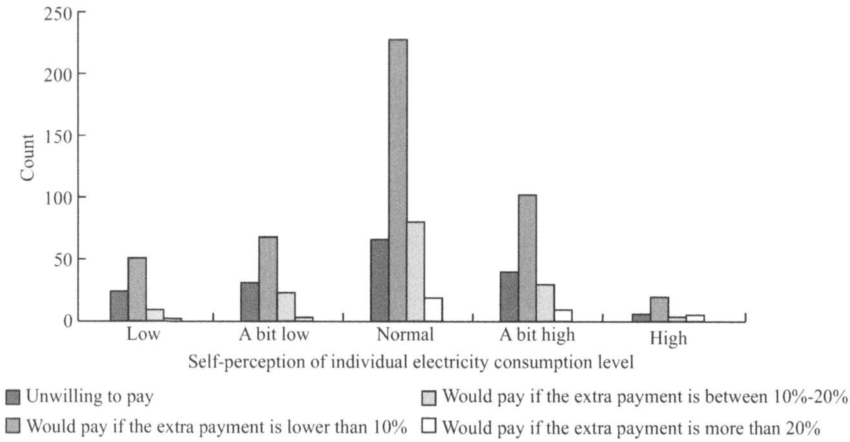

Fig. 2.4 Crosstab between household willingness to pay extra money for electricity saving appliances and self-perception of individual electricity consumption level

2.1.4 Empirical Analysis for Determinants of Household Electricity Saving Behavior

2.1.4.1 The Logistic Regression Analysis

Table 2.2 presents the results of logistic regression analysis. To assess the overall fitness of the model, Hosemer-Lemeshow goodness-of-fit test was examined, since it is usually considered as a critical statistic to detect incorrect model specification such as non-linearity in the predictors or missing predictors. The output corresponding to the Hosmer–Lemeshow statistic together with likelihood ratio values (Table 2.2) indicates that it is reasonable to reject the null hypotheses that the independent variables are not associated with the dependent variable. Cox and Snell R^2 and Nagelkerke R^2 are 0.147 and 0.2, respectively. The empirical results in terms of estimated coefficients and corresponding Wald-test values are displayed in Table 2.2. Wald statistic in the outcome shows that the coefficients are significantly different from zero, so we can assume that the predictors are making a significant contribution to the prediction of the outcome. Multi-collinearity was further checked among independent variables. Variance inflation factor (VIF) for all independent variables range from 1.024 to 1.401, which are all well below the maximum level of 10.0 suggested by Mason and Perreault (1991). This means multi-collinearity should not be a serious concern in our regression. These results support the sound explanatory power and validity of an integral estimate.

Table 2.2 Parameter estimates for the logistic regression model

Independent variables	β	S.E.	Wald	*df*	Sig.	Exp (β)	VIF
Economic benefits	0.758	0.140	29.169	1	0.000	2.134	1.192
Policy and social norms	0.301	0.147	4.226	1	0.040	1.352	1.319
Past experience	0.221	0.125	3.133	1	0.077	1.247	1.284
Perceived inconvenience	−0.363	0.127	8.209	1	0.004	0.696	1.081
Social interaction	−0.098	0.135	0.531	1	0.466	0.906	1.357
House area	−0.177	0.080	4.851	1	0.028	0.838	1.024
Age	0.386	0.074	27.142	1	0.000	1.471	1.054
Information	0.304	0.136	5.027	1	0.025	1.356	1.251
Environmental awareness	0.189	0.141	1.794	1	0.180	1.208	1.401
Constant	−3.793	0.616	37.864	1	0.000	0.023	
−2 Log-likelihood	951.139				0.000		
Hosmer and Lemeshow	6.791				0.559		

Note Estimation terminated at iteration number 4 because parameter estimates changed by less than 0.001

In general, the variables having statistically significant impacts on the stated willingness to save more electricity seem to be more or less the same as the theoretical framework built. It indicates that economic benefits, policy and social norms, perceived inconvenience and information, are statistically significant at the 5% significant level, and past experience is significant at a 10% level. However, environmental awareness and social interaction were not verified in our model. Individual differences in the socioeconomic background (such as level of education, income and gender) were initially included in the logistic regression estimations, but most of them had no significant impact on the willingness to reduce electricity consumption. The only exceptions were age and dwelling area, both of which were estimated both significant at a 5% level.

The results of the regression analysis demonstrated that economic benefit is one of the main antecedents of residents' electricity-saving behavior in Beijing. With the great improvement of living standards for Beijing residents, various appliances flow into the families and the electricity consumption accounts larger share in residents' daily consumption. Electricity saving has become one of the main channels to decrease household daily expenses. According to our survey, 80.2% of respondents would not accept the increasing electricity price, and respondents that recognized the financial pressure from electricity costs were also more reluctant to undertake further measures to save electricity.

Our results also support the hypothesis that policies and social norms play a significant role in promoting the electricity-saving behavior of residents in Beijing. The Chinese government has attached great importance to electricity saving and implemented many measures to promote residents' energy-saving behavior. For example, some subsidies for energy-saving appliances purchasing were put into effect to encourage electricity saving. About 57.5% of respondents said that they would choose electricity-saving products, on the condition of being subsidized by the government.

Moreover, people holding more information on electricity saving and with better knowledge of policies regarding electricity use are more willing to participate in electricity saving than those without. This is because this specific knowledge presupposes a high awareness of methods for efficient electricity saving. Adequate information normally results from the frequent promotion of energy saving and environment protection by media.

Furthermore, the residual effect was verified as a positive influence on residents' electricity-saving behavior. Respondents with past experience of energy-saving practices or electricity shortage are more willing to participate in electricity-saving activities. Because of the combination of rapidly rising industrial demand for electricity and high household power consumption, shortfalls in electricity supply are often announced in Beijing. Many respondents have experienced curtailed electricity use through rolling brownouts initiated by the government. As expected, electricity shortage has a positive correlation with residents' willingness to save electricity. More frequent electricity shortages might raise residents' awareness of the need to save electricity. This is consistent with our prior expectations and previous research studies (Longo et al. 2008; Carlsson and Martinsson 2008).

Inconvenience and discomfort caused by electricity saving have a significantly negative effect on respondents' willingness to reduce unnecessary electricity use. This observation points to the limitation of electricity-saving infrastructure and poor technologies in electricity efficiency projects in Beijing. It is worth mentioning that all these disadvantages have erected visible obstacles to residents' daily electricity-saving activities. According to our research, the stronger the discomfort the respondents feel when they take part in electricity saving, the less willing they are to save electricity.

Contrary to many previous research studies (e.g., Sardianou 2007; Linden and Carlsson-Kanyama 2006; Viklund 2004), the logistic regression results provide limited support for the environmental awareness. As is supposed in the theoretical frame, environment awareness might not be easily translated into direct pro-environment behavior. There is still a big gap in China between electricity-saving awareness and real action. Also, social interactions and knowledge of others' behavior in electricity saving were not significant in Beijing. One possible explanation might be that it is difficult to identify other people's electricity behavior, and the accelerating pace of life reduces the opportunities for residents to communicate with each other.

For the demographic variables, seniority in age exerts a positive effect on electricity saving while spaciousness of a dwelling works as a negative influence. The old respondents, especially the retired people, report a high willingness to reduce unnecessary electricity use, while respondents living in larger dwellings are less willing to take part in electricity saving. One plausible explanation for the former result is that old people experienced more electricity shortfalls, especially in the 1970s, a time when conservative energy behavior was fostered in China. This induced constant saving habits among Chinese households. For the latter result, the cause is that respondents who live in large dwellings are well off. The opportunity cost of spending time in electricity-saving activities is perceived high by those residents, because the time for them is a scarcer resource, better to be spent enjoying a convenient life.

2.1.4.2 Influencing Effect Analysis and Policy Implications

The results hereby presented demonstrate that household electricity-saving behavior lacks sufficient encouragement in Beijing, because the policy to address this issue is so far limited to financial incentives, infrastructure construction, and public acceptance. The design and implementation of any electricity saving policy should first identify the barriers to household electricity conservation. It follows that further policy measures have to be conducted with the determinants examined above taken into account.

First, more educational campaigns and social propaganda of energy crisis and household electricity-saving skills should be initiated with more government support. Meanwhile, excessive administrative interventions in the electricity market need to be phased out. This study points to the lack of awareness of the energy shortage among Beijing residents. Most residents perceive little burden imposed in their daily lives by the electricity shortage, a problem largely due to policies that ensure Beijing's precedence over other regions in energy use. As we know, the neighboring provinces (autonomous regions, municipalities) (e.g., Hebei and Shanxi province) provide more than 65% of electricity supply for Beijing. It is sometimes the case that administrative directives by Chinese central government even require these provinces (autonomous regions, municipalities) to meet the electricity demand of Beijing at the expense of their own electricity needs, especially during important activities (e.g., 2008 Beijing Olympic Games) or during the period of power consumption peak (e.g., Spring Festival). Fewer residents in Beijing, especially the young people, have ever experienced the electricity brownouts, which points to the necessity to strengthen residents' consciousness of electricity shortfall in an effort to improve household electricity efficiency. At the same time, more market measures should replace present mandatory directives in order to balance the demand and supply of electricity in Beijing. The price variation following the change in electricity demand and supply under market mechanisms can intensify residents' awareness of electricity scarcity and motivate their electricity-saving behavior.

Moreover, policies applied for enhancement of residents' electricity-saving awareness should be combined with cost saving and utility improvement approaches. Many researches increasingly highlight the residents' environmental awareness as a key factor to promote electricity-saving behavior, without an emphasis on how to transform such awareness into actual conservation behavior. As shown in the current study, however, the role of environmental awareness in electricity saving is very limited. There may be a strong consciousness of energy scarcity and environmental degradation in Beijing households. However, a significant inertia among Beijing residents in their electricity saving behavior is observed. This finding uncovers a challenge in translating environmental awareness into electricity-saving action. An important reason for this is that the lack of technologically feasible "cost-effective" approaches for reducing electricity use and other notable barriers such as management, retraining time and capital constraints are overlooked, while environment awareness is promoted. Inconveniences and discomfort arising from electricity-saving were not effectively mitigated. This impedes the residents from transforming their environment awareness into electricity-saving actions. Therefore, it is necessary to further encourage the improvement of electricity-saving technology with more policy support, and replace the backward household electricity infrastructure with a more energy-efficient one.

Furthermore, policy-makers should pay more attention to the increasing electricity consumption of residents living in big dwellings. As shown in this research, a small dwelling area could be beneficial for household electricity saving. However, the dwelling area of newly constructed dwellings in Beijing shows little indication of being smaller. Even the newly completed economically affordable dwellings (EAH), which provide housing for low-income residents in Beijing, reached 101.44 m^2 each on average in 2008 (Beijing Bureau of Statistics of China 2009). Given this reality, it is necessary to push forward tiered pricing for household electricity in Beijing. Two or three hierarchies with an incremental unit price for electricity in each hierarchy could be drawn up, excluding the basic household demand. The extra charge would be levied if the amount of electricity consumption exceeds the standard for each hierarchy. Residents, who pay more marginal cost of electricity due to their large dwellings, in turn have to pay more attention to their daily electricity consumption. Above all, policy tools with respect to household electricity saving would not only focus on residents' environmental awareness, but also incorporate financial benefits (e.g., financial subsidies and tax preference) and technological upgrading (e.g., to make the usage of electricity-efficient products more convenient). It is argued that there exists a relatively high awareness among some residents about the benefits associated with electricity conservation, but few awareness-related actions are implemented. Such a situation demonstrates that more attention should be paid to encourage the household to undertake saving measures. Wide spread information about electricity saving should also be part of effective policy to stimulate substantial reductions in household electricity consumption.

2.2 Rebound Effect of Residential Electricity Consumption

In order to reduce energy demand and CO_2 emissions, all regions and departments are seeking ways to improve energy efficiency. Many scholars also argue that enhancing energy efficiency can reduce energy consumption (von Weizsäcker et al. 1997; World Energy Council 2008; Bosseboeuf et al. 1997; Patterson 1996). However, can the improved energy efficiency reduce energy consumption? In reality, Chinese energy consumption still kept growing from 1996 to 2010, though energy efficiency was continuously improved during this period.

In fact, the rebound effect may arise due to the ignorance of the market reaction to energy efficiency, and thus the energy efficiency policies are not as effective as people expected. The rebound effect means that the improvement of energy efficiency does not necessarily reduce energy consumption and sometimes even makes energy consumption increase. As we know, the improved energy efficiency gives rise not only to a decrease in energy consumption, but also to a reduction in energy prices. Nevertheless, lower price will increase energy consumption to some extent, and thereby cause a rebound effect. The existence of this phenomenon triggers a sustained boom in the study on rebound effects. Many scholars have investigated the reasons for rebound effects in detail and put forward the corresponding policy and suggestion (Brookes 1978; Khazzoom 1980; Bentzen 2004). As China has been the biggest energy consumer and an important CO_2 emission country, most of the relevant literature has begun to investigate causality between energy consumption and economic growth as well as energy demand forecast. However, these studies overemphasized energy efficiency improvement, as a result of technological advancement and ignored other implications from the improvement, such as the rebound effect, which lead policy-makers to go on a tangent from the main topic. This research will empirically deal with the issue of the energy rebound effect.

Although the urban household accounted for about 10% of energy consumption during 1996–2010 in China, electricity use increased from 11% in 1996 to 26% in 2010. This rising trend will continue as people's living standards and quality of life gradually improve. Therefore, it is necessary to investigate the rebound effect of electricity consumption in urban China. This research will use the 1996–2010 China's 30 provincial government panel data to build a cointegration equation and a panel error correction model to analyze the direct rebound effect of urban residential electricity use. Finally, based on the empirical analysis, this research will also discuss ways to reduce energy consumption in theory.

2.2.1 Rebound Effect and Economic Mechanism

2.2.1.1 Definition of Direct Rebound Effect

On the basis of single energy service or a single-department energy service, there exist several prevalent definitions of the direct rebound effect. In order to quantitatively analyze and estimate the direct rebound effects of residential energy consumption, it is necessary to identify a feasible definition. Similarly, there are some proxy variables that can be used to measure the rebound effect, the practicality of which depends on certain conditions (Mizobuchi 2008). These definitions are summarized as follows.

Mizobuchi graphically illustrated direct rebound effects in Fig. 2.5.

$$\begin{aligned}\text{rebound effect (RE)} &= \frac{\text{rebound consumption}}{\text{expected savings}} \times 100\% \\ &= \frac{\text{expected savings} - \text{actual savings}}{\text{expected savings}} \times 100\% \\ &= \frac{E_2 - E_1}{E_0 - E_1} \times 100\% \\ &= 1 - \frac{E_0 - E_2}{E_0 - E_1} \times 100\%\end{aligned}$$

In Fig. 2.5, ε_0 and ε_1 ($\varepsilon_0 < \varepsilon_1$) represent different energy efficiency levels of energy service. RE of 10% means that 10% of the expected savings is offset by the increased consumption. Accordingly, the rebound effect is defined as follows (Haas and Biermayr 2000).

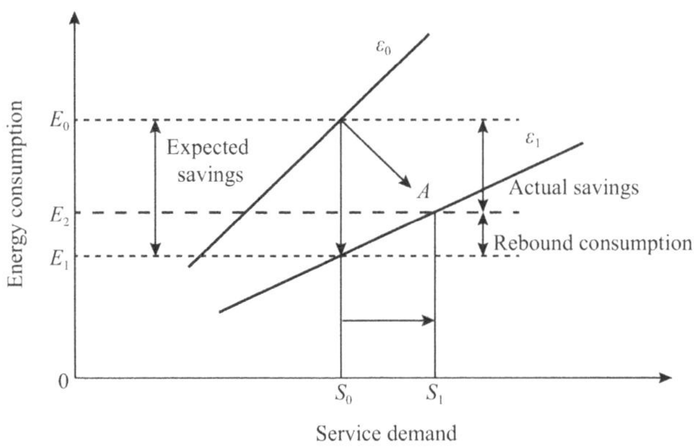

Fig. 2.5 Graphic of the direct rebound effect

By comparing energy demand before and after the improvement of energy efficiency, this method can estimate the change in energy consumption. Since the changes in other factors may influence the energy consumption demand, we have to control other variables during the calculation. However, it is always very difficult to control those variables in practice. Consequently, the results of this approach are potentially biased.

Another method stems from the household production function proposed by Becker (1965), in which the energy service is formulated as a function of energy consumption, time, capital, and other inputs, assuming that the effectiveness of individual family comes from the energy service, such as the comfortable space temperature. Based on this framework, Wirl proposed the economic definition of energy efficiency: $\varepsilon = S/E$, in which ε is defined as the ratio of useful energy service to energy inputs (Wirl 2009). That is, if the energy efficiency increases, the energy consumption will decrease. So, we get the price of the energy service: $P_S = P_E/\varepsilon$. By means of the above concepts, many scholars have presented the most common definition of rebound effect as follows (Brookes 1978; Berkhout et al. 2000):

$$\mathrm{RE} = \eta_\varepsilon(S) = 1 + \eta_\varepsilon(E) \tag{2.3}$$

where $\eta_\varepsilon(E)$ denotes the efficiency elasticity of energy demand, and $\eta_\varepsilon(S)$ denotes the efficiency elasticity of energy service. The energy saving caused by energy efficiency improvement is effective only when the efficiency elasticity of energy demand service is zero, that is, the efficiency elasticity of energy demand is equal to minus one. A positive rebound effect implies that $\eta_\varepsilon(S) > 0, |\eta_\varepsilon(E)| < 1$.

Saunders has done some specifications for the previous studies, and defined the rebound effect as follows (Saunders 1992, 2005):

If RE > 1, the rebound effect is called the backfire effect;

if RE = 1, the rebound effect is called the full rebound effect;

if RE < 1, the rebound effect is called the partial rebound effect;

if RE = 0, the rebound effect is called the zero rebound effect;

if RE < 0, the rebound effect is called the super conservation effect.

Since it is difficult to calculate ε, the energy rebound effect is often estimated from the price elasticity of energy service (Berkhout et al. 2000). That is:

$$\mathrm{RE} = \eta_\varepsilon(S) = -\eta_{P_S}(S) \tag{2.4}$$

Some other scholars have estimated the rebound effect in this way which is easier to implement than Eq. (2.3) (Khazzoom 1980; Binswanger 2001; Greene et al. 1999). However, this definition is mainly based on two hypotheses:

(1) Symmetry: consumers respond in the same way to energy price decline and energy efficiency improvement.

(2) Exogeneity: energy prices' change can't affect energy efficiency. That is,

$$\eta_{P_E}(\varepsilon) = \frac{\partial \ln \varepsilon}{\partial \ln P_E} = 0$$

In this method, symmetry plays a critical role. The direct changes in price have a greater effect on the improvement of production efficiency. When the time series is stationary, the decline in energy prices will not affect energy efficiency. However, the rising energy prices will promote technology progress. Exogeneity means that energy efficiency may affect energy costs, and vice versa. This can be proved by the empirical analysis of the cointegration relationship through the introduction of instrumental variables in the simulation model.

Due to data accessibility, Eq. (2.4) is employed more widely than Eq. (2.3) to estimate the direct rebound effect. Similarly, data on energy consumption are more available and more accurate than data on energy services. Since $P_S = P_E/\varepsilon$, if energy efficiency ε is regarded as a constant, then the definition of the rebound effect can be achieved by transforming the price elasticity of energy consumption (Saunders 2005; Sorrell et al. 2009; Frondel et al. 2008). That is,

$$\mathrm{RE} = \eta_{\varepsilon}(S) = -\eta_{P_E}(E) \tag{2.5}$$

Analogous to Eq. (2.4), Eq. (2.5) is also based on the symmetry and exogeneity hypotheses. That is, in order to get the rebound effect of residential energy consumption, we only need to calculate the elasticity of energy consumption with respect to the price of the energy.

Khazzoom argued that the empirical analysis of the direct rebound effect would be limited to microeconomic level (Khazzoom 1980). At this level, the technology progress will reduce energy consumption, and in turn, the rising price of energy will affect energy efficiency. Many economists use the price elasticity of energy consumption as instrumental variables to estimate the rebound effect. Haas and Biermayr argued that when the energy efficiency was unavailable, the price elasticity of energy consumption could be calculated by constant elastic demand dynamic standard function in the double-logarithmic form (Haas and Biermayr 2000). Due to the availability and accuracy of data, this research tends to use the price elasticity of energy consumption to estimate the energy rebound effect.

2.2.1.2 Economic Mechanism

Energy rebound effect is used to describe the paradox between energy consumption and energy efficiency. Based on utility theory, Jevons first proposed this issue in his research "The coal question", in which the author presented the idea that it was ridiculous to reduce fuel consumption through using it economically and instead the opposite was true. This is the so-called "Jevons' paradox".

The rebound effect raised widespread concern after the first oil crisis in the 1970s. Earlier research and debates about energy conservation focused on the fields of economics, environment, energy, engineering, etc. Specifically, since engineering technical experts argued that improving energy efficiency could significantly reduce energy consumption, energy conservationists and environmentalists claimed that improving energy efficiency was an effective approach to reduce energy consumption, and an energy efficiency improvement could reduce energy expenditure more or less. Therefore, energy conservation should rely mainly on the efforts of science and technology. However, more scholars took the opposite view. Khazzoom pointed out that improvements in energy efficiency could lead to an increase in the energy service demand (Khazzoom 1980). Therefore, the actual reduction in energy consumption would not change in proportion to the reduction in energy consumption per unit of energy service. Brookes (1978) argued that energy efficiency would lead to economic growth which would sequentially increase energy consumption. This is the famous Khazzoom–Brookes Postulate, that is, when energy price remains unchanged, energy efficiency improvements caused by technological advantages will increase rather than reduce the energy consumption.

Based on the perfect market mechanism hypothesis, especially the energy market mechanism, Greening et al. divided the rebound effect mechanism into the following four main categories (Greening et al. 2000):

(1) Direct effects: improved energy efficiency will decrease the effective price of energy service and therefore lead to an increase in consumption of that service. This will offset the reduction of energy consumption caused by advanced efficiency. The direct effect is mainly reflected in the substitution effect and income effect.
(2) Economy-wide effects or market-clearing price and quantity adjustments: the energy rebound effect on the overall economy.
(3) Transformation effects: technology progress will change consumer preferences, reform the social system and reconstruct the production organization.
(4) Secondary effects: the impact on other production/service. Energy efficiency improvement will reduce the cost of the manufacturing sector; therefore, the prices of their products will come down, which will promote consumption. So, the manufacturing sector will increase energy demand.

Transformation effects and secondary effects are collectively called the indirect rebound effects. Due to the complexity of indirect and economy-wide effects, most of the literature studied the direct effect. For the same reason, this research will study the direct rebound effect based on residential energy consumption.

2.2.2 *Methodology and Data for Residential Electricity Consumption*

2.2.2.1 Econometric Model and Variables

This research aims to estimate the long-term and short-term rebound effects of Chinese urban residential electricity consumption. Based on the work by Haas and Biermayr (2000), we add the population as a new variable P_{it} into the econometric model. The revised econometric model is written as follows:

$$\ln E_{it} = \alpha + \beta_1 \ln I_{it} + \beta_2 \ln P_{E_{it}} + \beta_3 \ln P_{it} + \beta_4 \ln \mathrm{DD}_{it} + \mu_{it} \tag{2.6}$$

in which α is a constant, $\beta_1 - \beta_4$ are the parameters to be estimated, and μ_{it} represents the random error term. A simple explanation of the variables in the model equation is given as follows.

E_{it}: the explanatory variable, indicating the electricity consumption of the province i in year t of the urban residents (unit: kW h).

I_{it}: per capita disposable real income of province i in year t of the urban residents (unit: yuan). It is calculated through numerical manipulation in which the nominal per capita disposable income value is reduced according to the urban consumer price index (1995 as the base period) of the corresponding year in the region and the regional price difference. Price difference is described by the purchasing power parity of all the provinces (autonomous regions, municipalities) of China in the period from 1996 to 2010, which refers to Brandt and Holz's research results of the regional purchasing power parity of China in 2000 (Brandt and Holz 2006). Generally, the real per capita disposable income reflects the residents' living standard in the region. The improvement of living standard has brought an increase in the number of new families' power consuming appliances and the rise of household electricity consumption. On the other hand, it promotes the process of replacing the old electric equipment with low energy efficiency and greatly reduces the residents' electricity consumption.

$P_{E_{it}}$: residential electricity prices of province i in year t. The residential electricity price is calculated according to the urban consumer price index, and the purchasing power parity of Chinese provinces (autonomous regions, municipalities) derived by following the logic of Brandt and Holz (2006). Because it is difficult to obtain residential electricity prices in various regions of China, this study selects the civil electricity price of all capital cities as a replacement. There is much more substitutability existing in household energy consumption. Intuitively, the rise of the residential electricity price has a negative impact on residential electricity consumption.

P_{it}: the population of province i in year t. Since the human being is the main body of energy consumption, the greater the population is, the greater the need for electrical equipment and electricity consumption is.

DD_{it}: the degree day value of province i in year t. Due to the diversities of latitude and geographical conditions in various regions of China, the climate is very different, which has a large effect on the electricity consumption of a region especially in winter and summer. Generally, the regions with a larger degree day value have a bigger demand for electricity consumption. In the early 1950s, Thom explored the relationship between energy consumption and temperature using the degree day method for the first time (Thom 1952). The degree day of a day refers to the actual deviation between daily average temperature and the prescribed base temperature. The degree day can be divided into two types, the heating degree day and the cooling degree day (Kadioğlu and Şen 1999). The annual heating degree days refers to the quantity of accumulated temperature when the daily average temperature is lower than the base temperature in a year. The annual cooling degree days refers to the quantity of accumulated temperature when the daily average temperature is higher than the base temperature in a year. The corresponding equation is given by

$$\begin{cases} \mathrm{HDD} = \sum\limits_{m=1}^{12} (1 - \mathrm{rd})(T_{b1} - T_m) \times M \\ \mathrm{CDD} = \sum\limits_{m=1}^{12} \mathrm{rd}(T_m - T_{b2}) \times M \end{cases} \tag{2.7}$$

in which HDD is the value of heating degree days of a particular year; CDD is the value of cooling degree days of a particular year; n is the number of days of a particular year; T_i represents the daily average temperature; T_{b1} is the base temperature of the heating degree days; T_{b2} is the base temperature of the cooling degree days; rd is a variable which indicates the daily average temperature. if the daily average temperature is higher than the base temperature, then rd = 1, otherwise, rd = 0 (Sailor 2001). Because this method demands a heavy workload and the required data are not readily available, the study improves the method proposed by Wayne by calculating the degree days of a year as follows:

$$\begin{cases} \mathrm{HDD} = \sum\limits_{i=1}^{n} (1 - \mathrm{rd})(T_{b1} - T_i) \\ \mathrm{CDD} = \sum\limits_{i=1}^{n} \mathrm{rd}(T_i - T_{b2}) \end{cases} \tag{2.8}$$

in which T_m represents the monthly average temperature; m is the number of days in a month; rd = 1, if the monthly average temperature is higher than the base temperature, and otherwise, rd = 0. The degree days' value in a particular year is: DD = HDD + CDD.

Because it is hard to acquire the data of daily average temperature in various regions of China, this study selects the temperatures in the capital cities as a replacement. By doing so, besides the availability of the data, this method considers the other two reasons: (1) within the same provincial unit, latitude, and geographical conditions are usually similar, and climate differences and temperature changes are small; (2) capital city is usually the largest city of each provincial unit, and has the largest population.

When we use the above method to estimate price elasticity, the premise is a price decline. However, the price is fluctuating. In order to solve this problem, Dargay and Gately put forward a method which can reflect the rise and fall of energy prices (Dargay 1992; Gately 1993), which the price change is broken down into three parts: $P_{E_{it}}^{\max}$ (the highest price in history), $P_{E_{it}}^{\mathrm{cut}}$ (prices fall), and $P_{E_{it}}^{\mathrm{rec}}$ (price recovery). In order to use this method in the logarithmic function, Haas and Biermayr (2000) made some changes from Dargay and Gately's decomposition method, and the pertinent definition is as follows:

$$P_{E_{it}} = P_{E_{it}}^{\max} \times P_{E_{it}}^{\mathrm{cut}} \times P_{E_{it}}^{\mathrm{rec}} \tag{2.9}$$

where: $P_{E_{it}}^{\max} = \max\{P_{E_{i1}}, P_{E_{i2}}, \ldots, P_{E_{it}}\}$

$$P_{E_{it}}^{\mathrm{cut}} = \prod_{m=0}^{t} \min\{1, \frac{P_{E_{im-1}}^{\max}/P_{E_{im-1}}}{P_{E_{im}}^{\max}/P_{E_{im}}}\}$$

$$P_{E_{it}}^{\mathrm{rec}} = \prod_{m=0}^{t} \max\{1, \frac{P_{E_{im-1}}^{\max}/P_{E_{im-1}}}{P_{E_{im}}^{\max}/P_{E_{im}}}\}$$

Taking logarithmic on both sides of Eq. (2.9) yields the following:

$$\ln P_{E_{it}} = \ln P_{E_{it}}^{\max} + \ln P_{E_{it}}^{\mathrm{cut}} + \ln P_{E_{it}}^{\mathrm{rec}} \tag{2.10}$$

Inserting Eq. (2.10) into Eq. (2.6) gives

$$\begin{aligned} \ln E_{it} = \alpha + \beta_1 \ln I_{it} + \beta_2^{\max} \ln P_{E_{it}}^{\max} + \beta_2^{\mathrm{cut}} \ln P_{E_{it}}^{\mathrm{cut}} + \beta_2^{\mathrm{rec}} \ln P_{E_{it}}^{\mathrm{rec}} \\ + \beta_3 \ln P_{it} + \beta_4 \ln \mathrm{DD}_{it} + \mu_{it} \end{aligned}, \tag{2.11}$$

In which β_2^{cut} (the coefficient of $\ln P_{E_{it}}^{\mathrm{cut}}$) stands for the long-term rebound effect.

2.2.2.2 Data Sources

Under the framework of Eq. (2.11), we use panel data to estimate the direct rebound effect of urban residential electricity consumption. Considering the consistency and availability of statistical data, Xizang, Taiwan, Hong Kong, and Macao are not included in this study. The data include the urban residential electricity consumption, the per capita disposable income of urban residents, the average temperature of winter and summer, the residential electricity price and urban population.

Table 2.3 Descriptive statistics of the variables

Indicators	$\ln E$	$\ln I$	$\ln P_E^{max}$	$\ln P_E^{cut}$	$\ln P_E^{rec}$	$\ln P$	$\ln DD$
Mean	21.972	8.927	−0.870	−0.110	0.063	16.394	6.240
Media	22.061	8.907	−0.865	−0.081	0.042	16.479	6.316
Maximum	24.232	9.943	−0.562	0.000	0.256	17.966	7.895
Minimum	19.035	8.107	−1.534	−0.494	0.000	14.158	2.518
Std. Dev.	0.943	0.412	0.184	0.096	0.060	0.778	1.014
Skewness	−0.581	0.186	−0.608	−1.164	1.267	−0.920	−1.502
Kurtosis	3.284	2.282	2.986	4.304	4.195	3.845	5.989
Observations	450	450	450	450	450	450	450
Cross sections	30	30	30	30	30	30	30

Specifically, the data of the per capita disposable income of urban residents and the average temperature of winter and summer in various provinces (autonomous regions, municipalities) are collected from China Statistical Yearbooks (1997–2011) (NBSC, 1997a, 1998b, 1999a, 2000c, 2001c, 2002c, 2003c, 2004c, 2005c, 2006c, 2007c, 2008c, 2009c, 2010c, 2011c). The data of the residential electricity price in various provinces (autonomous regions, municipalities) are collected from China Price Yearbooks (1997–2011) (NBSC, 1997b, 1998c, 1999b, 2000d, 2001d, 2002d, 2003d, 2004d, 2005d, 2006d, 2007d, 2008d, 2009d, 2010d, 2011d). China Energy Statistical Yearbooks (1997–2011) (NBSC, 1997c, 1998a, 1999c, 2000a, 2001a, 2002a, 2003a, 2004a, 2005a, 2006a, 2007a, 2008a, 2009a, 2010a, 2011a) provides the data of the urban residential electricity consumption, and the data of urban population in various provinces (autonomous regions, municipalities) come from China Population Statistical Yearbooks (1997–2011) (NBSC, 1997d, 1998d, 1999d, 2000e, 2001e, 2002e, 2003e, 2004e, 2005e, 2006e, 2007e, 2008e, 2009e, 2010e, 2011e). We described the variables of this study in Table 2.3.

2.2.3 *Econometric Analysis of Residential Electricity Consumption*

2.2.3.1 Panel Unit Root Test

There are many methods for the panel unit root test and each of them has its own uniqueness. Therefore, it is difficult to get the same conclusion with different test methods. To guarantee the results' robustness and improve the credibility of the conclusion, this research uses LLC, IPS, and Fisher-type methods to test the panel unit root of each variable (Results are shown in Table 2.4). Except for the $\ln P_E^{max}$ and $\ln DD$, the other variables do not refuse the null hypothesis of the existing panel unit root, but there does not exist panel unit roots for the first-order difference of all variables. The result of the three-unit root tests show that all variables of the regression model are integrated order 1.

Table 2.4 Results of panel unit root tests

Variables	LLC test	IPS test	Fisher-type test	
			ADF-Fisher	PP-Fisher
$\ln E$	2.49782 (0.9938)	7.10510 (1.0000)	19.7053 (1.0000)	25.2762 (1.0000)
$\Delta \ln E$	−17.1743 (0.0000)	−15.3238 (0.0000)	306.603 (0.0000)	407.616 (0.0000)
$\ln I$	3.90261 (1.0000)	10.6106 (1.0000)	8.13534 (1.0000)	14.7752 (1.0000)
$\Delta \ln I$	−12.3338 (0.0000)	−10.4638 (0.0000)	212.986 (0.0000)	268.107 (0.0000)
$\ln P_E^{\max}$	−2.65676 (0.0039)	−10.3790 (0.0000)	150.293 (0.0000)	195.449 (0.0000)
$\Delta \ln P_E^{\max}$	−9.03794 (0.0000)	−13.0427 (0.0000)	206.791 (0.0000)	186.239 (0.0000)
$\ln P_E^{\text{cut}}$	6.41257 (1.0000)	11.0828 (1.0000)	8.48944 (1.0000)	11.2667 (1.0000)
$\Delta \ln P_E^{\text{cut}}$	−13.8670 (0.0000)	−9.80560 (0.0000)	212.933 (0.0000)	253.711 (0.0000)
$\ln P_E^{\text{rec}}$	−5.29377 (0.0000)	0.73324 (0.7683)	51.634 (0.7706)	84.0129 (0.0221)
$\Delta \ln P_E^{\text{rec}}$	−18.6206 (0.0000)	−13.5418 (0.0000)	265.528 (0.0000)	310.433 (0.0000)
$\ln P$	−2.73878 (0.0031)	5.22259 (1.0000)	38.6405 (0.9855)	69.0028 (0.1993)
$\Delta \ln P$	−9.08932 (0.0000)	−7.56879 (0.0000)	166.179 (0.0000)	164.466 (0.0000)
ln DD	−12.1265 (0.0000)	−11.3941 (0.0000)	241.233 (0.0000)	299.799 (0.0000)
Δ ln DD	−22.2490 (0.0000)	−20.4152 (0.0000)	395.457 (0.0000)	640.295 (0.0000)

Notes Δ represents the first-order difference; the value in the brackets is P value

2.2.3.2 Panel Cointegration Test

In applied economics, if the regression models are not cointegrated, the parameter estimates and corresponding test statistics would be biased and inconsistent (Bilgili et al. 2011). From the panel unit root tests in Table 2.4, all variables in the model are single integrated order series, fulfill the requirements of the panel cointegration test and can continue the panel cointegration test. Panel PP, Panel ADF, Group PP, and Group ADF refuse the null hypothesis of the existence of no cointegration relationship at the 5% level, while Panel v, Panel rho and Group rho can't refuse the null hypothesis, see Table 2.5. Considering the sample period in this empirical study was only 15 years, we argue that there exist cointegration relationships among the variables. The Kao test also refuses the null hypothesis of no cointegration (Kao 1999), which further supports the conclusion of the existence of

Table 2.5 Results of panel cointegration tests

Tests	Statistics	With no trend	With trend
Pedroni test	Panel v-statistic	−2.541749 (0.9945)	−4.775044 (1.0000)
	Panel rho-statistic	4.865800 (1.0000)	5.647109 (1.0000)
	Panel PP-statistic	−10.88599 (0.0000)	−9.423157 (0.0000)
	Panel ADF-statistic	−10.66998 (0.0000)	−6.905713 (0.0000)
	Group rho-statistic	7.539132 (1.0000)	
	Group PP-statistic	−19.45867 (0.0000)	
	Group ADF-statistic	−8.516581 (0.0000)	
Kao test	ADF-statistic	−4.614029 (0.0000)	

cointegration relationships among the variables. Then, this study takes the Engle and Granger two-step method to estimate the long-term equilibrium equation (the cointegration equation) (Liang and Gao 2007). According to correlated random effects (Table 2.6) and redundant fixed effects test (Table 2.7), we argue that the cross-sectional fixed effect model is the optimal estimation model, so we do a cross-sectional fixed effect regression. Finally, through the least-squares estimation, we get the long-term equilibrium equation. The results of cross-section fixed effects regression are shown in Table 2.8.

According to the value of adjusted *R*-squared and *F*-statistics, we can conclude the whole long-term cointegrating model fits very well. DW of 1.591550 also means there is no self-correlativity for residuals series. We further test the unit root

Table 2.6 Correlated random effects—Hausman test

Test summary	Chi-Sq. statistic	Chi-Sq. *df*	Prob.
Cross-section random effect	72.770794	6	0.0000

Table 2.7 Redundant fixed effects test

Tests	Statistic	*df*	Prob.
Cross-section *F*	13.298820	(29,414)	0.0000
Cross-section χ^2	296.247559	29	0.0000

Table 2.8 Estimation of long cointegration equation and unit root test of residual series

Estimation model				Unit root test of residual series		
	$\ln E_{it} = \alpha + \beta_1 \ln I_{it} + \beta_2^{\max} \ln P_{E_{it}}^{\max} + \beta_2^{\text{cut}} \ln P_{E_{it}}^{\text{cut}} + \beta_2^{\text{rec}} \ln P_{E_{it}}^{\text{rec}} + \beta_3 \ln P_{it} + \beta_4 \ln \text{DD}_{it} + \mu_{it}$					
Variable	Coefficient	t-Statistic	Prob.	Test method	Statistic	Prob.
$\ln I_{it}$	0.947757	20.40914	0.0000	LLC test	−4.745*	0.0000
$\ln P_{E_{it}}^{\max}$	−0.079360	−0.853581	0.3938	IPS test	−4.097*	0.0000
$\ln P_{E_{it}}^{\text{cut}}$	−0.741594	−5.009917	0.0000	ADF-Fisher test	119.251*	0.0000
$\ln P_{E_{it}}^{\text{rec}}$	0.670163	3.346687	0.0009	PP-Fisher test	124.998*	0.0000
$\ln P_{it}$	0.261263	4.574507	0.0000			
$\ln \text{DD}_{it}$	0.138698	5.669505	0.0000			
c	8.170993	9.977485	0.0000			
Adjusted R-squared	0.984765					
Durbin–Watson stat	1.591550					
F-statistic	830.2236					

Note * means refusing the null hypothesis of the existing panel unit roots

of the residuals series by the four kinds of testing methods, and all results refuse the null hypothesis of the existing panel unit root at the 5% level. This means that the origin of the residual series is stationary series, see Table 2.8. Therefore, the results from the Engle-Granger two-step test method also suggest that a cointegration relationship exists among the variables.

2.2.3.3 Testing the Exogeneity of the Price Variable

The microeconomic theory indicates that the price of any good is a function of the supply of this good in the marketplace (keeping demand constant). Furthermore, the demanded amount is determined by the market price and vice versa the good demanded also affects the market price. As a result, it is difficult to identify the direction of the causality for many goods, i.e., whether the demand affects the price or the price affects the demand. In such conditions, we can assume that the price is an endogenous explanatory variable, and thus we can use the Hausman test to demonstrate the exogeneity of the price in Eq. (2.6) (Hausman 1978). In this research we take the one-period lag logarithm of the price ($\ln P_{E_{it-1}}$) and its logarithm square ($\ln^2 P_{E_{it-1}}$) as the instrumental variables. Take the endogenous variable $\ln P_{E_{it}}$ as the dependent variable, and the other exogenous explanatory variables and instrumental variables of Eq. (2.6) as explanatory variables of the linear regression.

According to the test of correlation, the test of significance of linear regression and the test of significance of regression coefficients, the instrumental variables and endogenous variables are proved to be highly correlated (the value of the F-statistics is 93.08730). After finding the instrumental variables satisfying the dependency and the exogeneity, we make a linear regression for the other explanatory variables and the instrumental variables ($\ln P_{E_{it-1}}$, $\ln^2 P_{E_{it-1}}$). Then we extract the residual $\widehat{\mu}$, and do a new linear regression by putting $\widehat{\mu}$ into Eq. (2.6). The t value of $\widehat{\mu}$ is 3.842165 which is significant at the 5% level. Therefore, we confirm the price variable is endogenous (Wooldridge 2007).

2.2.3.4 Panel Error Correction Model (PECM)

The error correction model (ECM) is often used to estimate short-term elasticity. ECM is a specific econometrics model, which uses a long-term cointegration equation as an instrumental variable to solve the spurious regression problem. In this study, the short-term price elasticity of the ECM is the short-term energy rebound effect which we want to obtain.

According to the panel cointegration analysis, we find that there is a long-term equilibrium relationship between the dependent variable (urban residential electricity consumption in China) and each explanatory variable (e.g., the per capita disposable income of urban residents). In order to offset the deficiency of the long-term statistical model, we construct a short-term dynamic model to reflect the correction mechanism for the short-term equation deviating from the long-term one. So according to Eq. (2.11) and Table 2.8, we can obtain the residual series $\widehat{\mu_{it}}$ as follows:

$$\begin{aligned}\widehat{\mu_{it}} = \mathrm{ecm}_{it} = \ln E_{it} - \widehat{\alpha} - \widehat{\beta}_1 \ln I_{it} - \widehat{\beta_2^{\max}} \ln P_{E_{it}}^{\max} - \widehat{\beta_2^{\mathrm{cut}}} \ln P_{E_{it}}^{\mathrm{cut}} \\ - \widehat{\beta_2^{\mathrm{rec}}} \ln P_{E_{it}}^{\mathrm{rec}} - \widehat{\beta_3} \ln P_{it} - \widehat{\beta_4} \ln \mathrm{DD}_{it}\end{aligned},$$

and take it as an error correction. Then the short-term estimations are obtained through the error correction model (ECM) as depicted by Eq. (2.12):

$$\begin{aligned}\Delta \ln E_{it} = \gamma_1 \Delta \ln I_{it} + \gamma_2 \Delta \ln P_{E_{it}}^{\max} + \gamma_3 \Delta \ln P_{E_{it}}^{\mathrm{cut}} + \gamma_4 \Delta \ln P_{E_{it}}^{\mathrm{rec}} \\ + \gamma_5 \Delta \ln P_{it} + \gamma_6 \Delta \ln \mathrm{DD}_{it} + \gamma_7 \Delta \ln E_{it-1} + \gamma \mathrm{ecm}_{it-1} + \varepsilon_{it}\end{aligned} \quad (2.12)$$

in which the value of γ_3 (coefficient of $\Delta \ln P_{E_{it}}^{\mathrm{cut}}$) is the short-term rebound effect, and ε_{it} is random error. Equation (2.12) shows that the short-term volatility of urban residential electricity consumption in China not only depends on various factors' short-term changes but also is influenced by the deviation from long-term equilibrium in the previous period (ecm_{it-1}). In addition, the difference series reflects the volatility of the variable; for example, $\Delta \ln I_{it}$ shows the volatility of the per capita disposable income of urban residents, and $\Delta \ln P_{E_{it}}$ shows the volatility of the household electricity price. The coefficient of difference series means short-term elasticity.

The long-term cointegration panel model estimated with ordinary least-squares (OLS) method is

$$\ln E_{it} = 8.17 + 0.95 \ln I_{it} - 0.08 \ln P_{E_{it}}^{\max} - 0.74 \ln P_{E_{it}}^{\mathrm{cut}} + 0.67 \ln P_{E_{it}}^{\mathrm{rec}} + 0.26 \ln P_{it} + 0.14 \ln \mathrm{DD}_{it} + \mu_{it}.$$

After establishing the model, we can use the error correction model to estimate the short-term rebound effect of urban residential electricity use in China. We firstly do a cointegration regression by using $\ln E_{it}$, $\ln I_{it}$, $\ln P_{E_{it}}^{\max}$, $\ln P_{E_{it}}^{\mathrm{cut}}$, $\ln P_{E_{it}}^{\mathrm{rec}}$, $\ln P_{it}$, and $\ln \mathrm{DD}_{it}$, and then take the stationary residual error series ecm_{it} as the error correction series establishing the error correction model as follows (Table 2.9):

$$\Delta \ln E_{it} = 0.87\Delta \ln I_{it} + 0.34\Delta \ln P_{E_{it}}^{\max} - 0.72\Delta \ln P_{E_{it}}^{\mathrm{cut}} + 0.16\Delta \ln P_{E_{it}}^{\mathrm{rec}} + 0.19\Delta \ln P_{it} + 0.08\Delta \ln \mathrm{DD}_{it} + 0.03\Delta \ln E_{i,t-1} - 0.40\mathrm{ecm}_{it-1} + \varepsilon_{it}$$

The coefficient value of ecm_{it-1} shows the speed of adjustment to reach the cointegration equilibrium at the current period. The coefficient value of ecm_{it-1} is −0.40. The adjustment value has the statistical significance at the 1% level. These results lead to estimations of the direct rebound effect of 72% in the short term and of 74% in the long term for all energy services that use electricity in households. An increase in the energy efficiency of these energy services potentially would bring about savings in electricity consumption of 10. This actually produces savings of 2.8 in the short term and of 2.6 in the long term. The results are more than 0 but less

Table 2.9 Results of the panel error correction model estimation

Variable	Coefficient	Std. error	t-Statistic	Prob.
$\Delta \ln I_{it}$	0.870367	0.077794	11.18808	0.0000
$\Delta \ln P_{E_{it}}^{\max}$	0.338731	0.104247	3.249313	0.0013
$\Delta \ln P_{E_{it}}^{\mathrm{cut}}$	−0.724889	0.171297	−4.231771	0.0000
$\Delta \ln P_{E_{it}}^{\mathrm{rec}}$	0.161346	0.214104	0.753587	0.4516
$\Delta \ln P_{it}$	0.187654	0.077922	2.408241	0.0165
$\Delta \ln \mathrm{DD}_{it}$	0.080002	0.013760	5.813943	0.0000
$\Delta \ln E_{i,t-1}$	0.031320	0.043956	0.712531	0.4766
ecm_{it-1}	−0.401446	0.040157	−9.996833	0.0000
Weighted statistics				
R^2	0.188561	Mean dependent var	0.170509	
Adjusted R^2	0.173691	S.D. dependent var	0.177143	
S.E. of regression	0.146741	Sum squared resid	8.225615	
F-statistic	3.673759	Durbin–Watson stat	2.049597	
Prob (F-statistic)	0.000000			

than 1, which means that urban residential electricity use in China exists a partial rebound effect.

2.2.3.5 Direct Rebound Effect Analysis and Policy Implication

The main objective of this research is to provide an estimation of the direct rebound effect for all energy services that consume electricity in urban households in China. Our empirical results reveal the existence of direct rebound for residential electricity use in urban China, with its long-term rebound effect being 0.74 and short-term rebound effect being 0.72. There are no backfire effects, but only partial rebound effects. This means that, though the expected results were not achieved when urban residents improved the efficiency of electricity consumption, it can still be a correct policy. If there was no energy efficiency improvement, more energy would be spent. From the energy-saving viewpoint, since the technology and policy of improving energy efficiency are not as effective as we expected in theory, they cannot always be regarded as the only means to implement energy saving or solve the energy problems. Therefore, we cannot take technological progress as the only approach to improve energy efficiency, to achieve results in energy saving, or to solve energy problems. In addition, the empirical results indicate that, regardless of long term or short term, the elasticity of energy consumption with respect to price drop is greater than that with respect to the price rise, which is totally different from the results from Dargay and Gately (1995) and Haas and Schipper (1998). This could be caused by the rapid growth of China's economy, the rise in urban disposable incomes, and the cheaper electricity in China. People hope to get a more comfortable way of living through electricity consumption, and the frequent use of home appliances will increase. This may lead people to respond to the electricity price to decline more quickly and effectively than the electricity price rise.

The direct rebound effect in this research may be overestimated or underestimated. Overestimation is due to two reasons: One is the exogenous hypothesis of energy efficiency (i.e., energy efficiency is unaffected by energy prices). However, according to symmetry, consumers react in the same way to the decline in the energy electricity price as to the improvement of energy efficiency. This hypothesis is necessary for estimating energy direct rebound effect through energy price elasticity of energy demand. On the other hand, since China is in the high-speed economic development stage, the acceleration of urbanization and residents' disposable income will speed up the rising demand for energy, which may offset the energy saving from the improvement of energy efficiency. Underestimation can be attributed to the relationship between the direct rebound effect and capital cost. If the cost of new efficient equipment is lower than that of the old inefficient one, it may enlarge the rebound effect (Dimitropoulos and Sorrell 2006).

To sum up, the magnitude of the rebound effect should be an alterative criterion for the efficiency estimation of a single energy service. Compared with developed countries, the direct rebound effect of urban residential electricity use in China is significantly higher. The main reason is that China is in the accelerating urbanization process and upgrading resident consumption stage. When the government makes energy policy, attention should be placed not only on the increase in energy efficiency, but also on the technology subsidies to the enterprises improving the household appliance products' energy efficiency, which will help cut residential energy consumption and make energy conservation and emission reduction more fruitful. Additionally, the Chinese government should get rid of the obstruction from middle-income earners to implement the tiered electricity price (TEP) reform and make consumers fully aware of the electricity cost and the scarcity of available resources (Wang et al. 2012). Only in this way can we jump out of the "prisoner's dilemma" in the situation of low present energy price and low energy efficiency, and lead the lifestyle in a sustainable direction. Moreover, the relationship between energy prices controlled by the government, the external cost of energy consumption and the residential energy consumption needs to be discussed in further research.

2.3 Summary

This chapter focuses on the features and determinants of electricity consumption. The research sets out to explore the possibilities for further savings in household electricity consumption and employee electricity consumption through empirical studies of the people's willingness to save electricity and behavioral characteristics in electricity saving, as applied within a Chinese context. Moreover, the direct rebound effect of urban residential electricity consumption is measured, evaluating the implementation effect of energy efficiency improvement project, and offering references for policy-making.

From the household electricity consumption perspective, we use logit regression model to analyze residents' electricity consumption behaviors through a sample of 816 randomly selected residents in Beijing. We identify that economic benefits, policy, social norms, and past experience affect electricity-saving behavior positively, while the discomfort caused by electricity-saving activities affects electricity-saving behavior negatively. The results provide references for designing future informative policy measures in dwelling electricity saving. Both economic motives and technologically feasible approaches (conducted to, for instance, avoid the inconvenience caused by electricity saving) are vital for the reduction of unnecessary electricity use in Beijing. Educational campaigns which put a stronger emphasis on disseminating information about electricity-saving measures to residents should be promoted and initiated by interested organizations, authorities, and residential communities. Furthermore, although there is a comparatively high environmental awareness among Beijing residents, an effective policy management system coordinating with the dissemination of energy-saving information and financial incentives needs to be constructed to translate the pro-environment consciousness into the electricity-saving action.

Though lots of energy policies and energy-efficient technologies for residential electricity consumption are produced, the rebound effect caused by technology progress negatively affects the effectiveness of energy efficiency policies. This research empirically investigates the direct rebound effect of urban residential electricity use in China. With China's 30 provincial government panel data from 1996 to 2010, we build a cointegration equation and a panel error correction model to analyze the direct rebound effect. The results indicate that there exists an obvious rebound effect in the Chinese urban residential electricity consumption. Specifically, the long-term rebound effect is 0.74, while the short-term rebound effect is 0.72. Therefore, the rebound effect significantly impairs the functioning of energy efficiency policies. For this reason, the Chinese government should take the rebound effect into consideration when formulating energy policies.

References

Abrahamse, W., and L. Steg. 2009. How do socio-demographic and psychological factors relate to households' direct and indirect energy use and savings? *Journal of Economic Psychology* 30 (5): 711–720.

Ajzen, I. 1991. The theory of planned behavior. *Organizational Behavior and Human Decision Processes* 50 (2): 179–211.

Andrews-Speed, P. 2009. China's ongoing energy efficiency drive: Origins progress and prospects. *Energy Policy* 37: 1331–1344.

Bamberg, S., and P. Schmidt. 2003. Incentives, morality, or habit? Predicting students' car use for university routes with models of Ajzen, Schwartz, and Triandis. *Environment and Behavior* 35 (2): 264–285.

Banfi, S., M. Farsi, M. Filippini, et al. 2008. Willingness to pay for energy-saving measures in residential buildings. *Energy Economics* 30 (2): 503–516.

Becker, G.S. 1965. A theory of the allocation of time. *Economic Journal* 75 (299): 493–517.

Becker, L.J., C. Seligman, R.H. Fazio, et al. 1981. Relating attitudes to residential energy use. *Environment and Behavior* 13 (5): 590–609.

Beijing Bureau of Statistics of China. 2009. *Beijing statistical yearbook 2009*. China Statistical Yearbook Press.

Bentzen, J. 2004. Estimating the rebound effect in U.S manufacturing energy consumption. *Energy Economics* 26: 123–134.

Berkhout, P.H.G., J.C. Muskens, and J.W. Velthuijsen. 2000. Defining the rebound effect. *Energy Policy* 28 (6–7): 425–432.

Bilgili, F., Y. Pamuk, and N.S.H. Tülüce. 2011. Short run and long run dynamics of residential electricity consumption: Homogeneous and heterogeneous panel estimations for OECD. *Economic Computation and Economic Cybernetics Studies and Research* 3: 113–126.

Binswanger, M. 2001. Technological progress and sustainable development: What about the rebound effect? *Ecological Economics* 36 (1): 119–132.

Borsch-Supan, A. 1990. On the compatibility of nested logit models with utility maximization. *Journal of Econometrics* 43 (3): 373–388.

Bosseboeuf, D., B. Chateau, and B. Lapillonne. 1997. Cross-country comparison on energy efficiency indicators: The on-going European effort towards a common methodology. *Energy Policy* 25 (7–9): 673–682.

Brandt, L., and C.A. Holz. 2006. Spatial price differences in China: Estimates and implications. *Economic Development and Cultural Change* 55 (1): 43–86.

Brookes, L.G. 1978. Energy policy, the energy price fallacy and the role of nuclear energy in the U.K. *Energy Policy* 6 (2): 94–106.
Bryman, A., and E. Bell. 2007. *Business research methods*, revised ed. Oxford: Oxford University Press.
Carlsson, F., and P. Martinsson. 2008. Does it matter when a power outage occurs? A choice experiment study on the willingness to pay to avoid power outages. *Energy Economics* 30 (3): 1232–1245.
Dargay, J.M. 1992. *Are price and income elasticities of demand constant? The UK experience*. Oxford: Oxford Institute for Energy Studies.
Dargay, J.M., and D. Gately. 1995. The imperfect price irreversibility of non-transportation of all demand in the OECD. *Energy Economics* 17 (1): 59–71.
Dimitropoulos, J., and S. Sorrell. 2006. The rebound effect: Microeconomic definitions, extensions and limitations. In *Proceedings of the 29th IAEE international conference*, Potsdam, Germany.
Ek, K., and P. Söderholm. 2010. The devil is in the details: Household electricity saving behavior and the role of information. *Energy Policy* 38 (3): 1578–1587.
Feng, D.S., B.K. Sovacool, and Minh K. Vu. 2010. The barriers to energy efficiency in China: Assessing household electricity savings and consumer behavior in Liaoning Province. *Energy Policy* 38 (2): 1202–1209.
Frondel, M., J. Peters, and C. Vance. 2008. Identifying the rebound: Evidence from a German household panel. *The Energy Journal* 29: 145–163.
Gately, D. 1993. The imperfect price reversibility of world oil demand. *Energy Journal* 14 (4): 163–182.
Gatersleben, B., L. Steg, and C. Vlek. 2002. Measurement and determinants of environmentally significant consumer behavior. *Environment and Behavior* 34: 335–362.
Greene, D.L., J.R. Kahn, and R.C. Gibson. 1999. Fuel economy rebound effect for U.S. household vehicles. *The Energy Journal* 20 (3): 1–31.
Greening, L.A., D.L. Greene, and C. Difiglio. 2000. Energy efficiency and consumption—The rebound effect—A survey. *Energy Policy* 28 (6–7): 389–401.
Haas, R., and P. Biermayr. 2000. The rebound effect for space heating: Empirical evidence from Austria. *Energy Policy* 28: 403–410.
Haas, R., and L. Schipper. 1998. Residential energy demand in OECD-countries and the role of irreversible efficiency improvements. *Energy Economics* 20 (4): 421–442.
Hansla, A., A. Gamble, A. Juliusson, et al. 2008. Psychological determinants of attitude towards and willingness to pay for green electricity. *Energy Policy* 36 (2): 768–774.
Harland, P., and H. Staats. 1999. Explaining pro-environmental intention and behavior by personal norms and the theory of planned behavior. *Journal of Applied Social Psychology* 29: 505–528.
Hausman, J. 1978. *Specification testes in econometrics. Econometrics* 46: 1251–1271.
Heath, Y., and R. Gifford. 2002. Extending the theory of planned behavior: Predicting the use of public transportation. *Journal of Applied Social Psychology* 32: 2154–2189.
Kadioğlu, M., and Z. Şen. 1999. Degree-day formulations and application in Turkey. *Journal of Applied Meteorology* 38 (6): 837–846.
Kao, C. 1999. Spurious regression and residual-based tests for cointegration in panel data. *Journal of Econometrics* 90: 1–44.
Khazzoom, J.D. 1980. Economic implications of mandated efficiency in standards for household appliances. *Energy Journal* 1 (4): 21–40.
Liang, Y.F., and T.M. Gao. 2007. Empirical analysis on real estate price fluctuation in different provinces of China. *Economic Research Journal* 8: 133–142.
Linden, A.L., and A. Carlsson-Kanyama. 2006. Efficient and inefficient aspects of residential energy behavior: What are the policy instrument for change? *Energy Policy* 34: 1927–1981.
Lindenberg, S., and L. Steg. 2007. Normative, gain and hedonic goal frames guiding environmental behavior. *Journal of Social Issues* 63: 117–137.
Longo, A., A. Markandya, and M. Petrucci. 2008. The internationalization of externalities in the production of electricity: Willingness to pay for the attributes of a policy for renewable energy. *Ecological Economics*, 67 (1): 140–152.

Lu, W. 2006. Potential energy savings and environmental impact by implementing energy efficiency standard for household refrigerators in China. *Energy Policy* 34: 1583–1589.

Mason, C., and W.D. Perreault. 1991. Collinearity, power, and interpretation of multiple regression analysis. *Journal of Marketing Research* 28 (3): 268–280.

McKenzie-Mohr, D., L.S. Nemiroff, L. Beers, et al. 1995. Determinants of responsible environmental behavior. *Journal of Social Issues* 51: 139–156.

Meijnders, A., C. Midden, and H. Wilke. 2001. Communications about environmental risks and risk-reducing behavior: The impact of fear on information processing 1. *Journal of Applied Social Psychology* 31 (4): 754–777.

Min, G.F., and E. Mills. 1997. Energy-efficient lighting in China: Problems and prospects. *Energy Policy* 25: 77–83.

Mizobuchi, K. 2008. An empirical study on the rebound effect considering capital costs. *Energy Economics* 30: 2486–2516.

Moll, H.C., K.J. Noorman, R. Kok, et al. 2005. Pursuing more sustainable consumption by analyzing household metabolism in European countries and cities. *Journal of Industrial Ecology* 9 (1): 259–276.

Murata, A., Y. Kondou, M. Hailin, et al. 2008. Electricity demand in the Chinese urban household-sector. *Applied Energy* 85 (12): 1113–1125.

NBSC. 2011. *China Energy Statistical Yearbook 2010*. Beijing: China Statistics Press.

Patterson, M.G. 1996. What is energy efficiency? Concepts, indicators and methodology issues. *Energy Policy* 24 (5): 377–390.

Price, L., and J. Sinton. 2002. Energy use and carbon dioxide emissions from steel production in China. *Energy Policy* 27: 429–446.

Saidur, R., H.H. Masjuki, and M.Y. Jamaluddin. 2007. An application of energy and exergy analysis in residential sector of Malaysia. *Energy Policy* 35: 1050–1063.

Sailor, D.J. 2001. Relating residential and commercial sector electricity loads to climate-evaluating state level sensitivities and vulnerabilities. *Energy* 26: 645–657.

Samuelson, C.D. 1990. Energy conservation: A social dilemma approach. *Social Behavior* 5: 207–230.

Sardianou, E. 2007. Estimating energy conservation patterns of Greek households. *Energy Policy* 35: 3778–3791.

Saunders, H.D. 1992. The Khazzoom-Brookes postulate and neoclassical growth. *The Energy Journal* 13 (4): 131.

Saunders, H.D. 2005. A calculator for energy consumption changes arising from new technologies. *Topics in Economic Analysis and Policy* 5 (1): 1–31.

Scarpa, R., and K. Willis. 2010. Willingness-to-pay for renewable energy: Primary and discretionary choice of British households' for micro-generation technologies. *Energy Economics* 32: 129–136.

Sorrell, S., J. Dimitropoulos, and M. Sommerville. 2009. Empirical estimates of the direct rebound effect: A review. *Energy Policy* 37: 1356–1371.

Stern, P.C., and S. Oskamp. 1987. Managing scarce environmental resources. In *Handbook of environmental psychology*, ed. D. Stokols, and I. Altman, 1043–1088. New York: Wiley.

Thom, H.C. 1952. Seasonal degree day statistics for the United States. *Monthly Weather Review* 80 (2): 143–149.

Viklund, M. 2004. Energy policy options—From the perspective of public attitudes and risk perceptions. *Energy Policy* 32: 1159–1171.

von Weizsäcker, E., A.B. Lovins, and L.H. Lovins. 1997. *Factor four: Doubling wealth, halving resource use*. London.

Wang, Z., B. Zhang, and Y. Zhang. 2012. Determinants of public acceptance of tiered electricity price reform in China: Evidence from four urban cities. *Applied Energy* 91 (1): 235–244.

Widen, J., M. Lundh, I. Vassileva, et al. 2009. Constructing load profiles for household electricity and hot water from time-use data—Modelling approach and validation. *Energy and Buildings* 41 (7): 753–768.

Wirl, F. 2009. *The economics of conservation programs*. London: Kluwer Academic Publishers.

Wooldridge, J.M. 2007. *Introductory econometrics: A modern approach*. Beijing: China Renmin University Press (in Chinese).

Zeng, S. 2005. *Chinese style management, 2*, 161–243. China Social Sciences Press (in Chinese).

Zografakis, N., E. Sifaki, M. Pagalou, et al. 2010. Assessment of public acceptance and willingness to pay for renewable energy sources in Crete. *Renewable and Sustainable Energy Reviews* 14 (3): 1088–1095.

Chapter 3
Low-Carbon Transportation for the Residential Sector in China

According to the report from IEA (2011), energy consumption from the transportation division has been growing by 9.3% per year from 1971 to 2010. To date, approximately 30% of global energy consumption and 25% of CO_2 emissions are from the transportation sector and energy consumption in this sector will double by 2050 (IEA 2011). Now, it is extremely urgent to slash down the transportation associated with CO_2 emissions. The process of urbanization and industrialization have been stepped up in a rapid development period due to the continuous development of the domestic economy, which consequently resulted in enhancing the number of trucks in China. As a result, the demand for transportation services and energy is rapidly increasing, especially petroleum product consumption.

3.1 Characteristics of CO_2 Emissions from Residential Transportation

Recently, there has been much interest in sustainable transport. However, few studies have examined the driving forces for household transportation emissions from the perspective of individual travel characteristics. This research examines the features and driving factors of CO_2 emissions from the household daily travel in Beijing from 2000 to 2012. It first investigates the changes in personal travel

---

This chapter takes the following literature for reference:

Wang Z, He W. 2017. CO_2 emission efficiency and marginal abatement costs of the regional transportation sectors in China. Transportation Research Part D: Transport and Environment, 50: 83–97.

Wang Z, Lu M. 2014. An empirical study of direct rebound effect for road freight transport in China. Applied Energy, 133(6): 274–281.

Wang Z, Liu W. 2015. Determinants of CO_2 emissions from household daily travel in Beijing, China: individual travel characteristic perspectives. Applied Energy, 158: 292–299.

Z. Wang and B. Zhang, *Low-Carbon Consumption in China: Residential Behavior, Corporate Practices and Policy Implication*,
https://doi.org/10.1007/978-981-15-2792-0_3

characteristics and CO_2 emissions, and then discusses the effects of population, economic activity, transport capacity, vehicle emission intensity, and the individual travel characteristic, which includes the effects of transport intensity, the transport mode share, and vehicle-use intensity on CO_2 emissions based on the decomposition analysis.

3.1.1 Environmental Concerns Raised by the Transport Sector

Global concentrations of carbon dioxide, methane, and nitrous oxide—three of the most notable greenhouse gases—have increased significantly over the past 250 years due to a direct result of human activities (IEA 2011). More and more people realized the need to adopt a more sustainable lifestyle to reduce the consumption of natural resources and the emission of pollutants. Many countries have established sustainability-related performance standards aiming at reducing the energy use and CO_2 emissions, such as promoting the energy efficiency in household appliances, the effectiveness of home insulation, the fuel economy in motor vehicles. Since the economic reforms of the late 1970s, China has been transformed into an economic giant due to the rapid economic growth, which has dramatically improved the living standard of the netizens (Tao and Yu 2011). However, rapid economic growth in China caused intensive investments in the development of manufacturing and heavy industries, which led to the rapid growth in energy consumption and CO_2 emission. The population is also experiencing a dramatic shift in lifestyle, as well as a significant increment in the demand for energy. Therefore, it is essential for China to promote sustainable consumption patterns to reduce energy consumption and CO_2 emissions, while meeting the domestic demand for energy consumption.

The transport sector plays a curial role in daily activities around the world. However, transport has been considered as a major contributor to greenhouse gas (GHG) emissions, which accounts for more than 60% of the global oil consumption and 23% of the CO_2 emissions in 2007 (Saboori et al. 2014). With the acceleration of urbanization and increasing levels of motorization in China, CO_2 emissions from the transport sector have dramatically increased in the past few years, especially in big cities, such as Beijing. According to the recent studies, the term of stainable transportation includes three aspects: (1) offering safe and efficient transport modes to meet residents' mobility needs; (2) using renewable resources and energy to minimize the pollutant and waste; (3) promoting equity within and between successive generations (Connolly et al. 2014; Liu et al. 2013; Mathiesen et al. 2008).

Recently, several studies analyzed the energy consumption and emissions reduction patterns in the transport sector (Brand et al. 2013; Becky and Linna 2012; Pongthanaisawan and Sorapipatana 2013). Some studies focused on a specific country or region (Ong et al. 2012; Al-Ghandoor 2013). For example, Brantley examined the consumption-driven environmental impacts by the transport sector in OECD countries (Liddle 2012). Mraihi et al. (2013) analyzed the driving factors behind the energy consumption change for road-based modes in Tunisian cities. There are also many studies analyzing the energy consumption and CO_2 emissions by the transport sector in the context of China (Lin and Xie 2014; Wang and Lu 2014; Cai et al. 2012). Furthermore, some researchers paid attention to different transport modes (Zhang et al. 2011). Chèze et al. (2013) researched the emissions from air transportation. Zhang et al. (2014) measured fuel consumption and CO_2 emissions for urban buses and provided scientific support for China's national fuel economy standard for buses to be established in the future. Hoffrichter et al. (2012) calculated the energy efficiency and CO_2 emissions for railway vehicles on a well-to-wheel basis. Okada assessed road transport emissions in Japan (Okada 2012). As for research methods, there are five kinds of basic methods: firstly the bottom-up sector-based analysis (Morán and González 2007; Zondag and de Jong 2011; Duffy and Crawford 2013; Sider et al. 2013); secondly, quantitative methods (Saide et al. 2009; Roxana et al. 2014; Poumanyvong et al. 2012; Wei et al. 2013); thirdly, econometric models (Török and Török 2014); fourthly, system optimization methods (Si et al. 2012; Kanzian et al. 2013; Szendro and Török 2012); and fifthly, the decomposition methods, which has been widely used to study the factors affecting the change of aggregate energy consumption and CO_2 emissions from the transport sector over time (Li et al. 2013; Andreoni and Galmarini 2012; Jennings et al. 2013; Phillip and Lee 2013; Brand et al. 2012; Chandran and Tang 2013). Two main decomposition techniques, namely, SDA and index decomposition analysis (IDA) are used to analyze these driving forces. There are similarities as well as dissimilarities between two approaches in terms of the study scope, the method formulation, and data requirements (Su and Ang 2012). For example, the SDA method is based on input–output coefficients and the final demand from an input–output framework, while the IDA approach uses an index number framework and requires sector-level data. The SDA method is often characterized by time periods of up to ten years since the input–output tables are not available annually for many countries. The IDA approach studies a yearly time period, because aggregated data at the sector level are often available (Cellura et al. 2012). Although there is no consensus among researchers regarding the preferred decomposition method between SDA and IDA, the IDA approach is more suitable for time series modeling in this study. There are three IDA methodologies: the Laspeyres index decomposition approach, the Arithmetic Mean Divisia Index (AMDI) method, and the LMDI approach. The LMDI approach, which was introduced by Ang (2005), has advantages in its theoretical foundation, adaptability, consistency of results decomposed by both multiplicative and additive methods, ease of use, and in the interpretation of its results: it also has the ability to perform a perfect decomposition and to accommodate zero values in the data set (Wang et al. 2015). Thus, this study constructs a decomposition analysis model by applying LMDI to investigate the major factors that may affect changes in Beijing's household daily travel CO_2 emissions.

In previous studies of CO_2 emissions from the transport sector, researchers mainly used decomposition methods to investigate factors influencing energy consumption and emissions (Jou and Chen 2014). For instance, Li et al. (2012) found that older people, especially older women, are more heavily dependent on car use than younger people are. Timilsina and Shrestha (2009) analyzed the effects of changes in the fuel mix, the modal shift, the economic growth, the emission coefficients, and the transportation energy intensity on the growth of the transport sector's CO_2 emissions in 20 Latin American and Caribbean countries. Chung et al. (2013) investigated the influence of the energy intensity, the regional shift, the energy mix, and transportation activity effects on the energy consumption and efficiency of China's transport sector during 2003–2009. These studies investigated the impact of different factors on the increased CO_2 emissions in the transport sector. However, few studies have examined the driving forces for household transport CO_2 emissions from the perspective of individual travel characteristics. Therefore, this study aims to fill this gap and develop a comprehensive picture of the driving forces behind changing CO_2 emissions, related to household daily travel, from a systemic point of view. Thus, taking Beijing as an example, this study first calculates the CO_2 emissions from household daily transportation from 2000 to 2012, and then constructs an SDA model to examine the main factors that influence the changes in emissions by using LMDI.

3.1.2 Methodology and Data

3.1.2.1 Estimation of CO_2 Emissions

To improve the air quality and mitigate the vehicle emissions in Beijing, we need to understand accurate emissions inventories and changing trends of emissions emitted by residents during their daily traveling. In general, two methods are used to calculate CO_2 emissions from the transport sector: the fuel-based method and the distance-based method (Greenhouse Gas Protocol 2005). In the fuel-based method, emissions are calculated by multiplying the fuel consumption by its CO_2 emission coefficient. This method is relatively authoritative but has difficulties in identifying the fuel type and fuel consumption of different transport modes. In addition, the transportation statistics in China mainly focus on operational transportation. Thus, this method is not suitable to estimate household daily travel emissions. In the distance-based method, emissions can be estimated by using distance-based emission factors and transport activity data, such as vehicle-kilometers or person-kilometers traveled by different vehicle types. Due to data limitations, this study uses an improved distance-based method to estimate CO_2 emissions during household daily travel in Beijing.

The urban passenger transport modes include public transport (the bus, the taxi, the subway, etc.), motorized transport (the privately owned vehicle, the motorcycle, the ferry, etc.) and non-motorized transport (bicycles and pedestrians that are deemed emission free). Here, we consider four main transport modes in Beijing: the taxi, the bus, the subway, and the private car. Hence, household daily travel CO_2 emissions can be calculated as follows:

$$C = \sum_i VKT_i \times EF_i \tag{3.1}$$

where C is household daily travel CO_2 emissions (in million tons, Mt), VKT_i is vehicle-kilometers traveled by different vehicle types (in km), EF_i is a distance-based emission factor (in g/km) and i represents each different passenger transport mode.

The vehicle-kilometers traveled by private car and subway are unavailable in the form of statistics. However, the annual average travel distances covered by private cars can be obtained from the survey launched by the former Beijing Transportation Research Centre (Beijing Municipal Committee of Transport 2007), and thus we can estimate the total vehicle-kilometers by multiplying it with the number of private cars. The vehicle-kilometers traveled by subway is calculated by multiplying the length of subway lines with the number of trains required for each line, that can be obtained from *Annual Reports on Development of Beijing Transport 2012*. As shown in Table 3.1, the distance-based emission factors in 2006 are summarized from several relevant studies such as the TREMOVE model work carried out by IIASA (International Institute for Applied Systems Analysis) (Herbruggen and Logghe 2005), the correlative study worked by Xiao et al. (2011), and comprehensive survey on transportation in Beijing conducted by Beijing Municipal Commission of Transport.

3.1.2.2 LMDI Decomposition Model

To define the driving forces of household daily travel CO_2 emissions, we choose demographic characteristics, economic characteristics, vehicle emission intensity characteristics, and individual travel characteristics which can be decomposed into transport intensity, transport mode shares, and vehicle-use intensity. According to the Kaya identity, the household daily travel CO_2 emissions in Beijing (C) can be decomposed into

Table 3.1 The distance-based CO_2 emission factor in 2006 (Unit: kg CO_2/km)

Emission factor	Private car	Taxi	Bus	Subway
CO_2 emission factor	0.23	0.25	1.07	2.11

$$C = \sum_i P \times \frac{INC}{P} \times \frac{TD}{INC} \times \frac{TD_i}{TD} \times \frac{RL_i}{TD_i} \times \frac{VP_i}{RL_i} \times \frac{C_i}{VP_i} \\ = \sum_i P \times I \times TI \times TS_i \times TC_i \times UI_i \times EI_i \tag{3.2}$$

where P is the population (in Million), and I is the per capita disposable income (in yuan). TI is the transport intensity (in person times/yuan), which is calculated as the passenger turnover volume (TD, in million person times) divided by the residential disposable income (INC, in million yuan), measuring the relative expenditure on daily travel. TS is the transport mode share (%), which refers to the ratio of the passenger turnover volume of transport mode i to the total passenger turnover volume, measuring the resident travel choice preference. TC is the transport capacity (in km/person times), which is calculated as the length of roads (RL, in Million km) divided by the passenger turnover volume, measuring the improvement of transportation infrastructure. UI is the vehicle-use intensity (in L/km), which is calculated as the vehicle population (VP, in Million) divided by the length of road (RL), measuring the usage efficiency of different transport modes. In addition, EI is the vehicle emission intensity (in tons), which is calculated as CO_2 emissions divided by the vehicle population, measuring the emission efficiency of different transport modes.

Then, following the LMDI method, the variation of C in the time period (0, T) can be decomposed as

$$\Delta C = C^T - C^0 = \Delta C_P + \Delta C_I + \Delta C_{TI} + \Delta C_{TS} + \Delta C_{TC} + \Delta C_{UI} + \Delta C_{EI} \\ = \sum WIn(\frac{P^T}{P^0}) + \sum WIn(\frac{I^T}{I^0}) + \sum WIn(\frac{TI^T}{TI^0}) + \sum WIn(\frac{TS^T}{TS^0}) \\ + \sum WIn(\frac{TC^T}{TC^0}) + \sum WIn(\frac{UI^T}{UI^0}) + \sum WIn(\frac{EI^T}{EI^0}) \tag{3.3}$$

where

$$W = \frac{P^T I^T TI^T TS^T TC^T UI^T EI^T - P^0 I^0 TI^0 TS^0 TC^0 UI^0 EI^0}{In(P^T I^T TI^T TS^T TC^T UI^T EI^T) - In(P^0 I^0 TI^0 TS^0 TC^0 UI^0 EI^0)}$$

Thus, changes in household daily travel CO_2 emissions in Beijing during the study period can be attributed to effects of: population scale (ΔC_P), economic activity (ΔC_I), transport intensity (ΔC_{TI}), transport mode share (ΔC_{TS}), transport capacity (ΔC_{TC}), vehicle-use intensity (ΔC_{UI}), and vehicle emission intensity effects (ΔC_{EI}).

3.1.2.3 Data Description

Passenger turnover volumes of different modes are estimated by multiplying the total passenger trip volume by the ratio of the four types of transport mode as collected from Annual Reports on Development of Beijing Transport 2012. The residential disposable income, population data, the length of roads, and the number of taxis, public buses, private cars as well as subway trains come from the Beijing Statistical Yearbooks for the appropriate period. However, due to the implementation of the end-number license plate policy in Beijing in October 2008, according to that private cars will be not allowed to enter inside of the fifth ring road zone for one day per week (except weekends) by means of grouping by the end number of the car license plate. Thus, the actual number of on-road private cars is a little less than the vehicle population reported by Statistical data. For convenience, we multiply the official data by 0.8 to get the number of private cars for decomposition in this study.

3.1.3 Household Daily Travel CO_2 Emissions in Beijing

We calculated the household daily travel CO_2 emissions by different transport modes in Beijing from 2000 to 2012. We chose this period before the shared bicycle and the shared traveling to avoid obscured results. Results show that household daily travel CO_2 emission continued to rise during the study period, which increased more than three times, from 4.34 Mt in 2000 to 18.58 Mt in 2012, and the annual growth rate was 13%. However, the growth rates in 2009 had slowed down due to the reduced daily trip frequency induced by the new odd–even traffic control measures launched in 2009. In addition, Beijing's government introduced hybrid electric–diesel buses in 2009, which produced much lower CO_2 emissions than traditional buses. Meanwhile, the CO_2 emissions from travel by subway and private car present the same increasing trend as total household daily travel CO_2 emissions, while emissions from taxis and buses take on a fluctuation, yet displaying generally increasing trend.

In terms of different transport modes, emissions from private car transport have increased dramatically, with an annual growth rate of 15%. The growth in emissions from taxis and the growth in emissions from buses are relatively slow in the same study period, with an annual growth rate of 7% and 5%, respectively. Due to the rapid construction and development of the Beijing subway in recent years, the emissions from subway travel increased from 0.017 Mt in 2000 to 0.13 Mt in 2012, with an annual growth rate of 18%. Public transport in Beijing has developed significantly from 2010 to 2012. For example, the subway in Beijing is the oldest metro system in China and has undergone rapid expansion since 2002. By the end of 2019, Beijing has 23 lines, 405 stations, and 699.3 km of track in operation, making it the second-longest subway system in the world after Shanghai. The rapid development of the Beijing Subway has resulted in a significant increase in

CO_2 emissions generated from the power consumption of train traction and subway station systems.

Results also show that private car transport is the biggest CO_2 emitter. The number of private cars in Beijing has grown quickly as urbanization and modernization continue to upsurge: this has caused an increase in passenger transport CO_2 emissions as well as severe congestion in some urban areas, especially at peak traffic flow hours. In 2012, approximately 585,500 new cars were sold in Beijing, which increased the city's total number of private cars to more than 3.97 million: the emissions from private cars reached 15.59 Mt, with the proportion increasing from 65% in 2000 to 84% in 2012. Private car transport is the dominant factor in the growth of CO_2 emissions from household daily travel in Beijing.

3.1.4 Driving Factors Behind Household Daily Travel CO_2 Emissions

By applying Eqs. (3.2) and (3.3), we analyze the driving factors affecting changes in household daily travel CO_2 emissions. The results are presented in Table 3.2, which shows that four factors exhibit positive effects on the growth of emissions during the study period: economic activity effect, vehicle-use intensity effect, population scale effect, and transport capacity effect. On the contrary, the transport intensity effect, vehicle emission intensity effect, and transport mode share effect contribute to the decreased CO_2 emissions. The accumulated effect meant that the CO_2 emissions increased by 14.24 Mt. According to the relative contributions of these factors, the economic activity effect is the major driver behind the growth of household daily travel CO_2 emission, whose contribution reached 85.18%. While the transport intensity effect has a significantly negative effect on emissions, and its contribution reached −46.21%. The following sections examine the impacts of these effects on household daily travel CO_2 emissions in Beijing.

Table 3.2 Factor decomposition for household daily travel CO_2 emission during 2000–2012

Driving factors	Changes in driving factors/%	Contribution value/Mt	Contribution ratio/%
ΔC_P	51.71	4.02	28.23
ΔC_I	252.37	12.13	85.18
ΔC_{TI}	−49.50	−6.58	−46.21
ΔC_{TS}	40.79	−0.88	−6.18
ΔC_{TC}	14.48	1.22	8.57
ΔC_{UI}	61.72	5.61	39.40
ΔC_{EI}	25.71	−1.28	−8.99
Total		14.24	100

3.1.4.1 Economic Activity Effect

Decomposition results show that the economic activity effect is the leading factor contributing to the increase in household daily travel CO_2 emissions in Beijing. It causes the emissions to increase by 12.13 Mt in the study period, accounting for 85.18% of overall emissions. The per capita disposable income factor reflects the quality of life as well as economic growth. In Beijing, the per capita disposable income has increased almost three folds from 10,300 yuan in 2000 to 36,500 yuan in 2012.

As their economic status improves, people with a higher living standard will pursue a higher quality of life, so they might shift from public transportation to faster and more comfortable traveling modes, such as private cars and taxis. In addition, residential transportation demand is closely linked to economic growth. High-income households tend to own more private cars and thus use each of these vehicles slightly more extensively than low-income households. Moreover many people consider private cars as a status symbol. In general, higher levels of car ownership imply higher travel rates, increased car use, and increased travel distances. Further, higher incomes generate greater demand for social connectivity, which entails a higher frequency of leisure travels. Therefore, increased incomes have inevitably led to excessive demand for private cars, which resulted in the rapid growth of CO_2 emissions.

3.1.4.2 Vehicle-Use Intensity Effect

The vehicle-use intensity effect is the second most important factor leading to the growth of household daily travel CO_2 emissions. The accumulated effect is an increase of 5.61 Mt, which accounts for 39.40% of the total emissions change. Among the four transport modes, the most significant effect on the growth of emissions comes from the private car use, which reaches 4.95 Mt, while taxi and bus use decrease emissions by 0.59 Mt and 0.19 Mt, respectively.

The change of vehicle-use intensity to some extent indicates the situation of the overall vehicle population. From the perspective of the vehicle population for each transport mode, the number of buses and the number of subway vehicles have increased by 4% and 17%, respectively. However, the number of private cars in Beijing has increased more than six folds in this study period, which caused a significant change in residential transport behavior. For one thing, the increased number of private cars makes people choose a private car instead of a public transport mode. For example, the share of travel by private cars out of the total passenger traffic volume in Beijing increased from 23% in 2000 to 39% in 2012. In addition, behavioral change influences not only the choices of transport modes, but also their usage efficiency. With the expansion of residents' ability to own private cars, the load factor for private car journeys in Beijing declined from 1.57 in 2000 to 1.2 in 2012, which reflects the decrease of vehicle usage efficiency in the past decades. The explosive growth in the number of private cars and increasing transportation demand can be considered as the main factors, which have a positive effect on the growth of household daily travel CO_2 emissions. Moreover, increasing travel time and travel distance has reinforced the growth of CO_2 emissions. During the period of 2000–2005, the average travel distance per journey increased from 8 km to 9.3 km, and the average travel time grew at a rate of 5%. Thus, in order to inhibit the rapid growth of daily travel CO_2 emissions, it is necessary to take measures to restrain the excessive growth in private cars as well as promote sustainable consumption patterns in a broader sense.

3.1.4.3 Population Scale Effect

The population scale effect is another factor, which increases Beijing's household daily travel CO_2 emissions from 2000 to 2012. The contribution of population effect reached 4.02 Mt, which accounts for 28.23% of the total increased emissions. During the study period, the population of Beijing increased from 13.64 million in 2000 to 20.69 million in 2012, with an annual growth rate of 3.5%. The population expansion with this pace will enhance the demand for transport accordingly. In addition, demographic change also accompanies the growth of urbanization. The newly urbanized population has a higher demand for passenger transport, which will lead to increased energy consumption and CO_2 emissions.

3.1.4.4 Transport Capacity Effect

The transport capacity effect also increases the household daily travel CO_2 emissions in Beijing: the contribution of this effect reached 1.22 Mt, which accounts for 8.57% of the total increased emissions. In recent years, Beijing has made a huge investment in the development of an intelligent transport system for public transportation. Despite that, traffic is the main problem in Beijing. To relieve traffic congestion, one of the direct solutions is to expand the capacity of road, such as adding more lanes, creating new routes. Therefore, the length of roads in Beijing has also increased dramatically during 2000 to 2012. The total operating mileage of buses and subways increased by 154% and 71% respectively, and the length of road in Beijing has increased more than seven folds in this study period. There is clear evidence that new or expanded roads can ease the traffic flows only for short term. As new road network capacity is taken up by the steady traffic growth, benefits from reduced congestion and shorter commute time are very limited. The expansion of the road network made it able to compete with other transport modes from the perspective of infrastructure spatial distribution. It is increasingly convenient to use the taxi and the private car for a trip, thus the average distance and travel times of road transportation is increasing. Our data show that the increasing degree of passenger turnover volume by private car, bus, and subway is higher than that of the length of roads. The net effect in combination with the new road is generally a considerable overall increase in traffic. The longer the road is, the more CO_2 is emitted.

3.1.4.5 Transport Intensity

The transport intensity effect displayed the leading factor contributing to decreasing the household daily travel CO_2 emissions in Beijing from 2000 to 2012. The accumulated effect of transport intensity effect displayed a decrease of 6.58 Mt, which accounts for 46.21% of the total change in emissions over the period. The transport intensity in Beijing presents a significant decrease during 2000–2012, which indicates that the relative expenditure on daily travel decreased with the increasing income. It is mainly due to various policy instruments introduced by Beijing's Government, such as Public Transport Priority, reduced public transport ticket prices, transfer benefits between rail transit and bus transit, which reduce the cost of traveling by public transportation and play an important role in achieving sustainable transport.

3.1.4.6 Vehicle Emission Intensity Effect

As shown in Table 3.2, the vehicle emission intensity effect plays a minor role in increasing the household daily travel CO_2 emissions. The accumulated effect displayed the 1.28 Mt decrease in emissions, which accounts for 8.99% of the total change in emissions over the period. This may be attributed to various effective measures and policies targeting emission intensity. For example, Beijing implemented a "license control" policy for light-duty gasoline vehicles (LDGVs), which will decrease CO_2 emissions of new LDGVs and improve their average speed. In addition, Beijing will impose increasingly stringent emission standards for new vehicles, adopt a series of traffic measures to improve driving conditions, further tighten the license control policy, and promote alternative fuel vehicles to address traffic problems. The effect of vehicle emission intensity on decreasing CO_2 emissions was much less than expected. Thus, simply lowering vehicle emission intensity is insufficient for reducing passenger transport emissions. It requires the integration of internal and external means that emphasize the economic, regulatory, and fiscal incentives, as well as changes in individual travel characteristics and technological innovation for reducing emissions.

3.1.4.7 Transport Mode Share Effect

Contrary to the results of other studies (Wang et al. 2011), we found that the transport mode share to some extent contributed to reducing the emissions, although its effect was very limited. The accumulated effect displayed a decrease of 0.88 Mt in emissions, and its contribution was only 6.18%. This may be attributed to the mode shifting, from more energy consumption intensive modes to the greener modes. According to the available data, the annual growth rate for share traveling by subway reached to 14% during the study period (from 3.6% in 2000 to 16.8% in 2012), which ranks highest among the other four transport modes. Although the subway has the highest CO_2 emission per hundred kilometers, its annual CO_2 emissions and CO_2 emissions per capita are far less than the other types of public transport mode. The reasons for this are as follows: firstly, the number of annual vehicle-kilometers traveled by subway is the lowest among other four transport modes during the study period; secondly, the subway has a large load factor so that the high CO_2 emissions per kilometer can be apportioned to each passenger.

3.1.5 Implications

By applying an LMDI method, CO_2 emissions due to passengers' daily travel are decomposed into the population scale, economic activity, transport intensity, transport mode share, transport capacity, vehicle-use intensity, and vehicle emission intensity effects. The decomposition results show that the economic activity effect explains most of the growth in CO_2 emissions, which is similar to the results of other studies (Becky and Linna 2012). Moreover, the vehicle-use intensity, especially the use intensity of private cars, is another main factor driving the growth of emission. The population and transport capacity effect also cause a rise in emissions. The results also identify that the transport intensity effect is primarily responsible for reducing the CO_2 emissions in Beijing. The vehicle emission intensity effect and transport mode share effect exhibit minor negative roles over the study period. However, the transport mode share effect plays a minor role over the study period, with attribution rates of 9% and 6%, respectively.

Firstly, results show that the growth of household daily travel CO_2 emissions is mainly due to private car travel: a series of effective policies that can result in the reduction of private car travel should urgently be implemented. Suggestions include imposing increasingly stringent emissions standard for new vehicles, further tightening the license control policy, and promoting alternative fuel vehicles. In addition, the government should improve the quality of mass transit by adding more bus stops, improving maintenance and adopting more air-conditioned buses to encourage people to drive less and make more use of public transport. Meanwhile, transport-pricing measures, such as the congestion charging, the fuel tax, and the vehicle purchase price tax, are another way of restraining the escalating demand for private cars (Mraihi et al. 2013).

The results also show that traveling by subway is an effective way to reduce traffic congestion and pollution in Beijing. However, with fare reductions and new lines drawing more riders to the network, Beijing's subway has experienced severe overcrowding in recent years. Though Beijing's subway has expanded rapidly since 2002, the existing network still cannot meet the city's mass transit needs adequately. The subway lines in Beijing, such as Lines 1, 13, and Batong are officially overloading, especially during the rush hour. The government can carry out segmented pricing strategies to promote its use by passengers whose travel distances are less than 10 km to transfer to the bus to reduce the risk of the large passenger flow in Beijing. Moreover, for alleviating peak passenger flow pressure, a time-sharing system, such as providing concessionary fares for passengers who travel during non-peak hours, can be proposed. Other modes of transportation, such as the biking, light-rail, and tram should be promoted to supplement the crowded subway systems.

According to the decomposition results, the effort to improve the fuel economy is almost offset by increasing vehicle-use intensity. The technological innovations alone cannot solve the problems related to the emissions reduction and climate change: emphasis should also be given on changing the individual behavior. Therefore, it is essential to take comprehensive measures to encourage the residents to adopt the behavioral changes introduced by the government.

3.2 Emission Efficiency and Marginal Abatement Cost for Transportation

Nowadays, the evaluation of CO_2 emission efficiency and its marginal abatement cost in the transportation sectors is a hot topic. However, while evaluating the CO_2 marginal abatement cost using the data envelopment analysis (DEA) approach, the weak disposability of CO_2 may imply the positive abatement cost, which undoubtedly violates our common sense. In order to obtain the nonpositive marginal abatement cost, CO_2 emissions should be treated as an input.

3.2.1 Productivity, CO_2 Emission Efficiency and Abatement Costs of China's Transport Sector

As the second-largest economy in the world, China has also become the largest energy consumer and carbon dioxide emitter in the world (IEA 2010). It has been proposed that by 2020 the CO_2 emissions per unit gross domestic product (GDP) in China should be reduced by 40–45% as that was in 2005. In addition, the proportion of nonfossil fuels consumed should be increased to 11.4% as was in 2015, and by 2015, the CO_2 emissions per unit GDP should be reduced by 17% compared to the 2010 figures. These policies have created much interest in exploring the renewable energy development and analyzing energy and environmental efficiency and CO_2 marginal abatement cost (Cong 2013; Cong and Shen 2014; Wu et al. 2014; Wang et al. 2013; Wang and Feng 2015; Zhou et al. 2010; Guo et al. 2010). For example, Cong and Shen (2014) explored how to develop renewable power in China. Wu et al. (2014) and He et al. (2013) concentrated on analyzing China's energy and environmental efficiency at the regional and industrial levels. However, few studies investigated the energy and environmental efficiency of regional transportation sectors in China.

Working as an important part of the service sectors, the transportation sectors emitted about 20% of the global CO_2 emissions in 2012. For developed countries like the United States, Japan, and others, their proportion of CO_2 emissions account for more than 20%. For China, the CO_2 emissions of the transportation sectors also account for 7–9% of total emissions during 2007–2012, while the added value accounts for 5% of GDP. (The data are calculated according to Word Bank, http://data.worldbank.org/?display=default.) In addition, the CO_2 emission intensity of China's transportation sectors is higher than the whole country.

3.2.2 *Methodology and Data*

3.2.2.1 Global Environmental Reference Technology

We suppose there are J DMUs in actual production. Each DMU employs the same N inputs, denoted as $x^t = (x_1^t, x_2^t, \ldots, x_N^t) \in \Re_+^N$ at time t, to produce M desirable outputs $y^t = (y_1^t, y_2^t, \ldots, y_M^t) \in \Re_+^M$, and Q undesirable outputs, $u^t = (u_1^t, u_2^t, \ldots, u_Q^t) \in \Re_+^Q$. The relationship between inputs and outputs at time t can be described as

$$P^t(x) = \left\{ (x^t, y^t, u^t) \in \Re_+^{N+M+Q} \mid x^t \in \Re_+^N \text{ can produce } (y^t, u^t) \in \Re_+^M \times \Re_+^Q \right\} \tag{3.4}$$

where $P^t(x)$ meets the convexity and cone conditions and, its inputs and desirable outputs meet strong disposability, which implies that for arbitrary $x' \geq x$, we have $(y, b) \in P^t(x')$, and for any $y' \leq y$, $(y', b) \in P^t(x)$ is also established. However, the undesirable outputs are weakly disposable (Färe and Grosskopf 2004), that is,

Weak disposability:

$$\textit{If } (y, b) \in P^t(x), \text{ then } \forall \theta \in (0, 1], \text{ we have } (\theta y, \theta b) \in P^t(x); \tag{3.5}$$

Null jointness:

$$\text{If } (y, b) \in P^t(x) \text{ and } b = 0, \text{ then } y = 0. \tag{3.6}$$

Proposition (I) means that the desirable outputs will decrease in the same proportion, while the undesirable outputs are reduced. And it also shows that reducing pollutant emissions will inevitably lead to economic loss in the actual production process. Proposition (II) implies that pollutants cannot be totally eliminated unless stopping the production process. According to Färe and Grosskopf (2004), with the assumption of variable returns-to-scale, the expression of $P^t(x)$, that meets convexity, cone, and weak disposability requirements can be formulated as

$$\begin{aligned} P^t(x) = \{(x^t, y^t, u^t)| \sum_{j=1}^{J} \lambda_j x_j^t \le x^t;\ \theta \sum_{j=1}^{J} \lambda_j y_j^t \ge y^t; \\ \theta \sum_{j=1}^{J} \lambda_j u_j^t = u^t;\ \sum_{j=1}^{J} \lambda_j = 1; \\ \lambda_j \ge 0;\ 0 \le \theta \le 1; j = 1, 2, \ldots J\}. \end{aligned} \tag{3.7}$$

where λ_j represents the intensity of the jth DMU; θ is an endogenous parameter.

As discussed in Sect. 3.2.1, the relative efficiency, evaluated by single-phase reference technology, is not stable. To overcome this shortcoming, a global reference technology is constructed as follows:

$$P^G = P^1 \cup P^2 \cup \ldots \cup P^G. \tag{3.8}$$

As with single-phase reference technology, global reference technology, which meets the conditions of variable returns-to-scale, can be expressed in the form:

$$\begin{aligned} P^G(x) = \{(x^t, y^t, u^t)| \sum_{t=1}^{T} \sum_{j=1}^{J} \lambda_j^t x_j^t \le x^t; \theta \sum_{t=1}^{T} \sum_{j=1}^{J} \lambda_j^t y_j^t \ge y^t; \\ \theta \sum_{t=1}^{T} \sum_{j=1}^{J} \lambda_j^t u_j^t = u^t; \sum_{t=1}^{T} \sum_{j=1}^{J} \lambda_j^t = 1; \\ \lambda_j^t \ge 0; 0 \le \theta < 1; j = 1, 2, \ldots, J, t = 1, 2, \ldots, T\}. \end{aligned} \tag{3.9}$$

where λ_j^t represents the intensity of the jth DMU in period t; θ is an endogenous parameter.

3.2.2.2 Efficiency Measurement Based on Non-radial Directional Distance Function Models

Chambers et al. (1996) proposed distance function, which aims to measure the maximum reduction in inputs when the outputs are fixed. Then, Chung et al. (1997) later established the directional distance function (DDF) model in the energy and environment field, which intended to radially maximize the desirable outputs and minimize the undesirable outputs spontaneously for a given direction. The radial DDF model by Chung et al. (1997) can be expressed as

$$D(x_{k'}, y_{k'}, u_{k'}; \vec{g}) = \sup\{\sigma : (x_{k'}, y_{k'} + \sigma g_y, u_{k'} - \sigma g_u) \in P^t(x)\}, \tag{3.10}$$

where $(x_{k'}, y_{k'}, u_{k'})$ is the input–output vector of the DMU, which is to be evaluated; $\vec{g} = (g_y, g_u)$ is a given direction vector, and different direction vectors lead to different directional distance functions; $P^t(x)$ refers to the reference technology in period t. However, as the relative efficiency evaluated by the single-phase reference technology is not stable and the efficiency scores in different periods cannot be compared directly, a global reference technology exhibiting variable returns-to-scale will be employed. Then the directional distance function in technology can be shown as:

$$D(x_{k'}, y_{k'}, u_{k'}; \vec{g}) = \sup\{\sigma : (x_{k'}, y_{k'} + \sigma g_y, u_{k'} - \sigma g_u) \in P^G(x)\} \tag{3.11}$$

where all the symbols in Model (3.11) have the same meaning with those in Model (3.10), except the reference technology $P^G(x)$. According to (3.9), the specific form of Model (3.8) can be expressed as

$$\begin{aligned}
& ND(x_{k'}^t, y_{k'}^t, u_{k'}^t; \vec{g}) = \max\ \delta \\
& \text{subject to} \\
& \sum_{t=1}^{T}\sum_{j=1}^{J} \lambda_j^t x_{jn}^t \le \theta x_{k'n}^{t'}, n = 1, 2, \ldots N; \\
& \sum_{t=1}^{T}\sum_{j=1}^{J} \lambda_j^t y_{jm}^t \ge y_{k'_m}^{t'} + \delta g_m^y;\ m = 1, 2, \ldots M; \\
& \sum_{t=1}^{T}\sum_{j=1}^{J} \lambda_j^t u_{jq}^t = u_{k'q}^{t'} - \delta g_q^u;\ q = 1, 2, \ldots Q; \\
& \sum_{t=1}^{T}\sum_{j=1}^{J} \lambda_j^t = \theta; \\
& \lambda_j^t \ge 0, \forall j; 0 \le \theta \le 1.
\end{aligned} \tag{3.12}$$

where $(x_{k'}, y_{k'}, u_{k'})$ is the input–output vector of the DMU, which is to be evaluated; $g = (g^y, g^u)$ represents the directional vector. Generally, we let $g = (y^{t'}_{k'm}, u^{t'}_{k'm})$. Also, δ represents the adjustment coefficient of the desirable and undesirable outputs, and λ^t_j is the intensity of the jth DMU in period t.

If the optimal solution to the mathematical programming [Eq. (3.11)] is denoted by $(\Lambda^*, \theta^*, \delta^*)$, then, according to Wang et al. (2013), the productivity of the k'th DMU at time t' can be expressed as

$$\text{Productivity} = 1 - \delta^* = 1 - \text{ND}(x^{t'}_{k'}, y^{t'}_{k'}, u^{t'}_{k'}, g). \tag{3.13}$$

The economic efficiency of the k'th DMU at time t' is defined as

$$\text{TFEE}^y_{k'} = \frac{1}{M}\sum_{m=1}^{M}\frac{y^{t'}_{k'm}}{y^{t'}_{k'm} + \delta^* g^y_m} \tag{3.14}$$

The pollutant emission efficiency of the k'th DMU at time t' is defined as

$$\text{TFEE}^u_{k'} = \frac{1}{Q}\sum_{q=1}^{Q}\frac{u^{t'}_{k'm} - \delta^* g^u_q}{u^{t'}_{k'm}} \tag{3.15}$$

Actually, economic efficiency measures the difference between target economic outputs and actual outputs. Less difference between target and actual economic outputs implies a higher economic efficiency. Pollutant emission efficiency also has a similar interpretation.

In addition, the economic efficiency and environmental efficiency of multiple areas, respectively, can be calculated according to the following expression:

$$\text{TFEE}^y_a = \frac{\sum_{i\in a} y^{t'}_{im}}{\sum_{i\in a} y^{t'}_{im} / \text{TFEE}^y_i} \tag{3.16}$$

$$\text{TFEE}^u_a = \frac{\sum_{i\in a} u^{t'}_{iq} * \text{TFEE}^u_i}{\sum_{i\in a} u^{t'}_{iq}} \tag{3.17}$$

where a represents the set of multiple decision-making units.

3.2.2.3 Calculation of CO_2 Marginal Abatement Cost

Reducing CO_2 emissions will undoubtedly give rise to economic loss because of the weak disposability of CO_2. The evaluation of the economic loss caused by CO_2 reduction is a hot topic for scholars. In academia, the economic loss caused by unit CO_2 reduction is called the “marginal abatement cost” or “hadow price” of the CO_2. Currently, much of the work has been done to measure the marginal abatement cost of the undesirable outputs at different levels (Mekaroonreung and Johnson 2012; Wei et al. 2013; Alphonse et al. 2015; Choi et al. 2012).

In the framework of nonparametric analysis, most scholars calculate marginal abatement cost by solving dual programming of the directional distance function or slack-based model. Many scholars also employed this method to construct the best production frontier (Chang et al. 2013; Choi et al. 2012; Mekaroonreung and Johnson 2012; Wang et al. 2013). In this chapter, we are dealing with CO_2 as the input, which could help to obtain the nonpositive CO_2 marginal abatement cost.[1]

To make up for the defect treating CO_2 as input, Leleu (2013) has proposed a hybrid model, which regards the undesirable outputs as inputs but gives the dual programming model an explicit economic interpretation through the transformation of Model (3.12). Based on the work of Leleu (2013), the equivalent model for Model (3.12), in the framework of global reference technology, is shown as

Prime programming

$$\begin{aligned}
& \mathrm{ND}(x_{k'}^{t'}, y_{k'}^{t'}, u_{k'}^{t'}; \vec{g}) = \max \delta \\
& \text{subject to} \\
& \sum_{t=1}^{T} \sum_{j=1}^{J} \lambda_j^t (x_{jn}^t - x_{k'n}^{t'}) \leq 0, n = 1, 2 \ldots N; \\
& \sum_{t=1}^{T} \sum_{j=1}^{J} \lambda_j^t (y_{jm}^t - y_{k'm}^{t'}) \geq \sigma y_{k'm}^{t'} + \delta g_m^y;\ m = 1, 2, \ldots M; \\
& \sum_{t=1}^{T} \sum_{j=1}^{J} \lambda_j^t (u_{jq}^t - u_{k'q}^{t'}) = \sigma u_{k'q}^{t'} - \delta g_q^u;\ q = 1, 2, \ldots Q; \\
& \sum_{t=1}^{T} \sum_{j=1}^{J} \lambda_j^t + \sigma = 1; \\
& \lambda_j^t \geq 0, \sigma > 0; \forall j, t.
\end{aligned} \tag{3.18}$$

Here, the economic meaning of each parameter is the same as Model (3.12). Now, in order to obtain the marginal abatement cost, we convert Model (3.18) into its dual programming form:

[1]Actually, nowadays, even though most scholars assume that undesirable outputs are weakly disposable, some scholars have challenged the equality constraint on undesirable outputs in environmental reference technology (Murty et al. 2012; Chen 2014). They argued that equality constraints will reduce the flexible solution to the DEA model, and it may have an impact on the results. On the contrary, dealing with undesirable output looks strange, it assures the completeness of production probability set and has no influence on the final results.

Dual programming

$$
\begin{aligned}
&\min \varphi \\
&\text{subject to} \\
&\Big(\sum_{m=1}^{M} \pi_m^y y_{jm}^t - \sum_{n=1}^{N} \pi_n^x x_{jn}^t - \sum_{q=1}^{Q} \pi_u^q u_{jq}^t\Big) \\
&-\Big(\sum_{m=1}^{M} \pi_y^m y_{k'm}^{t'} - \sum_{n=1}^{N} \pi_x^n x_{k'n}^{t'} - \sum_{q=1}^{Q} \pi_u^q u_{k'q}^{t'}\Big) \le \varphi; \\
&j = 1, 2, \ldots J; t = 1, 2, \ldots T; \\
&\sum_{m=1}^{M} \pi_y^m y_{k'm}^{t'} - \sum_{q=1}^{Q} \pi_u^q u_{k'q}^{t'} + \varphi \ge 0; \\
&\sum_{m=1}^{M} \pi_y^m y_{k'm}^{t'} + \sum_{q=1}^{Q} \pi_u^q u_{k'q}^{t'} = 1; \\
&\pi_x^n \ge 0, \forall n; \pi_y^m \ge 0, \forall m; \\
&\pi_u^q \ \text{unconstrained}, \forall q.
\end{aligned}
\tag{3.19}
$$

Here, π_n^x, π_m^y, and π_q^u, $(n = 1, 2, \ldots, N; m = 1, 2, \ldots, M; q = 1, 2, \ldots, Q)$ represent the dual variables of inputs, desirable outputs, and undesirable outputs, respectively. According to Wang and Feng (2015), these dual variables can be interpreted as the virtual price of the inputs/outputs. Because of the weak disposability of CO_2 in Model (3.18), the dual variable corresponding to CO_2, π_q^u, loses its nonnegative constraint, which may lead to positive abatement costs. According to the discussion in Leleu (2013), the dual model of Model (3.18) has explicit economic interpretation after the equivalent transformation from Model (3.12) to Model (3.18). The first $J \times T$ constraint aims to achieve the upper limits of the profit between the k'th DMU and any other DMUs. The $J \times T + 1$ constraint indicates that the profit of the k'th DMU, which is going to be assessed, should be greater than zero.

Based on the explicit economic interpretation of Model (3.17), we add a nonnegative constraint on π_q^u which implies that we regard the undesirable output as an input. This is to make sure that we obtain the expected sign for the marginal abatement cost. The specific model is expressed as

$$
\begin{aligned}
&\min \varphi \\
&\text{subject to} \\
&\Big(\sum_{m=1}^{M} \pi_m^y y_{jm}^t - \sum_{n=1}^{N} \pi_n^x x_{jn}^t - \sum_{q=1}^{Q} \pi_u^q u_{jq}^t\Big) - \\
&\Big(\sum_{m=1}^{M} \pi_y^m y_{k'm}^t - \sum_{n=1}^{N} \pi_x^n x_{k'n}^t - \sum_{q=1}^{Q} \pi_u^q u_{k'q}^t\Big) \le \varphi; \\
&\sum_{m=1}^{M} \pi_y^m y_{k'm}^{t'} - \sum_{q=1}^{Q} \pi_u^q u_{k'q}^{t'} + \varphi \ge 0; \\
&\sum_{m=1}^{M} \pi_y^m y_{k'm}^{t'} + \sum_{q=1}^{Q} \pi_u^q u_{k'q}^{t'} = 1; \\
&j = 1, 2, \ldots J; t = 1, 2, \ldots T; \\
&\pi_x^n \ge 0, \forall n; \pi_y^m \ge 0, \forall m; \\
&\pi_u^q \ge 0, \forall q.
\end{aligned}
\tag{3.20}
$$

If the optimized solution to the model in Model (3.18) is denoted by $(\pi_n^{x*}, \pi_m^{y*}, \pi_q^{u*}, \varphi^*)$, then, according to Wang et al. (2013), Wang and Feng (2015) and Choi et al. (2012), the marginal abatement cost of the qth undesirable output of the k'th DMU at time t' can generally be expressed as

$$
\text{Marginal abatement cost of the } k'\text{th DMU at time } t' = \frac{\pi_u^{q*}}{\pi_y^{m*}} \times 1\,\text{yuan} \tag{3.21}
$$

Since π_u^{q*} and π_y^{m*} are virtual price of undesirable and desirable outputs, we transform the virtual abatement cost of the undesirable output into factual price through Eq. (3.21).

3.2.2.4 Data Description

1. Division of the administrative regions in China

As of data availability issues, Hong Kong, Macao and Taiwan are not included. In accordance with their economic development level and geographical location, the 31 provinces (autonomous regions, municipalities) are divided into eight economic regions: Northeast areas (including Heilongjiang, Liaoning and Jilin); Middle Yellow River areas (including Inner Mongolia, Shanxi, Shaanxi and Henan); Lower Yellow River areas (including Beijing, Tianjin, Hebei and Shandong); Upper Yangtze River areas (including Sichuan, Chongqing, Guizhou and Yunnan); Middle Yangtze River areas (including Hunan, Hubei and Jiangxi); and Lower

Yangtze River areas (including Jiangsu, Shanghai, Zhejiang and Anhui); South Coastal areas (including Fujian, Guangdong, Guangxi and Hainan), and Northwest areas (including Xinjiang, Qinghai, Gansu and Ningxia).

2. Variables and data set

This chapter aims to study China's regional transportation sectors' productivity, economic efficiency, CO_2 emission efficiency, and CO_2 marginal abatement costs from 2007 to 2012, except for Xizang, due to data unavailability. Here, the item of "transportation sectors" are taken that consist of the transportation, storage, and postal services industry. According to the Cobb–Douglas production function theory (Shi et al. 2010; Choi et al. 2012), a DMU should employ capital and labor, which are regarded as non-energy inputs in the chapter, to produce outputs. Given that energy played a crucial role in economic development, we also regard energy consumption by transportation sectors as another input. Also, "desirable output"[2] denotes the added value of the regional transportation sectors. Recognizing the important role that transportation sectors played in global warming, we choose CO_2 emissions from transportation sectors as an undesirable output.

Data on all types of energy consumption (coal, gasoline, diesel, kerosene, natural gas, and power) are extracted from *China Energy Statistical Yearbooks 2008–2013* (NBSC, 2008a, 2009a, 2010a, 2011a). Here, the figures are converted into units of tens of thousands of tons of standard coal. The data related to the labor force and the industrial added value are taken from the *China Statistical Yearbook 2008–2013* (the industrial added value is measured at the 2004 constant price). Capital stock figures are estimated using the Perpetual Inventory Method, which can be expressed as $K_j^t = (1-\eta)K_j^{t-1} + I_j^t$, where K_j^t and K_j^{t-1} denote the DMU's capital stock in years t and $t-1$, respectively. The quantity I_j^t denotes the investment in the fixed assets in year j, and η denotes the depreciation rate. Here, we used the fixed assets investment in 2004 as the capital stock of the starting year. Each provincial unit's depreciation rate was obtained from *China Economic Census Yearbook 2004* and *China Statistical Yearbooks 2008–2013* (NBSC, 2008c, 2009c, 2010c, 2011c). We assume that the depreciation rate for each provincial unit (Table 3.3), is essentially unchanged during 2004–2012. The fixed assets and fixed asset investments for each provincial unit were both measured at 2004 constant price.

According to the guidelines on the calculation of national greenhouse gas emissions issued by the Intergovernmental Panel on Climate Change (IPCC 2006), CO_2 emissions CE can be calculated by

[2]Wu et al. (2015) also chose the passenger turnover volume and the freight turnover volume as desirable outputs. However, the added value of the transportation sector is a comprehensive index of the passenger turnover volume and the freight turnover volume. Therefore, both the added value of the transportation sectors and the passenger turnover volume and freight turnover volume can reflect the economic outputs of transportation sectors.

Table 3.3 The depreciation rate of Chinese provincial transportation sectors in 2004

Area	Depreciation rate/%	Area	Depreciation rate/%	Area	Depreciation rate/%
Beijing	3.52	Zhejiang	5.19	Hainan	4.29
Tianjin	4.78	Anhui	7.95	Chongqing	5.04
Hebei	4.64	Fujian	7.50	Sichuan	5.64
Shanxi	9.88	Jiangxi	3.22	Guizhou	6.54
Inner Mongolia	6.40	Shandong	7.93	Yunnan	5.27
Liaoning	8.15	Henan	7.45	Shanxi	5.36
Jilin	7.33	Hubei	9.76	Gansu	5.64
Heilongjiang	7.30	Hunan	7.81	Qinghai	8.14
Shanghai	4.41	Guangdong	5.83	Ningxia	4.62
Jiangsu	6.21	Guangxi	8.28	Xinjiang	8.11

$$CE = \sum_{i-1}^{I} A_i \times COF_i \times CV_i \times CCF_i \times \frac{44}{12} \tag{3.22}$$

where A_i, COF_i, CV_i, and CCF_i denote the consumption, carbon oxidation factor, calorific value, and carbon content factor of the ith fossil energy, respectively. The 44/12 is a coefficient converting carbon into CO_2.

Table 3.4 shows that both the standard deviations of energy consumption and CO_2 emissions are relatively large, with the explanation that different regional transportation sectors have widely different demands for energy and different CO_2 emissions. At the same time, the economic outputs they produced are also totally different. The existence of these large differences is one of the reasons we analyzed the economic efficiency and CO_2 emission in this chapter.

Table 3.4 The statistical description of input and output indices

Variables	Unit	Observations	Maximum	Minimum	Mean	Std.edv
Energy	Thousand tons	180	28668.4	693.7	8232.2	5722.6
Labor	Thousand people	180	618.0	28.0	213.0	124.0
Capital stock	Billion yuan	180	153.9	2.6	38.9	29.8
Added value	Billion yuan	180	318.3	3.9	78.4	64.8
CO_2 emission	Thousand tons	180	61566.4	1494.1	17564.3	12234.3

3.2.3 Productivity, Economic Efficiency, and CO_2 Emission Efficiency

According to the models in Sect. 3.2.2.3, and data in Sect. 3.2.2.4, we investigate the productivity, economic efficiency, and CO_2 emission efficiency of China's regional transportation sectors during 2007–2012. The regional transportation sectors' productivity is shown in Table 3.5.[3] It is clear from the figures presented herein that the productivity in different areas differs greatly. During the sampled period, some areas' productivity scores are more than 0.8 (Heilongjiang, Shandong, Jiangsu, Zhejiang, Qinghai, and so on). On the contrary, the productivity scores in other areas are relatively much lower (Chongqing, Shanghai, Gansu, Xinjiang, and so on). The data show that the provincial transportation systems' utilization efficiency of input factors, i.e., energy, capital, and labor, are vastly different depending on the provinces (autonomous regions, municipalities) involved. However, while the cars are congested on the road, they would still consume energy, and cause tear and wear. Undoubtedly, bad traffic problems will have a significant impact on the productivity of the transportation system. In terms of productivity change, many areas' productivity scores increase year-by-year, e.g., Liaoning. At the same time, the productivity in Guizhou, Ningxia, and Qinghai decreases throughout the whole of the sampling period.

After the productivity of regional transportation sectors is analyzed, the economic efficiency and CO_2 emission efficiency of the transportation systems of eight economic areas in China during 2007–2012 are shown in Table 3.6. From the table, we can observe that economic efficiency appears to be greater than CO_2 emission efficiency in each economic area.

Just considering the economic efficiency, the economy of the transportation sectors in the Lower Yellow River areas, Northeast areas, and South Coastal areas performed much better than other economic areas. During 2007–2012, all areas had relatively high economic efficiency with more than 80%; moreover, it increases year by year. In particular, their economic efficiency is more than 0.9 in 2012 (0.952 in the South Coastal areas). In stark contrast to the South Coastal areas, the economic efficiency of the transportation sectors in the Upper Yangtze River areas and Northwest areas are clearly much lower (less than 0.7 in the sampled period). The reason for this phenomenon is easily understood. The Lower Yellow River and South Coastal areas contain economically developed provincial units, e.g., Tianjin, Guangdong, Hainan, Fujian, etc., and these provincial units are ones that always lead China's economy. In these regions, the government investment and the introduction of advanced technology are relatively high. Therefore, the economic efficiency scores of their transportation sectors are also relatively high. However, the Upper Yangtze River and Northwest areas are less well developed. Because of

[3]Results were obtained using MATLAB 2012b.

Table 3.5 The productivity of transportation sectors in China during 2007–2012

Areas and provinces		Productivity					
		2007	2008	2009	2010	2011	2012
Northeast areas	Liaoning	0.758	0.817	0.867	0.914	0.940	0.957
	Jilin	0.597	0.580	0.616	0.628	0.652	0.695
	Heilongjiang	1.000	1.000	0.995	1.000	0.958	0.947
Middle Yellow River areas	Shanxi	0.506	0.335	0.364	0.394	0.418	0.466
	Inner Mongolia	0.496	0.419	0.410	0.387	0.392	0.378
	Henan	0.830	0.848	0.869	0.868	0.884	0.903
	Shaanxi	0.596	0.545	0.524	0.526	0.537	0.589
Lower Yellow River areas	Beijing	0.412	0.409	0.431	0.439	0.449	0.483
	Tianjin	0.781	1.000	0.807	0.829	0.911	0.923
	Hebei	0.845	0.900	0.960	0.937	0.949	1.000
	Shandong	0.806	0.880	0.916	0.967	0.990	1.000
Upper Yangtze River areas	Chongqing	0.397	0.382	0.451	0.418	0.424	0.444
	Sichuan	0.563	0.542	0.531	0.602	0.703	0.749
	Guizhou	0.429	0.376	0.405	0.388	0.412	0.380
	Yunnan	0.395	0.405	0.429	0.396	0.408	0.418
Middle Yangtze River areas	Hubei	0.414	0.423	0.489	0.521	0.502	0.567
	Hunan	0.579	0.727	0.675	0.678	0.695	0.799
	Jiangxi	0.755	0.800	0.830	0.771	0.774	0.808
Lower Yangtze River areas	Shanghai	0.311	0.335	0.371	0.382	0.411	0.446
	Jiangsu	0.820	0.853	0.915	0.946	0.984	1.000
	Zhejiang	0.811	0.831	0.872	0.915	0.930	0.968
	Anhui	0.701	0.724	0.754	0.741	0.720	0.655
South Coastal areas	Guangdong	0.724	0.768	0.839	0.890	0.946	1.000
	Guangxi	0.560	1.000	0.609	0.620	0.633	0.637
	Hainan	0.648	0.576	0.613	0.721	0.702	0.860
	Fujian	1.000	0.945	0.966	1.000	0.980	1.000
Northwest areas	Gansu	0.439	0.450	0.455	0.445	0.453	0.463
	Qinghai	1.000	1.000	0.950	0.991	0.902	1.000
	Ningxia	1.000	1.000	0.866	1.000	0.814	0.695
	Xinjiang	0.379	0.388	0.422	0.429	0.439	0.460

their poor technology, the economy of the transportation sectors in these areas performs worse (Table 3.6).

The trend of the CO_2 emission efficiency of the eight economic areas is similar to economic efficiency. The Lower Yellow River areas, Northeast areas, and South Coastal areas have relatively high CO_2 emission efficiency with the tendency to increase year by year. However, CO_2 emission efficiency of the Upper Yangtze River and Northwest areas are relatively low (almost less than 0.6). This again demonstrates the result of the imbalanced development of China's national economy.

Table 3.6 The economic efficiency and CO_2 emission efficiency of transportation sectors in China during 2007–2012

Areas	Index	Time					
		2007	2008	2009	2010	2011	2012
Northeast areas	Economic efficiency	0.832	0.851	0.873	0.896	0.901	0.913
Middle Yellow River areas		0.755	0.731	0.734	0.734	0.743	0.756
Lower Yellow River areas		0.811	0.865	0.875	0.892	0.914	0.936
Upper Yangtze River areas		0.660	0.652	0.658	0.667	0.692	0.704
Middle Yangtze River areas		0.696	0.732	0.738	0.737	0.737	0.778
Lower Yangtze River areas		0.776	0.792	0.823	0.845	0.862	0.874
South Coastal areas		0.817	0.861	0.860	0.893	0.916	0.952
Northwest areas		0.682	0.686	0.685	0.694	0.682	0.688
Northeast areas	CO_2 emission efficiency	0.788	0.811	0.846	0.875	0.897	0.912
Middle Yellow River areas		0.611	0.522	0.526	0.527	0.544	0.564
Lower Yellow River areas		0.737	0.798	0.817	0.845	0.872	0.900
Upper Yangtze River areas		0.461	0.446	0.470	0.469	0.500	0.508
Middle Yangtze River areas		0.523	0.565	0.604	0.619	0.610	0.684
Lower Yangtze River areas		0.586	0.620	0.666	0.699	0.728	0.750
South Coastal areas		0.730	0.800	0.804	0.849	0.881	0.926
Northwest areas		0.536	0.544	0.547	0.571	0.542	0.541

Figure 3.1 presents the kernel density curve for the economic efficiency and CO_2 emission efficiency of China's regional transportation sectors (i.e., the distribution of economic efficiency and CO_2 emission efficiency). According to Fig. 3.1, it can be found that the economic efficiency distribution is fairly concentrated between 0.6 and 1, which indicates that the economic efficiency score of the provincial transportation systems is pretty good and the difference between different areas is also comparatively small. On the contrary, the CO_2 emission efficiency distribution is dispersed more widely and mainly falls in the range between 0.4 and 0.9. This implies that there is a large difference between CO_2 emission efficiency of different areas.

The economic efficiency is much greater than the productivity and CO_2 emission efficiency. In addition, all variables present an upward trend during 2007–2012. In

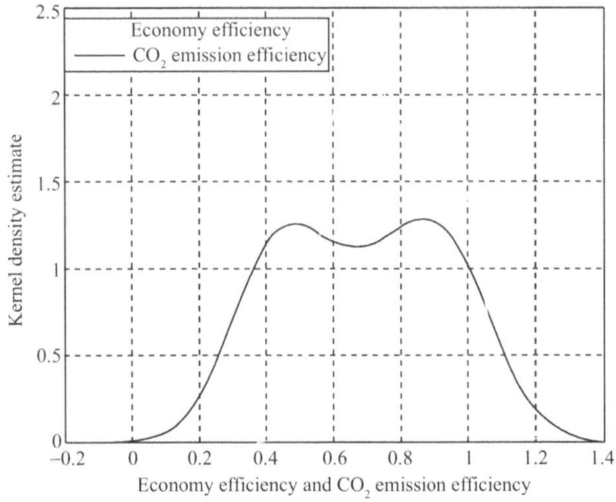

Fig. 3.1 The kernel density estimate of economic efficiency and CO_2 emission efficiency

2007, the economic efficiency of China's transportation system was 0.77, and this rose to 0.854 in 2012. The average value of the economic efficiency in the sampling period exceeds 0.8. When it comes to CO_2 emission efficiency, its value was 0.65 in 2007 and 0.76 in 2012. The average growth rate in economic efficiency during 2007–2012 was 1.67%. However, for economic efficiency, the equivalent figure was 1.8% and for productivity, it was 1.42%. Therefore, the CO_2 emission efficiency of China's transportation sectors is growing at a slightly higher rate than its productivity and economic efficiency.

3.2.4 CO_2 Marginal Abatement Cost of the Provincial Transportation System

In Sect. 3.2.3, we analyzed China's regional transportation sectors' productivity, economic efficiency, and CO_2 emission efficiency. From the results, we observed that many regions display relatively low CO_2 emission efficiency, which implies that the gap between the target and actual CO_2 reduction is relatively large. Therefore more attention should be paid on the transportation sector of these regions to reduce the CO_2 emissions.

Table 3.7 presents the CO_2 marginal abatement costs of the transportation sectors during 2007–2012. From the table, we find that CO_2 marginal abatement cost of the northeast areas is the lowest during 2007–2012, with an average value of only 1,980 yuan per ton (i.e., its economic output decreases by 1,980 yuan, when

Table 3.7 The CO_2 marginal abatement cost of transportation sectors among eight economic areas (Unit: 10^3 yuan/t)

Areas	Time						Mean
	2007	2008	2009	2010	2011	2012	
Northeast areas	1.97	1.97	2.18	2.20	2.05	1.51	1.98
Middle Yellow River areas	6.49	7.44	7.44	6.04	6.33	5.61	6.56
Lower Yellow River areas	7.30	4.99	7.33	6.65	5.21	4.19	5.94
Upper Yangtze River areas	7.79	8.28	7.71	7.80	8.72	7.80	8.02
Middle Yangtze River areas	8.34	8.34	7.33	7.33	7.28	7.28	7.65
Lower Yangtze River areas	7.38	5.75	4.93	4.93	5.32	3.85	5.36
South Coastal areas	3.68	5.56	3.87	1.96	2.45	1.33	3.14
Northwest areas	3.49	3.77	4.84	4.34	4.79	4.90	4.35
The whole country	5.85	5.80	5.77	5.18	5.31	4.57	5.41

the amount of CO_2 emissions is reduced by one ton). However, during the sample period, at first its CO_2 marginal abatement cost shows an upward trend and then decline, to merely 1,510 yuan per ton in 2012. The second-lowest area is the South Coastal areas where an average CO_2 marginal abatement cost over the sampling period was 3,140 yuan per ton. Moreover, the abatement cost before 2010 appeared to be more than in 2010 for South Coastal areas. Thus, the marginal abatement cost in the south coastal areas presents a significant downward trend.

Some areas, including the Middle Yellow River areas and Upper and Middle Yangtze River areas, had much higher CO_2 marginal abatement costs, almost exceeding 6,000 yuan per ton. In particular, the costs in the Upper and Middle Yangtze River areas remained at a high level during 2007–2012, which indicates that any reduction in CO_2 emission has a significant impact on the local economy.

The marginal abatement costs derived in this chapter are higher than the results published in Wang et al. (2013), which focuses on the CO_2 marginal abatement cost of the industrial sectors. This probably maybe because of the higher carbon intensity associated with the transportation sectors compared to the industrial sectors. The industrial added value of the transportation sectors only accounted for 5% of the whole GDP, whereas its CO_2 emissions accounted for 6–7% (during 2007–2012). However, the large gaps between the figures for different areas may be caused by a multitude of reasons. In order to further explore the specific relationship between the marginal abatement cost and CO_2 emission efficiency, a scatter plot was constructed, which is shown in Fig. 3.2.

From Fig. 3.2, we can find out that the CO_2 marginal abatement cost has a close relationship with CO_2 emission efficiency. Generally speaking, the marginal abatement cost and emission efficiency are negatively correlated. That is, the areas with lower CO_2 emission efficiency probably have higher abatement costs, and vice versa. In order to obtain a more quantitative measure regarding this relationship, regression analysis was established as follows:

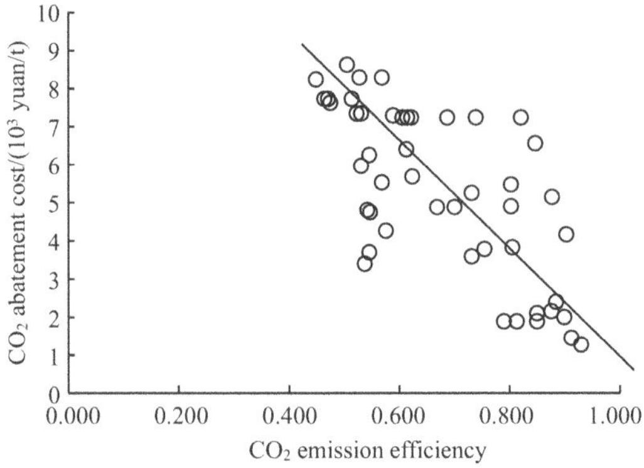

Fig. 3.2 The scatter plot between the CO_2 marginal abatement cost and CO_2 emission efficiency

Table 3.8 The regression result of the abatement cost and CO_2 emission efficiency

Variable	Coefficient	Std. error	*t*-Statistic	Prob.
C	12223.75	1060.170	11.52999	0.0000
Efficiency	−10217.74	1544.804	−6.614262	0.0000
R^2: 0.487456 *F*-statistic: 43.74846				

$$\text{Marginal abatement cost} = C + b \times \text{efficiency} + \varepsilon \tag{3.23}$$

where C is a constant; b is the regression coefficient, and ε is the residual. The regression results are presented in Table 3.8.[4]

Table 3.8 shows that the results are statistically significant, i.e., CO_2 emission efficiency has a significant impact on the marginal abatement cost. The specific regression expression is as given below:

$$\begin{aligned} \text{abatement cost} = &\ 12223.75 - 10217.74 \times \text{efficiency} \\ &\ (11.53) \qquad (-6.61) \end{aligned} \tag{3.24}$$

It implies that if the CO_2 emission efficiency increases by 1%, then the marginal abatement cost decreases by 102 yuan per ton. This significant relationship may be useful in seeking new routes and methods to reduce the marginal abatement cost. For example, the Upper Yangtze River areas, which need to reduce their abatement costs, should pay more attention to the improvement of their CO_2 emission efficiency. At the same time, according to Zhou et al. (2010) and Wang et al. (2016), technical progress was the main contributor to the improvement of CO_2 emission efficiency. Hence, these regions should also emphasize the improvement of their technology.

[4]The results were obtained using EViews (v8.0) software, and the significant level is 5%.

3.2.5 Implications

The results in Sect. 3.2.4 indicate that the economic efficiency, productivity, and CO_2 emission efficiency in the South Coastal areas, Lower Yellow River areas, and Northeast areas are all higher than other areas, e.g., Upper Yangtze River and Northwest areas.

Aiming at measuring the economic efficiency and CO_2 emission efficiency at the same time, we utilized the DEA method to evaluate abatement costs. There are some limitations to this approach. As discussed, DEA is used to evaluate the abatement cost by solving the dual model of the directional distance function model. However, when we transform the DDF model to its dual programming, the dual variables corresponding to the undesirable output lose their nonnegative constraints, which can lead to positive abatement costs. In order to obtain non-positive abatement costs, we treat the undesirable outputs as inputs, and revise the equality constraints of the undesirable output to inequality constraints. This approach may cause happening of deviation.

When estimating the capital stock, we assume that the depreciation rate of each transportation sector during 2007–2012 is constant and equals to that in 2004. This means that the capital stock estimates are not particularly accurate. Moreover, in this chapter, we do not study the energy efficiency of the transportation sectors, as the model involved variable returns-to-scale. If we estimate the energy efficiency, then model becomes nonlinear, which is difficult to be solved, but we would like to deal with these difficulties in future work.

In this chapter, we employed the global data envelopment analysis method based on Leleu (2013) to investigate the productivity, economic efficiency, CO_2 emission efficiency, and CO_2 marginal abatement costs of regional transportation sectors in China during 2007–2012. The chapter makes up the fact that few literature systemically examine these aspects of China's transportation sectors and also obtains some valuable results.

Firstly, there is an obvious difference between the productivity of transportation sectors in different areas. Specifically, the productivity of Heilongjiang, Shandong, Jiangsu, Zhejiang regions is significantly higher than Chongqing, Shanghai, Gansu, Xinjiang, etc. Besides, the productivity of transportation in Liaoning presents an upward trend during the sample periods; however, it decreases year by year in Qinghai, Ningxia, and Guizhou.

For economic efficiency, the transportation sector of South Coastal areas, Lower Yangtze River areas, and Northeast areas perform better than the other areas and their economic efficiency scores is more than 0.8 over the sampled period, thus show a progressively upward trend. However, the economic efficiency of the Upper Yangtze River and Northwest areas is relatively lower, fluctuating around 0.7. For CO_2 emission efficiency, the situation is similar to that of economic efficiency. The CO_2 emission efficiency of the transportation system in the South Coastal areas and Northeast areas are relatively high and increase year by year, reaching 0.9 in 2012. But for the Upper Yangtze River areas, Northwest areas, and Lower Yellow River areas, CO_2 emission efficiency almost drop down below 0.6 during 2007–2012, which implies that the transportation of these areas reduces their total CO_2 emissions by 40%.

In addition, the CO_2 marginal abatement costs of the transportation sector in different regions differ greatly. The marginal abatement costs in South Coastal areas and Northwest areas are the lowest, with an average value of only 1,980 yuan per ton of CO_2 reduction during 2007–2012. However, the average abatement cost in the Upper Yangtze River areas, are almost higher than 7,000 yuan. Moreover, we find that CO_2 emission efficiency and marginal abatement costs are negatively correlated. Specifically, a 1% increase in CO_2 emission efficiency results in a 102 yuan reduction in the marginal abatement cost. Improving CO_2 emission efficiency provides an important route to reduce the CO_2 marginal abatement cost.

3.3 Rebound Effect for Transportation

Improving energy efficiency has been considered as a major approach to reduce transportation energy consumption, whereas its effectiveness that reduces the transportation cost may be incremental to energy use, i.e., a "direct rebound effect". The existence of the rebound effect and its magnitude will directly affect the realization of energy saving and carbon reduction targets.

3.3.1 Rebound Effect from Transportation

3.3.1.1 Definition of Energy Rebound Effect

Energy rebound effect is used to describe the paradox between energy consumption and energy efficiency. In 1865, William Stanley Jevons first proposed this issue in his research "The Coal Question", in which he presented the idea that "it was ridiculous to reduce fuel consumption through using it economically and instead the opposite was true." This is the so-called "Jevons' paradox". (The graph of Jevons Paradox is shown in Fig. 3.3). Improving energy efficiency could reduce energy consumption which is not as effective as people expected due to the changes in consumer behavior. When energy efficiency was improved, the costs of unit energy service was decreased, thus will stimulate consumers to get more energy services,

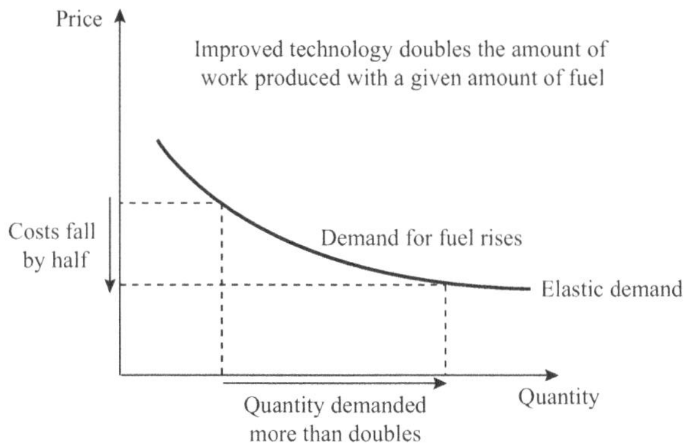

Fig. 3.3 Graphic of Jevons paradox

which will lead to a decline in energy consumption inconsistent with the increased energy efficiency. These behaviors will consume more fuel.

Based on previous researches and perfect market mechanism hypothesis, especially energy market mechanism, Greening et al. (2000) divided the rebound effect mechanism into four main categories: (1) Direct effect: improving energy efficiency will decrease the effective price of an energy service and therefore lead to an increase in consumption of that service. This will offset the reduction of energy consumption caused by advanced efficiency. The direct effect is mainly reflected in the substitution effect and the income effect; (2) Economy-wide effect or market-clearing price and quantity adjustments: the energy rebound effect impacts on the overall economy; (3) Transformation effect: technology progress will change consumer preferences, reform the social system and reconstruct the production organization; (4) Secondary effect: the impact on other production/service. Energy efficiency improvement will reduce the cost of the manufacturing sector; therefore, the prices of their products will come down, which will promote consumption. Based on Greening's research (Greening et al. 2000), Sorrel referred (3) and (4) as the indirect rebound effect, and economy-wide effect referred to as the synthesis of direct effect and indirect effect. Due to the complexity of the indirect rebound effect and economy-wide rebound effect research, scholars rarely involve the two forms of studies (Turner 2009; Turner and Hanley 2010). Most researches focus on the direct rebound effect.

The concept of direct rebound effect was first proposed by Khazzoom (1980), and he argued that improving energy efficiency could not reduce energy services. On the contrary, it could increase the demand for energy services. Therefore, the actual reduction in energy consumption would not change in proportion to the reduction in energy consumption per unit of energy service. Brookes (1978) also argued that improvements in energy efficiency would lead to economic growth, and economic growth would also increase energy consumption. This is famous Khazzoom–Brookes Postulate, which states that, when energy price remains unchanged, the energy efficiency improvements caused by technological advantages will increase rather than reducing energy consumption.

Since the theory of the energy rebound effect is proposed, there are two different ways to estimate the magnitude of the direct rebound effect. One is to measure it directly through the initial data of investigation, like Mizobuchi (2008) described in Fig. 2.5. He assumed ε_0 and $\varepsilon_1 (\varepsilon_0 < \varepsilon_1)$ as different energy efficiency levels of energy service (ε_0 is the initial energy efficiency and ε_1 is the improved energy efficiency). Haas and Biermayr (2000) defined the rebound effect by a simple equation (2.5).

RE of 10% means that 10% of the expected savings are offset by the increased consumption. Saunders (2005) made a specification for the rebound effect according to some previous researches.[5] This method estimates the energy rebound effect by comparing the energy demands before and after the improvement of energy efficiency. Since changes in other factors may influence the energy consumption demand, we also have to control other variables during the calculation. Yet in practice, most researchers only focus on comparing the energy consumption before and after energy efficiency is changed, but they fail to control other necessary variables, which leads to the results that are potentially biased (Frondel and Schmidt 2005).

Another method is the econometric model which applies some proxy variables for estimation and the most common definition for estimation of the direct rebound effect is as follows (Khazzoom 1980; Berkhout 2000; Sorrell and Dimitropoulos 2008):

$$RE = \eta_\varepsilon(S) = \frac{\partial \ln S}{\partial \ln \varepsilon} = 1 + \eta_\varepsilon(E) \tag{3.25}$$

in which ε is the efficiency elasticity of energy demand, and $\eta_\varepsilon(S)$ is the efficiency elasticity of an energy service. The energy saving caused by energy efficiency improvement is only effective when the efficiency elasticity of demand for energy is zero, that is, the efficiency elasticity of demand for energy is equal to minus one. A positive rebound effect implies that $\eta_\varepsilon(S) > 0, |\eta_\varepsilon(E)| < 1$. In addition, energy

[5]If RE > 1, the rebound effect is called backfire effect;
If RE = 1, the rebound effect is called full rebound effect;
If 0 < RE < 1, the rebound effect is called partial rebound effect;
If RE = 0, the rebound effect is called zero rebound effect;
If RE < 0, the rebound effect is called super conservation effect.

efficiency $\varepsilon(\varepsilon = S/E)$ is defined as the ratio of useful energy service to energy input. If energy efficiency improves, the energy consumption will decline. According to the definition of energy efficiency, we get the price (cost) of the energy service as follows: $P_S = P_E/\varepsilon$, which is proportional to the energy price P_E.

However, energy efficiency ε is difficult to acquire in many empirical studies. Therefore, researchers set some premises to avoid the ε variable.[6] They used the negative price elasticity of energy service to estimate the direct rebound effect (Berkhout 2000; Binswanger 2001).

That is,

$$RE = \eta_{\varepsilon}(S) = -\eta_{P_S}(S) = -\frac{\partial \ln S}{\partial \ln P_S} \tag{3.26}$$

Because this, data of P_S and S can be acquired more easily, Eq. (3.26) is used more often than Eq. (3.25). However, this definition needs rigorous data of energy service price, the energy service price depends on the changes in energy price and energy efficiency ($P_S = P_E/\varepsilon$). If we consider ε as a constant, the rebound effect is equal to the price elasticity of energy consumption ($\eta_{P_E}(E)$) (Saunders 2005; Sorrell and Dimitropoulos 2008; Frondel et al. 2008).

$$\mathrm{RE} = \eta_{\varepsilon}(S) = -\eta_{P_E}(E) = -\frac{\partial \ln E}{\partial \ln P_E} \tag{3.27}$$

In Eq. (3.26), we assume that energy price is as an exogenous variable, and energy service demand is only affected by the price of energy services. If energy efficiency ε is constant, then Eq. (3.26) is equal to Eq. (3.27) (Frondel et al. 2008).

However, as in many cases, the data of energy consumption and energy efficiency is difficult to obtain, then we can use the energy price elasticity of energy service to estimate the direct rebound effects (Matos and Silva 2011).

$$\mathrm{RE} = \eta_{\varepsilon}(S) = -\eta_{P_E}(S) = -\frac{\partial \ln S}{\partial \ln P_E} \tag{3.28}$$

Like Eq. (3.26), Eq. (3.28) also must satisfy the symmetry and exogeneity assumptions to get the same effectiveness of Eq. (3.25). The Matos and Silva's (Matos and Silva 2011) empirical study demonstrated the estimation results of the direct rebound effect of Portugal's road freight transport and used the same Eq. (3.26) with Eqs. (3.25) and (3.26).

---

[6](1) Exogeneity: energy prices' change can't affect energy efficiency. That is, $\eta_{P_E}(\varepsilon) = \frac{\partial \ln \varepsilon}{\partial \ln P_E} = 0$.
(2) Symmetry: Energy services are only affected by the cost of energy services, consumers respond in the same way to energy price decline and energy efficiency improvement.

As discussed above, most former researches used Eq. (3.26) and Eq. (3.27) to estimate the direct rebound effect. But, none of these studies identified here directly measured the rebound effect of increased fuel consumption due to efficiency improvements in road freight transport [Eqs. (3.26) and (3.27)]. However, these studies used different variables of "useful work" (energy service) to estimate the rebound effect. Because of the lack of detail data for energy consumption in transportation division (road transport, sea transport, urban transport, rail transport, air transport, and other transport modes), we will use the data of energy service and Eq. (3.28) to estimate China's direct rebound effects of road freight transport.

3.3.1.2 Evidence of Direct Rebound Effect from Transportation Department

Empirical studies of direct rebound effect mostly focus on passenger transportation, household heating, household cooling, and other energy services (Sorrell and Dimitropoulos 2008; Bentzen 2004; Freire 2010; Missemer 2012). Among them, the energy rebound effect of the transport division has received much attention. Blair et al. (1984) used the monthly data of motor transport of Florida from 1967 to 1976, therein short-term rebound effect of motor transport was estimated between 25 and 40%, and long-term rebound effect is found 25–40%. Mayo and Mathis (1988) used the yearly data of motor transport of the US from 1958 to 1984 to estimate the short-term and long-term rebound effect of motor transport, which is and found as 22%, and 26%, respectively. The other similar studies' estimation results regarding the direct rebound effect of motor transport neither the short term nor the long term were less than 40% (Greene 1992; Jones 1993).

Since the twenty-first century, the energy rebound effect of the transport sector has received much attention from the researchers. With the panel data of US states from 1966 to 2001, Small and van Dender (Small and Dender 2005; Small and Dender 2007) took into account the endogeneity of energy efficiency, and used three-stage least squares method to estimate the long-term and short-term direct rebound effect of vehicle energy consumption which consequently reckoned at 4.7% and 22%, respectively. It is observed that the rebound effect decreased with an increase in people's income. Based on the cross section and time serial data of the US from 1966 to 2004, Hymel et al. (2010) calculated that the direct rebound effect of road transport was to be about 0.16. The results of the aforementioned study were slightly lower than the former research results but this further confirmed the previous conclusion that "the rebound effect decreases with an increase of people's income, increases with the rise of energy costs, increases with the higher congestion level." The rebound effect analysis of cars and light trucks by Greene (2007) also confirmed the relationship between rebound effect and people's income. Frondel et al. (2012) and Su (2012) estimated the direct rebound of road transport effect in Germany and America with a quantile regression method. Their study found that income's influence on the rebound effect is positive, and decreases with the rise of road mileage. Wang et al. (2012) are earliest reporters, who investigated the direct rebound effect of passenger transport in developing countries, and found that the direct rebound effect of passenger transport in urban China was 96%, which means a majority of the expected reduction from efficiency improvement could be offset due to the existence of the rebound effect.

Until now, the current researches rarely involve freight transportation. Borger and Mulalic (2012) argued that the rising fuel prices would promote freight energy efficiency improvement, and then the rebound effect would be decreased. Matos and Silva (2011) reported that the road freight direct rebound effect in Portugal is more stimulating to improve the operating efficiency instead of increasing the energy efficiency (fuel economy) to enhance the total efficiency. As the same with passenger transport, in the road freight transport sector, when energy efficiency (fuel economy) improves, the unit cost of truck traveling will reduce. Further, if the road freight transport industry is a perfect competitive market structure, then it will lead to a lower freight charge. Thus the demand for trucking services will be stimulated, and fuel consumption will increase as well. Consequently, this partially offsets the initial reduction in fuel use induced by higher fuel efficiency, thus at the end proportional response of total fuel use is less than exogenous changes in fuel efficiency. However, different from passenger transport, road freight transport is closely tied to economic activity and restricted by consumers' demand. Therefore, in the road freight transport division, the direct rebound effect not only involves the potential behavioral or operational changes of the owner-operators that may lead to the increased fuel demand, but also impacts of the external factors driving the road freight demand (James et al. 2012).

The road freight transportation energy rebound effect of road freight transport in developing countries is fundamentally different from the developed countries; its magnitude may be much higher compared with developed countries; the energy service demand in developing countries is also far away from saturation. Therefore, research in this area is significant to conserve the energy consumption in developing countries.

3.3.2 Methodology and Data

3.3.2.1 Econometric Model

The development of road freight transport is closely tied with economic activity and restricted by the consumers' demand. So this study adopts the freight turnover quantity (hundred million tons per kilometer) as the estimator of freight transportation demand, and based on the factors of China's road freight demand and road freight CO_2 emissions (Wang et al. 2011, 2012), we select the key indicators that affect the road freight demand, given as follows: total retail sales of consumer goods

(CONS) which reflect people's desire and capability of spending, vehicle numbers per kilometer, which reflect density of traffic (DT) and the price of fuel (P_E). Although there are various functional forms, the double-log model seems to be easy to specify and convenient to estimate. Besides this, its coefficients can be directly interpreted as constant elasticity. Therefore, in this study, the double-log model is adopted. We can obtain the following recursive equation for estimation of the direct rebound effect for road freight transport in China.

$$\ln TKM_{it} = \alpha + \alpha_1 \ln P_{E_{it}} + \alpha_2 \ln DT_{it} + \alpha_3 \ln CONS_{it} + \mu_{it} \tag{3.29}$$

in which α is a constant, α_1–α_3 are the parameters to be estimated, and μ_{it} represents the random error term. Then a simple explanation of the variables in the model equation is given as follows.

TKM_{it}: The explained variable, indicate the road freight demand of the province i in year t (unit: hundred million tons per kilometer)

$P_{E_{it}}$: Energy prices of the province i in year t. As diesel oil is taking 80% share in China's road transport fuel use for past twenty years, so this study uses retail price of 0# diesel oil as the energy price (at the constant price of the year 1998). Thus, the rise of diesel price has a negative impact on the road freight demand.

DT_{it}: Traffic density of the province i in year t. Traffic engineers take the phenomenon of input capability to exceed output as congestion, then this study uses the traffic density to estimate its level. For adjacent vehicles, the increased speed decreases the traffic density, while decreased speed increases the traffic density. Thus, it will cause the uncertainty of vehicle travel time; freight turnover of that day will also be affected. Congestion will cause frequent acceleration, which results in the consumption of more gasoline and diesel.

$CONS_{it}$: Total retail sales of consumer goods of the province i in year t (at the constant price of the year 1998, unit: 100 million yuan). The increase of total retail sales of consumers good directly causes the upsurge of road freight demand.

3.3.2.2 Data Description

Under the framework of Eq. (3.29), this study conducts the empirical analysis of the entire nation, and eastern, central, and western direct rebound effect of China. However, in order to maintain consistency and keep in view the availability of data, Taiwan, Hong Kong, and Macao are not included in this study. The data used include the road freight turnover, total retail sales of consumer goods, total civil cars, total road mileage, retail price of 0# diesel oil of China's 31 provinces (autonomous regions, municipalities) and autonomous regions from 1999 to 2011. Among these, the provincial data of total retail sales of consumer goods are from

China Statistical Yearbooks (1999–2012), the data of retail price of 0# diesel oil (1999–2006) are from *China Price Yearbooks* (2000–2007), and the data of retail price of 0# diesel oil (2006–2011) are from Wind Economic Database. We described the variables of this study in Table 3.9.

3.3.3 Econometric Results and Analysis

3.3.3.1 Panel Unit Root Test

Econometric theory indicates that most of the economic variables panel data are not stationary, the conventional econometrical method may lead to spurious regression, and this study adopts the stationary test to avoid this. There are many methods for panel unit root test and all have their own uniqueness. Therefore, it is difficult to harvest the same conclusion from the results coming from different test methods. In order to guarantee the robustness of the test results and improve the credibility of the conclusion, this study adopts the LLC (Levin–Lin–Chu) test, IPS (Im–Pesaran–Shin) test and CH (Ch-statistics) test method to test the panel unit root of each variable. The panel unit root test results are shown in Table 3.10.

From Table 3.10, we can conclude that, except for the lnP_E variables, the others do not refuse the null hypothesis of the existing panel unit root, but the first-order difference of all variables does not exist panel unit root.

Table 3.9 Descriptive statistics of the variables in 31 provinces (autonomous regions, municipalities)

Variables	ln TKM	ln P_E	ln DT	ln CONS
Mean	5.634493	1.214872	2.779429	7.199587
Media	5.608372	1.223254	2.583838	7.395082
Maximum	8.798515	1.820159	5.404569	9.521927
Minimum	2.602690	0.430984	1.296685	3.574019
Std. Dev.	1.176631	0.306298	0.780822	1.169984
Skewness	0.187462	−0.200579	1.184069	−0.780050
Kurtosis	3.008056	2.106945	4.259978	3.397114
Observations	403	403	403	403
Cross sections	31	31	31	31

3.3.3.2 Panel Cointegration Equilibration and Long-Term Rebound Effect Analysis

In applied economics, if the regression model was not cointegrated, the parameter estimates and corresponding test statistics would be biased and inconsistent. From the panel unit root test in Table 3.10, all variables in the model are in a single integrated order series, fulfill the requirements of the panel cointegration test and can continue the panel cointegration test. According to Eq. (3.29), we finish the correlated individual random effects with Hausman test and individual fixed effect F-test, and we argue that cross-sectional fixed effect model is the optimal estimation model, so this study established cross-sectional fixed effect regressions of the entire nation, and eastern, central, and western regions. The established long-term equilibrium equations of road freight demand are as follows:

$$\ln(\text{TKM}_{j,it}) = \alpha_{j0} + \alpha_{j1}\ln(P_{E_{j,it}}) + \alpha_{j2}\ln(\text{DT}_{j,it}) + \alpha_{j3}\ln(\text{CONS}_{j,it}) + \mu_{j,it} \quad (3.30)$$

where j = 0, 1, 2, 3 refers to the entire nation, and the eastern, central, and western regions, i is region number, t is interval time.

The results of the panel unit root test suggest that there is a cointegration relationship between variables. Therefore, we adopt the Engle-Granger two-step method and cross-sectional fixed effect regression to estimate the long-term equilibrium equation (i.e., cointegration equation). According to the value of adjusted R-squared and F-statistics, we can conclude t the completely long-term cointegrating model fit very well. The value of DW also means there is no self-correlativity for residual series. Further test for the unit root of the residuals series suggests that the four testing methods all refuse the null hypothesis of the existing panel unit root at the 5% level. This means that the origin of the residual series is a stationary series (Table 3.10). Therefore, the results of the Engle-Granger two-step test method also suggest that the cointegration relationship exists between the variables.

Table 3.10 Results of panel unit root tests

Variable	LLC test	IPS test	CH test	
			ADF-Fisher test	PP-Fisher test
$\ln TKM$	2.49782 (0.9938)	7.10510 (1.0000)	4.58490 (1.0000)	3.95271 (1.0000)
$\Delta \ln TKM$	−14.4134 (0.0000)	−8.50428 (0.0000)	177.017 (0.0000)	201.165 (0.0000)
$\ln P_E$	−4.60549 (0.0000)	2.56124 (0.9948)	26.0537 (1.0000)	29.7206 (0.9998)
$\Delta \ln P_E$	−17.4921 (0.0000)	−14.7531 (0.0000)	286.086 (0.0000)	330.785 (0.0000)
$\ln CONS$	−8.89460 (1.0000)	14.5055 (1.0000)	1.69101 (1.0000)	0.76913 (1.0000)
$\Delta \ln CONS$	−10.5096 (0.0000)	−6.30046 (0.0000)	138.113 (0.0000)	144.681 (0.0000)
$\ln DT$	1.76233 (0.9610)	4.62252 (1.0000)	30.2910 (0.9998)	23.7549 (1.0000)
$\Delta \ln DT$	−12.3552 (0.0000)	−8.47108 (0.0000)	176.466 (0.0000)	211.150 (0.0000)

Notes Δ means the first-order difference; the value in the brackets is P-value, and it means that it refuses the null hypothesis of existing panel unit root at the 5% level

Based on the estimated results of long-term equilibrium analysis in Table 3.11, we argue that China's long-term direct rebound effect of road freight transport (its magnitude is 0.84) is higher than OECD countries. Among these, the lowest rebound effect in eastern China is 0.52, and in central China it is recorded highest as 0.80, while in western China it is 0.78. However, the influence of total retail sales of consumer goods is just the opposite. In addition, the analysis of road traffic density revealed that the east China as a most seriously affected region by traffic congestion and its influence on the road freight transport demand is negative, which is precisely akin with our prediction. Its influence on the road freight transport demand in central China is higher than in western China, which is not akin to our prediction. Additionally, estimations of cross-sectional fixed effect varying-coefficient models show that the rebound effect decreased with decreasing the urbanization growth rate (Table 3.12).

3.3.3.3 Panel Error Correction Model and Short-Term Rebound Effect Analysis

Error correction model (3-ECM) is often used to estimate the short-term elasticity. ECM is a specific econometrics model, which adopts the long-term cointegration equation as an instrumental variable to solve the spurious regression problem. In this study, according to the definition of ECM and Eq. (3.28), the short-term price elasticity of the ECM is the short-term energy rebound effect, which we suppose to acquire, according to Eq. (3.30) and Table 3.13. We can acquire the residual series $\hat{\mu}_{j,it}$, and take it as an error correction term.

Table 3.11 Estimation of long cointegration equation

Estimates		National	Eastern	Central	Western
Constant term	α_0	-6.690219^{c} (−13.07746)	-5.691975^{c} (−4.970504)	-8.891121^{c} (−6.951024)	-5.400921^{c} (−8.235164)
Coefficient	α_1	-0.842634^{c} (−5.618644)	-0.516447^{b} (−2.006993)	-0.798643^{b} (−2.207092)	-0.782739^{c} (−3.124608)
	α_2	1.898152^{c} (19.85111)	1.671939^{c} (8.322379)	2.000602^{c} (8.781983)	1.758425^{c} (11.31041)
	α_3	-0.114234^{b} (−2.092929)	-0.3000498^{c} (−2.635123)	0.341402^{b} (2.312696)	0.167222^{a} (1.547823)
R^2		0.951316	0.957710	0.920607	0.935729
Adjusted R^2		0.946962	0.953448	0.912070	0.929348
DW statistics		0.956994	0.741901	0.994596	1.046615
F-statistics		218.5006	224.7196	107.8391	146.6313
Sample size		403	143	104	156

[a]Coefficient is significant at the 10% level; [b]Coefficient is significant at the 5% level; [c]Coefficient is significant at the 1% level; the value in the brackets is t value

$$\mathrm{ecm}_{j,it} = \hat{\mu}_{j,it} = \ln(TKM_{j,it}) - \hat{\alpha}_{j0} - \hat{\alpha}_{j1}\ln(P_{E_{j,it}}) - \hat{\alpha}_{j2}\ln(DT_{j,it}) - \hat{\alpha}_{j3}\ln(CONS_{j,it})$$

Then we can construct a panel error correction model as follows:

$$\Delta\ln(TKM_{j,it}) = \beta_{j0} + \beta_{j1}\Delta\ln(P_{E_{j,it}}) + \beta_{j2}\Delta\ln(DT_{j,it}) + \beta_{j3}\Delta\ln(CONS_{j,it}) + \beta_{j4}\Delta\ln(TKM_{j,it-1}) + \gamma_j ecm_{j,it-1} + \varepsilon_{j,it} \quad (3.31)$$

where the value of β_{j1} (the coefficient of $\Delta\ln(P_{E_{j,it}})$) is the short-term rebound effect and $\varepsilon_{j,it}$ is a random error term. Equation (3.31) shows that the short-term volatility of road freight transport demand in China not only depends on various factors' short-term changes but also is affected by deviation from the balanced trend ($\mathrm{ecm}_{j,it-1}$). In addition, the difference series reflects the volatility of the variable, for example, $\Delta\ln(P_{E_{j,it}})$ shows the volatility of the price of 0# diesel oil, $\Delta\ln(DT_{j,it})$ shows the volatility of the traffic density, and $\Delta\ln(CONS_{j,it})$ shows the volatility of total retail sales of consumer goods. The coefficient of difference series means short-term elasticity.

By analyzing the estimation results of ECM, we acquired the negative direct rebound effect on the road freight transport of China, which is a super conservative effect. That is very different from long-term rebound effect, in case of the short-term rebound effect, improving the truck's energy efficiency will further not only yield a proportional reduction in fuel consumption but also decrease the expected consumption. This shows that, in the short term, improving the energy efficiency will significantly reduce energy consumption. In addition, the density of road traffic has a positive impact on the road freight transport demand in the short term.

3.3.4 *Implications*

The empirical study indicates that: in aspect of long term, the partial rebound effect exists in China's energy consumption of road freight transport, and its magnitude of entire nation, eastern, central, and western region is 84%, 52%, 80%, and 78%, respectively; however, in the aspect of short term, its magnitude is negative, in other words, slight super conservative effect exists. In the aspect of long term, the majority of the expected energy reduction from efficiency improvement could be offset due to the existence of a rebound effect; improving the energy efficiency policy is not as effective as expected. The magnitude of the rebound effect should be an alternative criterion for the efficiency estimation of a single energy service.

In addition, we found that the direct rebound effect for road freight transport tends to decline with the decrease of the urbanization's growth rate. It is shown that improving energy efficiency can result in rebound effect; however, the effect of improving energy efficiency caused by scale is weakened. While, at present,

Table 3.12 Rebound effect and urbanization growth rate

Year	1999	2000	2001	2002	2003	2004	2005	2006	2007	2008	2009	2010	2011
Rebound effect	1.14[c] (−4.74)	1.07[c] (−4.89)	1.25[c] (−4.42)	1.10[c] (−3.95)	0.95[c] (−3.36)	0.90[c] (−3.19)	0.78[c] (−2.97)	0.76[c] (−3.04)	0.80[b] (−3.01)	0.36[a] (−1.30)	0.42[a] (−1.44)	0.31[a] (−1.10)	0.11[a] (−0.39)
Urbanization growth rate	0.043	0.041	0.040	0.038	0.037	0.030	0.029	0.031	0.034	0.024	0.028	0.033	0.026

[a]Coefficient is significant at the 10% level; [b]Coefficient is significant at the 5% level; [c]Coefficient is significant at the 1% level; the value in the brackets is *t* value

Table 3.13 Results of the panel error correction model estimation

Estimates		National	Eastern	Central	Western
Constant term	β_0	0.055258[c] (5.594282)	−0.026701[a] (−1.916812)	−0.062397[c] (−3.091048)	0.088239[c] (14.83503)
Coefficient	β_1	0.094873[c] (2.675754)	0.111939[a] (1.896865)	0.295508[c] (2.929211)	0.130678[c] (3.777887)
	β_2	0.484013[c] (0.072182)	0.865405[c] (7.378928)	1.343202[c] (9.333835)	0.189655[c] (4.726371)
	β_3	0.102669[c] (4.402975)	0.072391[c] (2.773107)	0.474530[c] (7.164038)	0.230371[c] (15.93622)
	β_4	0.139387[c] (12.47540)	0.267466[c] (4.510106)	0.182541[c] (2.967263)	0.164081[c] (32.51226)
	γ	−0.303449[c] (−20.189111)	−0.260256[c] (−8.031111)	−0.350990[c] (−7.719242)	−0.354672[c] (−59.77356)
R^2		0.698807	0.576557	0.721440	0.991437
Adjust R^2		0.694312	0.558146	0.704455	0.991098
DW statistics		1.944734	2.047098	2.153863	2.074875
F statistics		155.4489	31.31658	42.47429	2917.819
Sample size		341	121	88	132

[a]Coefficient is significant at the 10% level; [b]Coefficient is significant at the 5% level; [c]Coefficient is significant at the 1% level; the value in the brackets is t value

Chinese oil price with frequent fluctuation is decided by the market, and in short term, the demand of oil products have reduced by increasing the price, which has been risen 23 times in 34 price adjustments issued by National Development and Reform Commission (NDRC) from 2005 to 2012. Therefore, the magnitude of the short-term direct rebound effect may be measured lower. China may get out of "prisoner's dilemma" under the lower energy price and energy efficiency, and realize the way to develop the sustainable energy consumption mode for Chinese road freight transport.

References

Al-Ghandoor, A. 2013. An approach to energy savings and improved environmental impact through restructuring Jordan's transport sector. *Renewable and Sustainable Energy Reviews* 18: 31–42.

Alphonse, G.S., O.L. Alfons, and E. Grigorios. 2015. Estimating shadow price and efficiency analysis of productive inputs and pesticide use of vegetable production. *European Journal of Operational Research* 245 (1): 265–272.

Andreoni, V., and S. Galmarini. 2012. European CO_2 emission trends: a decomposition analysis for water and aviation transport sectors. *Energy* 45 (1): 595–602.

Ang, B.W. 2005. The LMDI approach to decomposition analysis: a practical guide. *Energy Policy* 33 (7): 867–871.

Becky, P.Y.L., and L. Linna. 2012. Carbon dioxide emissions from passenger transport in China since 1949: implications for developing sustainable transport. *Energy Policy* 50 (6): 464–476.

Beijing Municipal Committee of Transport. 2007. The third comprehensive survey report on transportation, Beijing Transportation Research Centre.

Bentzen, J. 2004. Estimating the rebound effect in U.S manufacturing energy consumption. *Energy Economics* 26: 123–134.

Berkhout, P.H.G., J.C. Muskens, and J.W. Velthuijsen. 2000. Defining the rebound effect. *Energy Policy* 28 (6–7): 425–432.

Binswanger, M. 2001. Technological progress and sustainable development: what about the rebound effect? *Ecological Economics* 36 (1): 119–132.

Blair, R.D., D.L. Kaserman, and R.C. Tepel. 1984. The impact of improved mileage on gasoline consumption. *Economic Inquiry* 22 (2): 209–217.

Borger, B.D., and I. Mulalic. 2012. The determinants of fuel use in the trucking industry—volume, fleet characteristics and the rebound effect. *Transport Policy* 24 (9): 284–295.

Brand, C., M. Tran, and J. Anable. 2012. The UK transport carbon model: an integrated life cycle approach to explore low carbon futures. *Energy Policy* 41: 107–124.

Brand, C., A. Goodman, H. Rutter, et al. 2013. Associations of individual, household and environmental characteristics with carbon dioxide emissions from motorised passenger travel. *Applied Energy* 104: 158–169.

Brookes, L. G. 1978. Energy policy, the energy price fallacy and the role of nuclear energy in the U.K. *Energy Policy* 6 (2): 94–106.

Cai, B.F., W.S. Yang, D. Cao, et al. 2012. Estimates of China's national and regional transport sector CO_2 emissions in 2007. *Energy Policy* 41: 474–483.

Cellura, M., S. Longo, and M. Mistretta. 2012. Application of the structural decomposition analysis to assess the indirect energy consumption and air emission changes related to Italian households consumption. *Renewable and Sustainable Energy Reviews* 16 (2): 1135–1145.

Chambers, R.G., Y. Chung, and R. Färe. 1996. Benefit and distance functions. *Journal of Economic Theory* 70 (2): 407–419.

Chandran, V.G.R., and C.F. Tang. 2013. The impacts of transport energy consumption, foreign direct investment and income on CO_2 emissions in ASEAN-5 economies. *Renewable and Sustainable Energy Reviews* 24 (8): 445–453.

Chang, Y.T., N. Zhang, D. Danao, et al. 2013. Environmental efficiency analysis of transportation system in China: a non-radial DEA approach. *Energy Policy* 58: 277–283.

Cheng, G. 2014. *Data envelopment analysis: method and MaxDEA software (in Chinese)*. Beijing: Intellectual Property Publishing House.

Chèze, B., J. Chevallier, and P. Gastineau. 2013. Will technological progress be sufficient to stabilize CO_2 emissions from air transport in the mid-term? *Transportation Research Part D: Transport and Environment* 18: 91–96.

Choi, Y., N. Zhang, and P. Zhou. 2012. Efficiency and abatement costs of energy-related CO_2 emission in China: a slacks-based efficiency measure. *Applied Energy* 98: 198–208.

Chung, Y.H., R. Färe, and S. Grosskopf. 1997. Productivity and undesirable outputs: a directional distance function approach. *Journal of Environmental Management* 51: 229–240.

Chung, W., G.H. Zhou, and I.M.H. Yeung. 2013. A study of energy efficiency of transport sector in China from 2003 to 2009. *Applied Energy* 112: 1066–1077.

Cong, R. 2013. An optimization model for renewable energy generation and its application in China: a perspective of maximum utilization. *Renewable and Sustainable Energy Reviews* 17: 94–103.

Cong, R.G., and S.C. Shen. 2014. How to develop renewable power in China? A cost-effective perspective. *The Scientific World Journal* (1): 946932.

Connolly, D., B.V. Mathiesen, and I. Ridjan. 2014. A comparison between renewable transport fuels that can supplement or replace biofuels in a 100% renewable energy system. *Energy* 73: 110–125.

Duffy, A., and R. Crawford. 2013. The effects of physical activity on greenhouse gas emissions for common transport modes in European countries. *Transportation Research Part D: Transport and Environment* 19: 13–19.

Färe, R., and S. Grosskopf. 2004. Modelling undesirable factors in efficiency evaluation: comments. *European Journal of Operational Research* 157: 242–245.

Freire, G.J. 2010. Empirical evidence of direct rebound effect in Catalonia. *Energy Policy* 38 (5): 2309–2314.

Frondel, M., and C.M. Schmidt. 2005. Evaluating environmental programs: the perspective of modern evaluation research. *Ecological Economics* 55 (4): 515–526.

Frondel, M., J. Peters, and C. Vance. 2008. Identifying the rebound: evidence from a German household panel. *The Energy Journal* 29: 145–163.

Frondel, M., N. Ritter, and C. Vance. 2012. Heterogeneity in the rebound effect: further evidence for Germany. *Energy Economics* 34: 461–467.

Greene, D.L. 1992. Vehicle use and fuel economy: how big is the rebound effect? *Energy Journal* 13 (1): 117–143.

Greene, D. 2007. Rebound: analysis of U.S. light-duty vehicle travel statistics. *Energy Policy* 41: 14–28.

Greenhouse Gas Protocol. 2005. Calculating CO_2 emissions from mobile sources: guidance to calculation worksheets. http://www.ghgprotocol.org/calculation-tools/service-sector [2011-12-18].

Greening, L.A., D.L. Greene, and C. Difiglio. 2000. Energy efficiency and consumption—the rebound effect—a survey. *Energy Policy* 28 (6–7): 389–401.

Guo, X.D., L. Zhu, Y. Fan, et al. 2010. Evaluation of potential reductions in carbon emission in Chinese provinces based on environmental DEA. *Energy Policy* 39: 2352–2360.

Haas, R., and P. Biermayr. 2000. The rebound effect for space heating: empirical evidence from Austria. *Energy Policy* 28: 403–410.

He, F., Q.Z. Zhang, J.S. Lei, et al. 2013. Energy efficiency and productivity change of China's iron and steel industry: accounting for undesirable outputs. *Energy Policy* 54: 204–213.

Herbruggen, B.V., and S. Logghe. 2005. Tremove version 2.3. Simulation model for European environmental transport policy analysis. In *14th International Conference. Transport and Air Pollution Graz.*

Hoffrichter, A., A.R. Miller, S. Hillmansen, et al. 2012. Well-to-wheel analysis for electric, diesel and hydrogen traction for railways. *Transportation Research Part D: Transport and Environment* 17 (1): 28–34.

Hymel, K.M., K.A. Small, and K.V. Dender. 2010. Induced demand and rebound effects in road transport. *Transportation Research B* 44 (10): 1220–1241.

IEA. 2010. CO_2 Emissions from Fuel Combustion Highlights. Paris: International Energy Agency.

IEA. 2011. CO_2 Emissions from Fuel Combustion. http://www.iea.org/co2highlights/CO2highlights.pdf [2012-11-11].

IPCC. 2006. IPCC guidelines for national greenhouse gas inventories. Hayama, Japan: The Institute for Global Environmental Strategies.

James, J.W., H.G. Erin, C. Bryan, et al. 2012. Estimating the direct rebound effect for on-road freight transportation. *Energy Policy* 48: 252–259.

Jennings, M., B.P. Gallachóir, and L. Schipper. 2013. Irish passenger transport: data refinements, international comparisons, and decomposition analysis. *Energy Policy* 56: 151–164.

Jones, C.T. 1993. Another look at U.S. passenger vehicle use and the rebound effect from improved fuel efficiency. *The Energy Journal* 14 (4): 99–110.

Jou, R.C., and T.Y. Chen. 2014. Factors affecting public transportation, car, and motorcycle usage. *Transportation Research Part A: Policy and Practice* 61: 186–198.

Kanzian, C., M. Kühmaier, J. Zazgornik, et al. 2013. Design of forest energy supply networks using multi-objective optimization. *Biomass and Bioenergy* 58: 294–302.

Khazzoom, J.D. 1980. Economic implications of mandated efficiency in standards for household appliances. *Energy Journal* 1 (4): 21–40.

Leleu, H. 2013. Shadow price of undesirable outputs in nonparametric analysis. *European Journal of Operational Research* 231 (2): 474–480.

Li, H., R. Raeside, T. Chen, et al. 2012. Population ageing, gender and the transportation system. *Research in Transportation Economics* 34 (1): 39–47.

Li, H.Q., Y. Lu, J. Zhang, et al. 2013. Trends in road freight transportation carbon dioxide emissions and policies in China. *Energy Policy* 57: 99–106.

Liddle, B. 2012. The systemic, long-run relation among gasoline demand, gasoline price, income, and vehicle ownership in OECD countries: evidence from panel cointegration and causality modeling. *Transportation Research Part D: Transport and Environment* 17 (4): 327–331.

Lin, B.Q., and C.P. Xie. 2014. Reduction potential of CO_2 emissions in China's transport industry. *Renewable and Sustainable Energy Reviews* 33: 689–700.

Liu, W., H. Lund, and B.V. Mathiesen. 2013. Modelling the transport system in China and evaluating the current strategies towards the sustainable transport development. *Energy Policy* 58: 347–357.

Mathiesen, B.V., H. Lund, and P. Nørgaard. 2008. Integrated transport and renewable energy systems. *Utilities Policy* 16: 107–116.

Matos, F.J.F., and F.J.F. Silva. 2011. The rebound effect on road freight transport: empirical evidence from Portugal. *Energy Policy* 39: 2833–2841.

Mayo, J.W., and J.E. Mathis. 1988. The effectiveness of mandatory fuel efficiency standards in reducing the demand for gasoline. *Applied Economics* 20 (2): 211–219.

Mekaroonreung, M., and A.L. Johnson. 2012. Estimating the shadow prices of SO_2 and NO_x for U.S. coal power plants: a convex nonparametric least squares approach. *Energy Economics* 34 (3): 723–732.

Missemer, A. 2012. William Stanley Jevons' The Coal Question (1865), beyond the rebound effect. *Ecological Economics* 82: 97–103.

Mizobuchi, K. 2008. An empirical study on the rebound effect considering capital costs. *Energy Economics* 30: 2486–2516.

Morán, M.Á.T., and P.R. González. 2007. Structural factors affecting land-transport CO_2 emissions: a European comparison. *Transportation Research Part D: Transport and Environment* 12 (4): 239–253.

Mraihi, R., K. Abdallah, and M. Abid. 2013. Road transport-related energy consumption: analysis of driving factors in Tunisia. *Energy Policy* 62: 247–253.

Murty, S., R.R. Russell, and S.B. Levkoff. 2012. On modeling pollution-generating technologies. *Journal of Environmental Economic and Management* 64: 117–135.

Okada, A. 2012. Is an increased elderly population related to decreased CO_2 emissions from road transportation? *Energy Policy* 45: 286–292.

Ong, H.C., T.M.I. Mahlia, and H.H. Masjuki. 2012. A review on energy pattern and policy for transportation sector in Malaysia. *Renewable and Sustainable Energy Reviews* 16: 532–542.

Phillip, Y.L., and S. Lee. 2013. Energy efficiency in the Japanese transport sector. *Energy Policy* 56: 248–258.

Pongthanaisawan, J., and C. Sorapipatana. 2013. Greenhouse gas emissions from Thailand's transport sector: trends and mitigation options. *Applied Energy* 101: 288–298.

Poumanyvong, P., S. Kaneko, and S. Dhakal. 2012. Impacts of urbanization on national transport and road energy use: evidence from low, middle and high income countries. *Energy Policy* 46: 268–277.

Roxana, J.J., N. Ali, and H. Katharine. 2014. Selection of CO_2 mitigation strategies for road transportation in the United States using a multi-criteria approach. *Renewable and Sustainable Energy Reviews* 38: 960–972.

Saboori, B., M. Sapri, and M. Baba. 2014. Economic growth, energy consumption and CO_2 emissions in OECD (Organization for Economic Co-operation and Development)'s transport sector: a fully modified bi-directional relationship approach. *Energy* 66: 150–161.

Saide, P., R. Zah, M. Osses, et al. 2009. Spatial disaggregation of traffic emission inventories in large cities using simplified top-down methods. *Atmospheric Environment* 43 (32): 4914–4923.

Saunders, H.D. 2005. A calculator for energy consumption changes arising from new technologies. *Topics in Economic Analysis and Policy* 5 (1): 1–31.
Shi, G.M., J. Bi, and J.N. Wang. 2010. Chinese regional industrial energy efficiency evaluation based on a DEA model of fixing non-energy inputs. *Energy Policy* 38: 6172–6179.
Si, B., M. Zhong, X. Yang, et al. 2012. Modeling the congestion cost and vehicle emission within multimodal traffic network under the condition of equilibrium. *Journal of Systems Science and Systems Engineering* 21 (4): 385–402.
Sider, T., A. Alam, M. Zukari, et al. 2013. Land-use and socio-economics as determinants of traffic emissions and individual exposure to air pollution. *Journal of Transport Geography* 33: 230–239.
Small, K.A., and K.V. Dender. 2005. The effect of improved fuel economy on vehicle miles traveled: estimating the rebound effect using U.S. State data, 1966–2001. University of California Energy Institute's Energy Policy and Economics, Working Paper Series.
Small, K.A., and K.V. Dender. 2007. Fuel efficiency and motor vehicle travel: the declining rebound effect. *Energy* 28 (1): 25–51.
Sorrell, S., and J. Dimitropoulos. 2008. The rebound effect: microeconomic definitions, limitations and extensions. *Ecological Economics* 65: 636–649.
Su, B., and B.W. Ang. 2012. Structural decomposition analysis applied to energy and emissions: some methodological developments. *Energy Economics* 34 (1): 177–188.
Szendro, G., and A. Török. 2014. Theoretical investigation of environmental development pathways in the road transport sector in the European region. *Transport* 29 (1): 12–17.
Tao, J., and S.R. Yu. 2011. Implementation of energy efficiency standards of household refrigerator/freezer in China: potential environmental and economic impacts. *Applied Energy* 88 (5): 1890–1905.
Timilsina, G.R., and A. Shrestha. 2009. Factors affecting transport sector CO_2 emissions growth in Latin American and Caribbean countries: an LMDI decomposition analysis. *International Journal of Energy Research* 33: 396–414.
Turner, K. 2009. Negative rebound and disinvestment effects in response to an improvement in energy efficiency in the UK economy. *Energy Economics* 31: 648–666.
Turner, K., and N. Hanley. 2010. Energy efficiency, rebound effects and the Environmental Kuznets curve. *Energy Economics* 33 (5): 709–720.
Török, Árpád, and Török Ádám. 2014. Macroeconomic analysis of road vehicles related environmental pollution in Hungary. *Central European Journal of Engineering* 4 (2): 186–191.
Wang, Z.H., and M.L. Lu. 2014. An empirical study of direct rebound effect for road freight transport in China. *Applied Energy* 133: 274–281.
Wang, Z.H., and C. Feng. 2015. Sources of production inefficiency and productivity growth in China: a global data envelopment analysis. *Energy Economics* 49: 380–389.
Wang, W.W., M. Zhang, and M. Zhou. 2011. Using LMDI method to analyze transport sector CO_2 emissions in China. *Energy* 36 (10): 5909–5915.
Wang, Z., B. Zhang, and Y. Zhang. 2012. Determinants of public acceptance of tiered electricity price reform in China: evidence from four urban cities. *Applied Energy* 91 (1): 235–244.
Wang, K., B. Lu, and Y.M. Wei. 2013. China's regional energy and environment efficiency: a range-adjusted measure based analysis. *Applied Energy* 112: 1403–1415.
Wang, Z.H., W. Liu, and J.H. Yin. 2015. Driving forces of indirect carbon emissions from household consumption in China: an input-output decomposition analysis. *Natural Hazards* 75 (2): 257–272.
Wang, Z.H., W.J. He, and K. Chen. 2016. The integrated efficiency of economic development and CO_2 emissions among Asia Pacific Economic Cooperation members. *Journal of Cleaner Production* 131: 765–772.
Wang, T.Y., H.Q. Li, J. Zhang, et al. 2012. Influencing factors of carbon emission in China's road freight transport. *Procedia—Social and Behavioral Sciences* 43: 54–64.
Wei, C., L. Andreas, B. Liu. 2013. An empirical analysis of the CO_2 shadow price in Chinese thermal power enterprises. *Energy Economics* 40: 22–31.

Wu, A.H., Y.Y. Cao, and B. Liu. 2014. Energy efficiency evaluation for regions in China: an application of DEA and Malmquist indices. *Energy Efficiency* 7: 429–439.

Wu, J., Q.Y. Zhu, J.F. Chu, et al. 2015. Measuring energy and environmental efficiency of transportation systems in China based on a parallel DEA approach. *Transportation Research Part D* 48: 460–472.

Xiao, Z.P., Y.W. Chai, and Z.L. Liu. 2011. Quantitative distribution and related factors for household daily travel CO_2 emissions in Beijing. *Urban Studies* 18 (9): 104–112.

Zhang, M., G. Li, H.L. Mu, et al. 2011. Energy and energy efficiencies in the Chinese transportation sector, 1980–2009. *Energy* 36: 770–776.

Zhang, S.J., Y. Wu, H. Liu, et al. 2014. Real-world fuel consumption and CO_2 emissions of urban public buses in Beijing. *Applied Energy* 113: 1645–1655.

Zhou, P., B.W. Ang, and J.Y. Han. 2010. Total factor carbon emission performance: a Malmquist index analysis. *Energy Economics* 32: 194–201.

Zondag, B., and G. de Jong. 2011. The development of the TIGRIS XL model: a bottom-up approach to transport, land-use and the economy. *Research in Transportation Economics* 31: 55–62.

Chapter 4
Residents' Willingness in Purchasing Low-Carbon Products

China overtook the United States as the world's biggest producer of carbon dioxide in 2007, and has become the largest energy consumer at the global level since 2010. By this, China seeks a "green and low-carbon development". Mode and the scheme of "13th Five-Year" comprehensive energy-saving emission reduction work was released in 2017. Low carbon production has aroused increasing concern. New energy vehicles and energy-efficient household appliances are two significant low-carbon products in China.

In this chapter, we conducted a nationwide survey in China to empirically investigate the features and determinants of residents' willingness in purchasing new energy vehicles and energy-efficient household appliances by applying multinomial logit regression and structure–function model. Furthermore, we also proposed some suggestions for policy-makers to promote the two low-carbon products.

4.1 Residents' Willingness and Determinants in Purchasing New Energy Vehicles

There is a rapid increase in car ownership in China, and this has brought significant pressure on fuel supply. China's oil consumption keeps rising; the CO_2 emission from the transportation industry accounts for 23% of total energy-related CO_2 emission (IEA 2012), and other harmful substances in automobile exhaust gas are also affecting health and the quality of life (Che et al. 2007; Courter et al. 2007;

This chapter takes the following literature for reference:

Wang Z, Wang X, Guo D. 2017. Policy implications of the purchase intention towards energy-efficient appliances among China's urban residents: do subsidies work? Energy Policy, 102: 430–439.

Wang Z, Dong X. 2016. Determinants and policy implications of residents' new energy vehicle purchases: the evidence from China. Natural Hazards, 82(1): 155–173.

Z. Wang and B. Zhang, *Low-Carbon Consumption in China: Residential Behavior, Corporate Practices and Policy Implication*,
https://doi.org/10.1007/978-981-15-2792-0_4

Møller et al. 2008). In this situation, the option of new energy vehicles (NEVs), is an alternative to traditional vehicles, in the automobile sector for environmental protection and energy-saving benefits (Arar 2010; Smith 2010). The government also views this as one of the 7th strategic emerging industries and therefore paying handsome subsidies. NEVs are not that popular, but China's NEV sales in 2012 were 12,800 units and in 2013 this rose to 17,600, for nearly 0.1% of all car sales.

For the promotion of NEVs, this research was conducted on the basis of the factors which stimulate and influence the purchase intentions of the peoples for NEVs. Mau et al. (2008) applied a discrete choice model to capture the dynamic changes in consumer preferences and found that the price of property and vehicle attributes has a great effect on consumer preferences for clean vehicle technology. Tran et al. (2013) applied Monte Carlo simulation to find that early adopters were influenced by economic interests, environmental attractiveness, new technologies, and automotive reliability to different extents. Li et al. (2013) determined the factors like energy security, environment, and accessibility of fuel that may affect consumer purchase intentions behind alternative fuel vehicles and hybrid vehicles. Carley et al. (2013) surveyed adult drivers in the US, and found that the disadvantages of electric cars in people's minds affected their willingness to buy plug-in hybrid vehicles. Wang et al. (2013) studied that the residents of Beijing and NEV purchase intentions, pointing out that perceived usefulness, perceived ease of use, subjective norms, and perceived behavioral control had a positive effect on purchase intention; their research also mentioned the effects of demographic variables in the model. According to the theory of planned behavior, Schniederjans and Starkey (2014) postulated the factors that influence consumer willingness to pay for green transportation and their findings revealed that consumer attitudes and peer pressure could positively influence the willingness to pay.

Research on NEVs purchase intentions is mainly based on the theory of planned behavior, rational choice theory, normative theories, environmental attitudes, symbols, self-identity, lifestyle, diffusion of innovations, and consumer innovativeness, as well as consumer emotion models (Cong 2013; Cong and Shen 2014; Rezvani et al. 2015). These theories determined the factors which influence NEVs purchase intentions. The main focus of these factors was on the cognition of NEVs (Schuitema et al. 2013), environmental protection awareness (Lane and Potter 2007; Egbue and Long 2012; Bockarjova and Steg 2014), the attitudes of people around them (Nayum and Klöckner 2014; Schniederjans and Starkey 2014), purchase costs (Lane and Potter 2007; Sovacool and Hirsh 2009), the purchasers' personal characteristics (Nayum and Klöckner 2014), government policies (Zhang et al. 2013), etc.

Preliminary studies indicated that families having personal cars have evidently higher intentions to purchase a NEV as compared to other families (Yin and Wang 2013). Whereas little literature exists on how the various factors influence the purchase intention of families with cars and few scholars investigated the effects of various factors on specific groups with different attitudes and behaviors. This research was conducted based on the improved theory of planned behavior, by taking an example of four major municipalities of China with urban residents to analyze each factor's effects on urban residents' purchase intentions toward NEVs and the moderating effects of perceived behavioral control. This study also empirically investigated the purchase intention determinants for residents with cars, which were measured using a multinomial logistic regression model (Zhang et al. 2013). The results showed that purchase intention determinants for NEVs were different for urban residents both with and without cars, and each factor had its specific domain of influence. The above results provide targeted suggestions for NEVs promotion policies.

4.1.1 Theoretical Framework for Determinants of Purchasing New Energy Vehicles

The theory of planned behavior (TPB) is widely spread and used to explore the relationship between attitudes and behaviors. It is a more comprehensive theory suitable for the study of NEVs purchase intentions, covering perceived attitudes toward NEVs, the perception of the surrounding environment, and the measurement of their perceived behavioral control. Hence, TPB was selected to provide a theoretical base to investigate how attitudes, subjective norms, and perceived behavioral control affect behavioral intentions. In this theory, intentions can be predicted highly accurately from attitudes toward the behavior, subjective norms, and perceived behavioral control; intentions and behavioral control account for considerable variances in actual behavior (Ajzen 1991). The structure of the theory is shown in Fig. 4.1.

As shown in Fig. 4.1, TPB postulates three independent conceptual determinants of intention, namely attitudes, subjective norms, and perceived behavioral control. Firstly, attitudes toward the behavior refer to the degree to which a person has a favorable or unfavorable evaluation or appraisal of the behavior in question. Secondly, subjective norms refer to the perceived social pressure to perform or not to perform the behavior. Thirdly the perceived behavioral control means the perceived ease or difficulty of performing the behavior (Ajzen 1991). Finally, a behavioral intention can find expression in behavior under behavioral control.

Purchase intentions, in this work, were assumed to capture the motivational factors to influence purchase behaviors, indicating how much people are willing and putting efforts to buy. Similarly, all other concepts in TPB which are referred to in this study correspond to new energy vehicle (NEV) purchase behaviors. To measure the attitude toward NEV purchase behavior conveniently, we valued it from the perspective of perceived usefulness and perceived ease of use and concepts borrowed from user acceptance of information technology. Perceived usefulness is defined as: "… the degree to which a person believes that using a

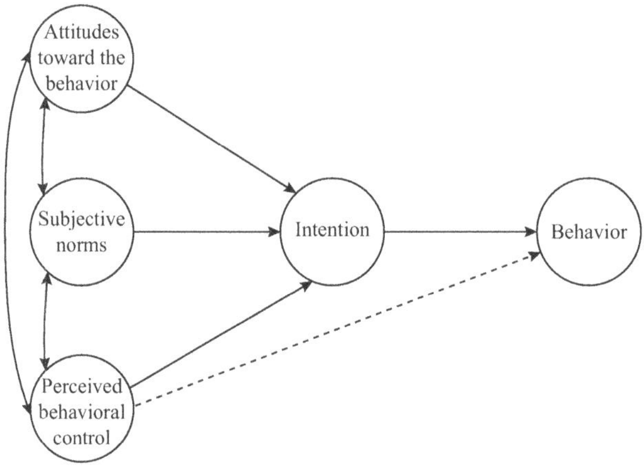

Fig. 4.1 The framework of TPB

particular system would enhance his or her job performance" while perceived ease of use refers to: "… the degree to which a person believes that using a particular system would be free of effort" (Davis 1989). As the objects and unit of analysis of this investigation are urban residents, it is pretty hard for us to know whether respondents carry out the purchase behavior or not: we mainly focused on the determinants of purchase intentions and did not involve the actual purchase behavior part in the theory (Schniederjans and Starkey 2014).

Figure 4.2 shows the theoretical framework used (Taylor and Todd 1995), which shows how perceived usefulness, perceived ease of use, subjective norms, and perceived behavioral control affect citizen's intentions when residents consider buying NEVs. The main focus of this study is urban residents with cars. This study explores the effect of variables and in the same way the moderating effect of perceived behavioral control: this is claimed as a novelty for the current study. In addition, we will discuss the effects of various factors on dissident groups, establishing a discrete choice model, which can provide more targeted advice for the promotion of NEVs.

The perceived attitudes here refer to the degree to which residents would like to purchase NEVs, whereas the higher perceived positive attitude is the more willing for residents to buy (Egbue and Long 2012). An NEV is an innovative product, and a lot of literature suggests that "perceived ease of use" and "perceived usefulness" have a significant impact on the perception of attitudes. Hence, we use "perceived ease of use" and "perceived usefulness" in a technology acceptance model to measure perceived attitudes (Davis 1989). "Perceived usefulness" is the perceived travel efficiency with NEVs. "Perceived ease of use" measures how easy it is to use NEVs. In general, people want to improve their travel efficiency to permit a potentially improved workplace performance and life experience, so if the new

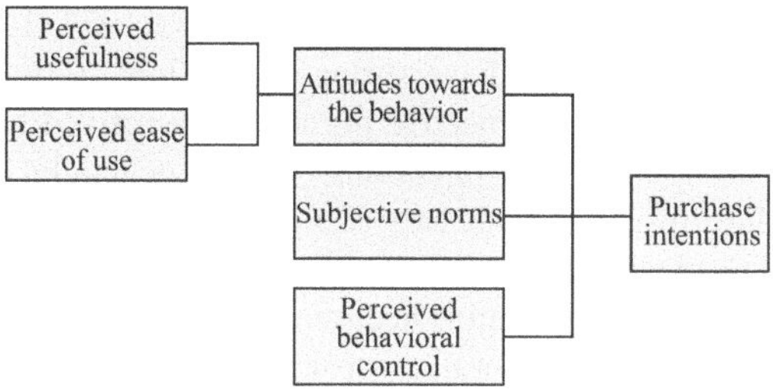

Fig. 4.2 The structure of the theory applied

product is effective in improving personal efficiency, people will be willing to buy it (Radner and Rothschild 1975; Pfeffer 1982). In addition, people's efforts are a finite resource, assigned to various activities for which they are responsible (Radner and Rothschild 1975). An easy-to-use product can save efforts, so products with perceived ease of use will be more easily accepted on the condition that others are the same (Davis 1989). From the above analysis, targeting on all urban residents and urban residents with cars, we made the following assumptions:

H4.1-1. For all urban residents, perceived usefulness significantly affects the intentions to purchase a NEV.
H4.1-2. For urban residents with cars, perceived usefulness significantly affects the intentions to purchase a NEV.
H4.2-1. For all urban residents, perceived ease of use significantly affects the intentions to purchase a NEV.
H4.2-2. For urban residents who have a car, perceived ease of use significantly affects the intentions to purchase a NEV.

According to TPB, the behavioral intention is positively affected by perceived attitudes, subjective norms, and perceived behavioral control (Beck and Ajzen 1991). The effects of perceived attitudes on purchase intentions have been mentioned above. Subjective norms here refer to the expectations of purchasing NEVs from their family, friends, and the people whom they respect. NEVs are considered innovative and environment friendly, which reflects the mainstream values of society and cause a positive impact on purchase intentions. Meanwhile, we can say that people will prefer the advice of family and friends to maintain good relationships and to make sound decision-making. On the basis of these assumptions we can say that subjective norms do matter. So it is assumed that all the residents will be significantly influenced by subjective norms. Perceived behavioral control is about how easy it is to take action, reflecting past experience and anticipated obstacles (Ajzen 1991; Cong and Wei 2010a, b). Personal behavior is strongly influenced by their confidence in taking action (Bandura 1977, 1982). As a result, the persons with more confidence have stronger intentions. From the above analysis, it is clear that the stronger subjective norms and perceived behavioral control are the cause of strong purchase intentions: this is generally applicable to all respondents including residents with cars (Coleman et al. 2011). Besides, in circumstances where residents can decide whether to buy NEVs by themselves, the effects of their environment are diminished, and vice versa, indicating the moderating effects of perceived behavioral control. From this analysis, initially, we assumed that the effects of determinants applied to all the residents:

H4.3-1. For all urban residents, subjective norms significantly affect the intention to purchase a NEV.
H4.3-2. For urban residents with cars, subjective norms significantly affect the intentions to purchase a NEV.
H4.4-1. For all urban residents, perceived behavioral control significantly affects the intentions to purchase a NEV.
H4.4-2. For urban residents with cars, perceived behavioral control significantly affects the intentions to purchase a NEV.
H4.5-1. For all urban residents, perceived behavioral control functions as a moderator of subjective norms' effects on purchase intentions.
H4.5-2. For urban residents with cars, perceived behavioral control functions as a moderator of subjective norms' effects on purchase intentions.

As the hypothesis framework reflects (Fig. 4.3), we empirically examined the impacts of perceived usefulness, perceived ease of use, and subjective norms, and perceived behavioral control on residents' purchase intentions toward NEVs. We also explored the moderate effect of perceived behavioral control on subjective norms, the different effects of determinants for urban residents with cars, and specific groups' effected by the variables.

4.1.2 Research Method

4.1.2.1 Samples and Data Collection

Samples were taken from urban households in the four major municipalities in China: Beijing, Shanghai, Tianjin, and Chongqing. A total of 390 questionnaires were floated, and 322 questionnaires were collected, with a recovery rate of 82.56%. We screened 240 questionnaires for analysis by checking their completeness and judged whether contradictory responses had been provided or not, with a valid response rate of 74.53%. Of the valid samples, 98 were male accounting for 40.83%, 105 were female, accounting for 43.75%, and 37 of them did not specify a gender, accounting for 15.42%. Questionnaire was referred to pure

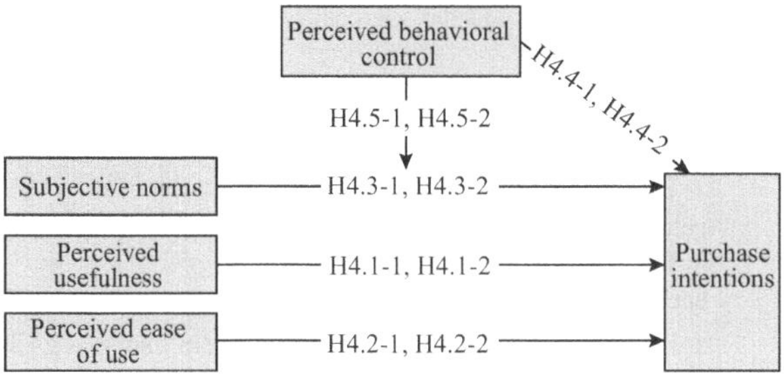

Fig. 4.3 The structure of this model produced

electric, full cell electric, and hybrid power (oil and gas, or oil and electric) vehicles. The vehicles with gasoline and diesel were excluded.

The questionnaire was designed according to TPB, where "purchase intentions" and "purchase intentions when I have owned a vehicle" were measured by "Yes", "No", and "Not Sure". Perceived usefulness, perceived ease of use, subjective norms, perceived behavioral control, and other factors were measured on a five-point Likert-type scale, ranging from "Strongly Disagree" to "Strongly Agree": the items used in the questionnaire are shown in Table 4.1.

Dependent variables in this research are the purchase intentions and purchase intentions when I have owned a vehicle. Purchase intention includes three possible discrete choices: "Willing to buy", "Not certain", and "Unwilling to buy", and is determined by the average value of the three corresponding items. Purchase intentions when I have owned vehicle is directly measured by the item "If I buy another vehicle, I will buy an NEV." Independent variables include perceived usefulness, perceived ease of use, subjective norms, perceived behavioral control, and the control variable are the other factors. To avoid subjectivity from artificial empowerment, we took the loading coefficients from factor analysis as weights to get the weighted average values of the measured values from corresponding items. In the end, we obtained five corresponding variable values and then conducted further research therewith.

4.1.2.2 Descriptive Analysis

Table 4.2 shows that the averages of most items ranged from 3 to 4 except for particular items on perceived ease of use, indicating that most of the residents consider NEVs useful, have positive purchase intentions towards an NEV, have behavioral control, and that their purchase intentions are supported by people around them. Items 14 and 15, measuring perceived ease of use, were less than 3,

Table 4.1 Items of the questionnaire and the sources

Determinants	Items	Sources
Purchase intentions	1. I plan to buy a new energy vehicle in the future 2. I am willing to buy a new energy vehicle in the future 3. I will make efforts to buy a new energy vehicle in the future	Ajzen (1991), Yoon (2011)
Purchase intentions when I have owned vehicles	4. If I buy another vehicle, I will buy a new energy vehicle	Yin and Wang (2013)
Perceived usefulness	5. On the whole, a new energy vehicle can improve the efficiency of the traveler 6. Compared to the traditional vehicle, a new energy vehicle can improve the efficiency of the traveler 7. Compared to having no vehicle, a new energy vehicle can improve the efficiency of the traveler 8. The new energy vehicle can improve my living quality	Davis (1989), Schniederjans and Starkey (2014)
Perceived ease of use	9. The charge time of pure electric vehicles can be accepted 10. It is easy for me to learn to drive new energy vehicles 11. It is convenient to drive new energy vehicles 12. I believe in the quality of new energy vehicle 13. I think the new energy vehicle is safe 14. The after-sales service of new energy vehicles is complete 15. There are some electric car charging infrastructures in my neighborhood	Davis (1989), Zhang et al. (2013)
Subjective norms	16. My friends encourage me to buy a new energy vehicle 17. My family approve me of buying a new energy vehicle 18. People I respect support my buying a new energy vehicle 19. Most of the people I consider their opinions support my buying new energy vehicle in the future	Schniederjans and Starkey (2014)

(continued)

Table 4.1 (continued)

Determinants	Items	Sources
Perceived behavioral control	20. I can decide whether to buy a new energy vehicle or not on my own 21. Whether I will buy a new energy vehicles is not influenced by others 22. I am confident that I can buy a new energy vehicle in the future if I want	Ajzen (1991), Yoon (2011)
Other factors	23. A single charge of electric vehicle could not satisfy my daily travel 24. The price of new energy vehicles is a little high 25. The life of new energy vehicles is short 26. I will not buy a new energy vehicle if the electricity price for new energy vehicles is higher than residential electricity	Sovacool and Hirsh (2009), Zhang et al. (2013), Schniederjans and Starkey (2014)

showing the residents are not satisfied with the after-sales service from NEVs and electric vehicle charging facilities. The standard deviation of item 15 was large, which means that charging facilities for NEVs in the four cities differ greatly. The means of items measuring "other factors" were all within the range of 3–4, indicating that the single charging mileage, the price, and service life of electric vehicles cannot meet the requirements of residents and that the prices of electric vehicles are effected by purchase intentions of residents.

According to the research results, we acquired the distribution of urban residents' purchase intentions, based on all samples. At the same time, we obtained the purchase intentions by the distribution of all the resident samples with cars.

The majority of urban residents are willing to purchase NEVs in the future, followed by uncertain residents, and unwilling residents. This shows that most residents have intentions to purchase a NEV in the future, and that NEVs will be popular in the future, which is consistent with the increasing trend in NEV sales in China. A large proportion of residents with cars hold uncertain attitudes and purchase intentions, possibly because they have had wider experiences and consider more factors, which indicates that the factors influencing uncertain residents should be paid greater heed.

Table 4.2 Descriptive statistics of samples

Determinants	Item	Mean	Mode	Std. deviation
Purchase intention	1	3.48	4	0.891
	2	3.82	4	0.763
	3	3.67	4	0.784
Purchase intention when I have owned a vehicle	4	3.69	4	0.795
Perceived usefulness	5	3.27	3	0.847
	6	3.13	3	0.883
	7	3.52	4	0.877
	8	3.32	3	0.815
Perceived ease of use	9	3.51	4	0.825
	10	3.44	4	0.872
	11	3.37	3	0.739
	12	3.28	3	0.809
	13	3.34	3	0.813
	14	2.98	3	0.884
	15	2.79	2	1.083
Subjective norms	16	3.46	3	0.891
	17	3.59	4	0.830
	18	3.62	3	0.799
	19	3.57	3	0.795
Perceived behavioral control	20	3.71	4	0.852
	21	3.59	4	0.904
	22	3.67	4	0.831
Other factors	23	3.43	4	0.866
	24	3.49	3	0.836
	25	3.42	3	0.802
	26	3.53	4	0.957

Table 4.3 KMO and Bartlett's test

Item		Value
Kaiser–Meyer–Olkin measure of sampling adequacy		0.804
Bartlett's test of sphericity	χ^2	1838.409
	df	231
	Sig.	0.000

4.1.2.3 Measurement Model

Cronbach's alpha value for the items in this questionnaire is more than 0.7, which is generally considered acceptable (Nunnally 1978). The construct validity of the questionnaire shows the degree of the questionnaire's ability to reflect the subject (Berry et al. 1988). The constructs in this research were considered to have high construct validity because the measuring items were derived from previous studies (Davis 1989; Ajzen 1991; Mourato et al. 2004; Ozaki and Sevastyanova 2011; Yoon 2011).

To simplify influencing factors and avoid multi-collinearity, the factor analysis method was applied to items 5–26. As shown in Table 4.3, the p-values of KMO and Bartlett's test were less than 0.05, which meant that factor analysis was applicable. Eventually, 22 items were classified into five categories which were: perceived usefulness, perceived ease of use, subjective norms, perceived behavioral control, and other factors, explaining a total of 59.137% of the variance. Whether measuring items should be retained or not depends on both loading coefficients and the contents measured. If its loading was 0.5, or above, and the differences between this loading and two other cross-loadings were more than 0.3, as listed in Table 4.4, this item would be retained while the item will be also retained with a definite meaning intended for the corresponding variable (Howell and Shea 2005). Then the weighted average values of five variables were obtained with the corresponding unitized factor loading coefficients as weights.

Table 4.4 Loadings and cross-loadings of the items

Items	Component				
	1	2	3	4	5
16	**0.762**	0.206	0.154	0.068	0.103
17	**0.718**	0.295	0.069	0.106	0.195
18	**0.799**	0.084	0.295	0.036	0.023
19	**0.706**	0.068	0.214	−0.070	0.222
8	0.276	**0.617**	0.216	0.193	−0.122
5	0.392	**0.745**	0.143	0.042	−0.064
6	0.236	**0.808**	0.102	−0.006	0.087
7	0.151	**0.635**	0.146	0.089	0.341
15	0.242	0.040	**0.588**	0.028	−0.240
9	0.005	−0.240	**0.568**	0.373	0.132
11	0.245	0.201	**0.661**	0.119	0.119
10	−0.104	0.274	**0.457**	0.365	−0.058
12	0.372	0.207	**0.585**	−0.052	0.082
13	0.129	0.400	**0.596**	−0.048	0.205
14	0.310	0.402	**0.514**	−0.123	0.117
23	0.334	0.024	0.056	**0.611**	−0.045
24	0.068	0.054	0.087	**0.754**	0.083
25	−0.053	0.136	−0.019	**0.782**	0.073
26	−0.097	−0.043	0.051	**0.644**	0.233
20	0.088	−0.030	0.051	0.134	**0.853**
21	0.153	0.020	−0.047	0.140	**0.747**
22	0.267	0.275	0.176	0.054	**0.604**

4.1.2.4 Regression Analysis

Purchase intention is a categorical variable made up of "willing to buy", "not certain", and "unwilling to buy". This study selected randomly 200 samples from the total 240 samples to study the variables' effects on purchase intentions, while the other 40 samples were used to examine the discrete choice model. Among the 200 samples selected, 86 samples were urban residents with cars while 21 of 40 test samples were urban residents with cars. In this study, with purchase intention as the dependent variable, the normalized values of the five factors were used as the independent variable for this regression. When ordinal variables were uniform, they were treated as continuous variables, so the five factors were treated as continuous variables here (Wen et al. 2005). Given the fact that purchase intention was a categorical variable out of order, we applied multinomial logistic regression to this research.

We established four models, from Eqs. (4.1) to (4.4), respectively: the first model without interaction for all urban residents, the second model without interaction for residents owning cars, the third model with interaction for all urban residents, and the fourth model with interaction for urban residents owning cars. Models (4.1) and (4.2) were used to test the effect of the variables on purchase intention.

$$\ln(p_i/p_3) = \beta_0 + \beta_1 X_n + \beta_2 X_u + \beta_3 X_e + \beta_4 X_o + \beta_5 X_c + \varepsilon \quad (i = 1, 2) \quad (4.1, 4.2)$$

Models (4.3) and (4.4) were used to examine the moderating effect of perceived behavioral control on the relationship between subjective norms and purchase intentions.

$$\ln(\hat{p}_i/\hat{p}_3) = \beta_0 + \beta_1 X_n + \beta_2 X_u + \beta_3 X_e + \beta_4 X_o + \beta_5 X_c + \beta_6 X_n * X_c + \varepsilon \quad (i = 1, 2) \quad (4.3, 4.4)$$

p_i is the probability of choosing situation i, p_1 denotes "unwilling to buy", p_2 denotes "not certain", and p_3 denotes "willing to buy"; X_n represents subjective norms; X_u represents perceived usefulness; X_e represents perceived ease of use; X_o represents other factors; and X_c represents perceived behavioral control. The results from the four models are shown in Table 4.5.

Empirical results reveal that, for all urban residents, subjective norms have a significant impact on the intentions to purchase an NEV, which is supporting Hypothesis 4.3-1, while the perceived ease of use and perceived behavioral control significantly influence the purchase intentions, supporting Hypotheses 4.2-1 and 4.4-1, respectively. However, the perceived usefulness of NEVs did not significantly influence the purchase intentions, at least not in accordance with Hypothesis 4.1-1. We investigated the influence of each variable upon purchase intention for residents with cars, and found that subjective norms have a significant impact on purchase intentions, in accordance with Hypothesis 4.3-2, while perceived

Table 4.5 Model verification results

Indicators		Model (4-1)	Model (4-2)	Model (4-3)	Model (4-4)
Model fitting information	Sig.	0.000	0.000	0.000	0.000
Goodness-of-fit	Pearson	0.000	0.934	0.268	0.439
	Deviance	1.000	1.000	1.000	1.000
Pseudo R^2	Cox and Snell	0.218	0.437	0.268	0.439
	Nagelkerke	0.284	0.539	0.350	0.542
	McFadden	0.169	0.346	0.214	0.348
p-value of likelihood ratio tests	X_n	0.007	0.000	0.004	0.000
	X_u	0.895	0.115	0.531	0.130
	X_e	0.026	0.774	0.013	0.847
	X_o	0.547	0.351	0.607	0.365
	X_c	0.056	0.144	0.099	0.153
	$X_n \times X_c$	–	–	0.003	0.848

usefulness, perceived ease of use, and perceived behavioral control did not influence the purchase intentions significantly, which was not consistent with Hypotheses 4.1-2, 4.2-2, and 4.4-2, respectively.

From these results, for both urban residents having and not having cars, perceived usefulness did not influence purchase intentions significantly, so, we divided the perceived usefulness into "perceived usefulness compared to conventional vehicles" and "perceived usefulness compared to the situation without cars". Hence, we used Items 6 and 7, and the purchase intentions to carry out logistic regression. Item 6 was "Compared to a traditional vehicle, a NEV can improve the efficiency of the traveler," referred to as "relative travel efficiency"; Item 7 was "Compared to having no vehicle, a NEV can improve the efficiency of the traveler," referred to as "absolute travel efficiency". As Tables 4.6 and 4.7 indicate, "relative travel efficiency" influenced purchase intentions significantly, while "absolute travel efficiency" had no significant effect thereon.

Table 4.6 Effect of perceived usefulness on purchase intentions for all urban residents

Indicators	Chi-square	*df*	Sig.
Intercept	144.042	2	0.000
Item 6	7.278	2	0.026
Item 7	1.338	2	0.512

Table 4.7 Effect of perceived usefulness on purchase intentions for residents owning cars

Indicators	Chi-square	*df*	Sig.
Intercept	41.956	2	0.000
Item 6	11.765	2	0.003
Item 7	1.364	2	0.506

Models (4.3) and (4.4) were developed by adding the interaction term of the subjective norms and perceived behavioral control to Models (4.1) and (4.2). The moderating effect was measured by the promotion of model fitting abilities and the coefficients' significance for each interaction term (Wen et al. 2005). Compared with Model (4.1), Model (4.3) offered a better fit, its Pearson test results were improved significantly, and the pseudo R^2 value also increased, indicating that the interaction term was necessary. Interaction term coefficients for subjective norms and perceived behavioral control in Model (4.3) were significant, which showed that perceived behavioral control had a moderating effect on subjective norms to purchase intention. Compared with the index in Model (4.2), the fitting index in Model (4.4) did not improve, and the coefficient of interaction was of no significance, implying that perceived behavioral control's moderating effect was insignificant in Model (4.4).

4.1.2.5 The Discrete Choice Models of Purchase Intentions

According to Model (4.3), a multinomial Logit model with reference to p_3 was established, and then we used estimations of the possibilities of purchase intentions across three possible situations. The Logit model form of Model (4.3) was as follows:

$$\begin{aligned}\ln(\hat{p}_1/\hat{p}_3) = {} & -3.182 - 0.284X_o - 0.649X_n - 1.490X_e + 0.129X_u - 0.771X_c \\ & - 0.751X_nX_c\end{aligned} \tag{4.5}$$

$$\begin{aligned}\ln(\hat{p}_2/\hat{p}_3) = {} & -1.719 - 0.178X_o - 0.987X_n - 0.103X_e - 0.391X_c \\ & - 0.276X_u + 0.734X_nX_c\end{aligned} \tag{4.6}$$

$\hat{p}_1$, $\hat{p}_2$, and $\hat{p}_3$ denote the possibilities of situations where purchase intentions are: "unwilling to buy", "not certain", or "willing to buy", in sequence. The significance of the coefficients in Eqs. (4.5) and (4.6) are shown in Table 4.12.

From the equations above, the probability distribution functions of the three situations were

$$\hat{p}_1 = \frac{\exp(\ln(\hat{p}_1/\hat{p}_3))}{1 + \exp(\ln(\hat{p}_1/\hat{p}_3)) + \exp(\ln(\hat{p}_2/\hat{p}_3))} \tag{4.7}$$

$$\hat{p}_2 = \frac{\exp(\ln(\hat{p}_2/\hat{p}_3))}{1 + \exp(\ln(\hat{p}_1/\hat{p}_3)) + \exp(\ln(\hat{p}_2/\hat{p}_3))} \tag{4.8}$$

$$\hat{p}_3 = \frac{1}{1 + \exp(\ln(\hat{p}_1/\hat{p}_3)) + \exp(\ln(\hat{p}_2/\hat{p}_3))} \tag{4.9}$$

With sample data applied to Model 4.3, the probabilities of each purchase intention were deduced, and the situation corresponding to the maximum probability was taken as the estimated result. With 200 samples used, the accuracy was 76.4%, as shown in Table 4.8. To verify the accuracy of the estimation model, the other 40 test samples were added to the model, and showed an accuracy of 72.5%. The estimation accuracy with these test samples was similar to the original results, showing good accuracy, and evince the stability of this multinomial Logit model.

In the same way, a purchase intention model for residents with cars, which was named Model (4.4), was built as

$$\begin{aligned}\ln(\hat{p}_{c1}/\hat{p}_{c3}) = & -3.894 - 0.526X_o - 1.250X_n - 0.538X_e \\ & - 1.249X_u + 0.228X_c + 0.503X_nX_c\end{aligned} \tag{4.10}$$

$$\begin{aligned}\ln(\hat{p}_{c2}/\hat{p}_{c3}) = & -0.893 + 0.372X_o - 2.072X_n - 0.116X_e + 0.161X_u \\ & - 0.742X_c + 0.283X_nX_c\end{aligned} \tag{4.11}$$

$$\hat{p}_{c1} = \frac{\exp(\ln(\hat{p}_{c1}/\hat{p}_{c3}))}{1 + \exp(\ln(\hat{p}_{c1}/\hat{p}_{c3})) + \exp(\ln(\hat{p}_{c2}/\hat{p}_{c3}))} \tag{4.12}$$

$$\hat{p}_{c2} = \frac{\exp(\ln(\hat{p}_{c2}/\hat{p}_{c3}))}{1 + \exp(\ln(\hat{p}_{c1}/\hat{p}_{c3})) + \exp(\ln(\hat{p}_{c1}/\hat{p}_{c3}))} \tag{4.13}$$

$$\hat{p}_{c3} = \frac{1}{1 + \exp(\ln(\hat{p}_{c1}/\hat{p}_{c3})) + \exp(\ln(\hat{p}_{c2}/\hat{p}_{c3}))} \tag{4.14}$$

where p_{ci} denotes the probability of the ith situation arising for the residents with cars, and Table 4.12 shows the significance of the coefficients in Eqs. (4.8) and (4.9). With 86 samples applied to Model (4.4), the accuracy rate was 75.6% (Table 4.9). To verify the accuracy of the estimation model, the other 21 samples were used, eventually getting an accuracy of 80.95%, evincing the accuracy and stability of Model (4.4).

Table 4.8 The test result of Model (4.3)

Observed value	Prediction value			
	0.00	1.00	2.00	Percentage (%)
0.00	3	0	9	25.0
1.00	0	10	24	29.4
2.00	0	8	120	93.8
Total percentage	1.7%	10.3%	87.9%	76.4

Table 4.9 The test results of Model 4

Observed value	Prediction value			
	0	1	2	Correct percentage (%)
0	2	2	0	50.0
1	1	17	10	60.7
2	0	6	40	87.0
Total percentage	3.8%	32.1%	64.1%	75.6

4.1.3 Analysis of Residents' Willingness and Determinants

For all urban residents, perceived ease of use, subjective norms, and perceived behavioral control could significantly affect their NEVs purchase intentions, which was consistent with the TPB (Ajzen 1991), while the influence of perceived usefulness was insignificant. The results show that these factors in TPB can also be applied to an analysis of residents' purchase intentions toward NEVs (Rezvani et al. 2015). To explore why perceived usefulness's effects were insignificant, the research analyzed the effects of "relative travel efficiency" and "absolute travel efficiency" on purchase intentions. For all urban residents, NEVs' relatively high efficiency compared to conventional cars could significantly affect the purchase intentions toward them, indicating that people make decisions about what kind of car should they buy rather than whether to buy a car or not, when considering NEVs. For goods subject to rigid demand, people are more concerned about relative characteristics, which is again another novel finding arising from this work.

For urban residents with cars, only the subjective norms could significantly influence purchase intentions toward NEVs, whereas the effects of perceived usefulness, perceived ease of use, and perceived behavioral control on purchase intention was insignificant. This shows that the purchase intentions of urban residents with cars are highly influenced by subjective norms, partly stemming from the influence of families and friends, and partly from the environmental awareness and social responsibility constraints (Sang and Bekhet 2015). We used the same way to analyze perceived usefulness, it was concluded that for urban residents with cars, "relative travel efficiency" rather than "absolute travel efficiency" significantly affects the purchase intentions, similar to the results for the whole urban residents. For residents with cars, the influence of perceived ease of use was no longer significant. The reason may be that residents with cars use different cars for specific purposes, and consequently they use NEVs within the city for short travel, caring less about ease of use. For urban residents with cars, perceived behavioral control did not exert a significant impact, which confirmed the other findings of (Schniederjans and Starkey 2014). The reason behind this is because people feel it difficult to conduct purchase behavior with plenty of factors around, even if they have purchase intentions (Yin and Wang 2013; Yuan et al. 2015). The moderating effect of perceived behavioral control on subjective norms was not significant, indicating that urban residents owning cars pay close attention to public views and social responsibilities (Marell et al. 1995), and this concern will not be influenced by perceived behavioral control (Nayum et al. 2013; Korcaj et al. 2015; Sang and Bekhet 2015).

Table 4.10 shows the variables' coefficients from Eqs. (4.5), (4.6), (4.10), and (4.11), with p_3, namely "willing to buy", as the reference. As Table 4.10 shows, in situation p_1, the coefficient of perceived ease of use was significant and negative, indicating that perceived ease of use had a higher effect on "willing to buy" residents than "unwilling to buy" residents. That is to say, perceived use of ease can positively influence purchase intentions of "unwilling to buy" residents to a large extent. In situation p_2, the effects of subjective norms and the moderating effects of perceived behavior control were significant. Coefficients of subjective norms were negative, indicating that subjective norms had a positive effect on purchase intentions, and coefficients of interaction terms were positive, indicating that the stronger perceived behavioral control was, the weaker the effects of the subjective norms were. In the same way, we dealt with samples of residents with cars. For residents with cars, in situation p_2, only subjective norms influenced purchase intentions effectively. A high proportion of residents with cars held uncertain attitudes towards the intentions to purchase an NEV, so we can promote their purchase intentions through subjective norms. This research used a discrete choice model to analyze the effects of each variable on people with specific purchase intentions, to know which group of residents is affected by these variables and to what extent so again, this was claimed as novelty of research.

Table 4.10 The parameters estimated of Model (4.3) and Model (4.4)

Coefficients		Model (4.3)			Model (4.4)		
		β	Sig.	Exp (β)	β	Sig.	Exp (β)
$p_1(p_{c1})$	Intercept	−3.182	0.000	0.042	−3.894	0.000	0.020
	Zcontrol	−0.771	0.082	0.463	0.228	0.846	1.256
	Zease	−1.490	0.006	0.225	−0.538	0.560	0.584
	Znorm	−0.649	0.207	0.523	−1.250	0.173	0.287
	Zusefulness	0.129	0.758	1.138	−1.249	0.097	0.287
	Zothers	−0.284	0.427	0.753	−0.526	0.533	0.591
	Znorm × Zcontrol	−0.751	0.139	0.472	0.503	0.611	1.654
$p_2(p_{c2})$	Intercept	−1.719	0.000	0.179	−0.893	0.008	0.409
	Zcontrol	−0.391	0.151	0.676	−0.742	0.073	0.476
	Zease	−0.103	0.757	0.902	−0.116	0.804	0.890
	Znorm	−0.987	0.003	0.373	−2.072	0.000	0.126
	Zusefulness	−0.276	0.312	0.759	0.161	0.697	1.175
	Zothers	−0.178	0.455	0.837	0.372	0.274	1.451
	Znorm × Zcontrol	0.734	0.007	2.083	0.283	0.629	1.327

Note All is with the reference to the $P_3(p_{c3})$ condition

4.1.4 Policy Implications

Analysis of the effects of factors on purchase intentions of residents with cars reflect that subjective norms influence purchase intentions towards NEVs significantly, but the moderating effect of perceived behavioral control is insignificant. The results indicate that subjective norms make a significant difference to the purchase intentions of residents with cars, regardless of their perceived ease of conducting the purchase. The survey shows that a high proportion of residents with cars hold uncertain attitudes toward purchasing NEVs. So manufacturers should pay close attention to public praise and family recommendations so as to introduce the second-household-vehicle market. For residents with cars, who are familiar with cars, they hold rational expectations of NEVs ease of use, and some of them only use NEVs within the city, so they are influenced by their ease of use slightly. On the basis of the aforementioned conclusions, residents with cars are probably the most likely potential purchasers of NEVs, especially in regions without complete NEVs infrastructure: this was considered to be an important breakthrough point, capable of improving NEV sales.

Although residents are sensitive and price conscious about the overall costs of NEVs, "other factors" do not influence purchase intentions towards NEVs significantly in this research: this is possibly due to current subsidy policies by the government. Discrete choice models were established in this research to measure the purchase intentions for NEVs and to analyze the effects of each variable on different groups of residents. This model was confirmed to be both accurate and robust. However, our research has some limitations as well like sample size is small, exploration of the effects of other factors is not particularly profound, and we do not measure the actual purchase behavior involved in this research. Thus these limitations may provide potential directions for our future research for scholars and policy-makers.

4.2 Residents' Willingness and Determinants in Purchasing Energy-Efficient Household Appliances

Energy consumption of appliances has aroused serious concerns in the current era. In 2004, the government proposed the establishment of a conservative-minded society and published a series of policies and regulations pertaining to energy savings. In August 2004, the Central Government issued an administrative measure that is termed as China Efficiency Energy Label and began to implement this energy efficiency labeling system in September 2005. Furthermore, the Government passed the Energy Conservation Law for the People's Republic of China in 2007. In 2009, to contribute and improve the expansion of domestic demand, especially consumer demand, and the energy efficiency of terminal products, the Chinese Government launched a project to promote energy-efficient products for the benefit of the people for the promotion of ten energy-efficient products, such as: air-conditioners, refrigerators, flat-screen computer monitors, washing machines, televisions, etc., whose energy efficiency grades are level 1 or level 2, through financial subsidies thereof. Particularly, in June 2012, during the implementation of this public-benefit project, the Government published detailed rules to promote five types of energy-efficient appliances (air-conditioners, flat-screen televisions, refrigerators, washing machines, and water heaters) through subsidy, with a tentative promotion period of one year.

Before China's project to promote energy-efficient products for the benefit of the people, the United States provided financial aid to energy-star products, such as subsidies for energy-saving washing machines in California at $75/set, a refrigerator subsidy at $75/set to $125/set; Korea and Australia have also used financial subsidies to promote energy-efficient products. In 2008, a subsidy worth ¥280 million was used for promoting energy-saving lamps in China. As a result, 62 million energy-saving lamps were sold which directly stimulated social consumption worth ¥650 million and saved electricity in the amount of 3.2 TW in that year. Both domestic and foreign experience shows that government subsidy plays an important role. It not only accelerates the promotion of energy-efficient products, but also eliminates market obstacles to their uptake project to promote energy-efficient products for the benefit of the people. The government achieved its policy targets almost perfectly from a macro point of view. Sales of color TVs, refrigerators, washing machines, air-conditioners, water heaters (the products winning the bidding) reached 50.13 million, 34.22 million, 33.17 million, 39.86 million, and 27.11 million units, respectively, according to the National survey of household appliances in China (by China Market Monitor Co., Ltd), during the period of implementing this subsidy policy for energy-efficient appliances from June 2012 to May 2013. In 2014, national total electricity consumption reached 5.52 trillion kW h, with a yearly growth of 3.8% and the growth rate falling by 3.8% points, which was the lowest level since 1998 (2.8%). In 2014, the growth in electricity consumption of urban and rural residents was 2.2% per anum basis with a fall of 6.7% points.

Under such "wonderful" macro data, the purchase intentions and behavior of Chinese residents towards energy-efficient appliances really changed because the government subsidy policies are still treating like a research question. Whether project to promote energy-efficient products for the benefit of the people subsidy policies have an impact on Chinese residents' willingness to pay for the energy-efficient appliances or not, what factors affect Chinese residents' purchase behavior of energy-efficient appliances remain key questions. In our study, we try to provide the solution for the aforementioned problems, we investigated the behavior of urban residents from 22 provinces (autonomous regions, municipalities) in China, covering seven major geographic regions, and established a model based on the method of a structural equation model.

4.2.1 Theoretical Framework for Determinants of Purchasing Energy-Efficient Household Appliances

Many scholars have postulated the purchase of energy-saving appliances. According to the main focus of this research, we examine the literature covering policy as a starting point. The review is linked with the theoretical framework, and keeping in mind the literature review and theoretical model we proposed the hypothesis accordingly.

According to Galarraga et al. (2016), the current academic studies on energy efficiency are seriously out of line with the problem of policy-making. To promote people's purchase behavior of energy-efficient appliances, many countries have introduced a lot of incentives, such as subsidies, taxes, reward and punishment systems, etc. Therefore, these studies are significant and linked with the policy implementation background of the country.

In terms of green and energy consumption, Yao et al. (2014) used Quantile Regression and the counterfactual Analysis to study the effect of subsidy policy and the inherent family property behavior on per capita electricity consumption. They concluded that the subsidy policy has a positive effect on the electricity consumption of urban and rural residents in general, but residents at two ends of the power consumption distribution are not sensitive to the subsidy policy. They also provided evidence that the subsidy policy has a rebound effect on the total electricity consumption of residents, and leads to an increase in power consumption. Yang and Zhao (2015) examined the effect of subsidy policy on high-income and low-income families and found that high-income families prefer green consumptions and green purchase behavior in comparison to low-income families.

The above facets are combined with the policy background, but other studies that combine relevant policies and energy-efficient appliance purchase behavior are rare. In the context of the end of China's energy-efficient appliances subsidy policy, this research regarding residents' purchase behavior toward energy-efficient appliances, aims to provide meaningful conclusions for policy-makers.

From the perspective of the object of an incentive policy, most people generally believe that the incentive policy is directly aimed at consumers. Actually, the incentive policy can also play other roles upstream of the entire product supply chain: for example, an incentive policy can encourage manufacturers to produce more energy-saving products (an "upstream incentive policy"), it can also encourage retailers to sell more energy-saving products (a "midstream incentive policy"). Similarly, incentive policies for consumers are called "downstream incentive policies". Upstream incentive policies include three types: technology procurement,[1] financial incentive, and subsidy. China's subsidy policy on

[1]Technology procurement is a means for the government to encourage manufacturers to carry out technological innovation. The specific method is that the government puts forward a standard of a personality of a product, such as energy saving, environmental protection, low cost, etc., and then

energy-efficient appliances is an upstream incentive policy. An upstream incentive policy has two advantages: it can control most of the market through a limited number of vendors and reduce the cost of delivery, and, because of the multiplier effect, it can cause a reduction in price for the manufacturers to pass on to the benefit of the consumer. Of course, there are drawbacks to such subsidies, such as: the subsidies cannot be open and transparent to consumers, and assessing manufacturing costs is difficult. Hence, it is necessary to improve the supervision and evaluation mechanism thereof (de la Rue Du Can et al. 2014).

de la Rue Du Can et al. (2014) also mentioned the evaluation method of incentive policies. They believe that the evaluation of an energy-saving policy has two points, energy saving and cost-effectiveness. Apart from these two points, we believe that there is a third evaluation point, and that is whether the policy has guided intentions and behaviors or not. In the first strand of this study, we mentioned that five categories of energy-efficient appliances enhanced in the retail market share and the growth rate of residents' electricity consumption dropped. However, we think that it is more important to assess the impact of subsidy policy on the purchase intention of Chinese residents at micro-level, which is also one of the critical issues in this research.

The theory of planned behavior (TPB) (Ajzen 1991) is commonly used to test environmental behavior from the perspective of psychology. Ajzen (1991) believed that all factors indirectly influence behavior via behavioral intentions: behavioral intentions are influenced by four factors, "attitude, perceived behavioral control, subjective norms and residual effect" (Fig. 4.1).

Attitude. It can be measured by the degree to which an individual has a favorable, or unfavorable evaluation of a particular behavior (Wang et al. 2011). In this research, attitude is people's environmental awareness (ENVA), and mostly people's environmental awareness depends on the degree of preference for energy-saving behavior and their product cognition (PROC) of energy-efficient appliances. Environmental awareness plays a significant role in energy use and -savings (Samuelson 1990). As Ek and Söderholm (2010) confirmed, residents' attitude to the environment is an important factor that predicts whether or not, they save electricity. Zografakis et al. (2010) argued this viewpoint and believed that people who have more energy-saving knowledge and a stronger concern for climate change are more likely to buy energy-saving products or new energy products, and they are more likely to participate in energy-saving activities. Moreover, people who have little knowledge of energy use can observe this in their behavior (Steg 2008). For example, when assessing the energy use of appliances, people tend to rely on a simple heuristic: they think that energy use is related to the size of an appliance. The larger the appliance, the more energy it is believed to use (Baird and Brier 1981). Obviously, this is not always correct. Moreover, people often

takes some similar approaches to the tender. Manufacturers who are able to create products which can meet the government's standards will receive a corresponding incentive or subsidy.

underestimate the consumption of energy. This also reflects that people do not have a strong awareness of using energy in an economical way.

Perceived behavioral control. This refers to the ease, or difficulty, of performing a particular behavior (Wang et al. 2011). It largely depends on the weighing of the costs and benefits in the process of performing a specific behavior, such as financial cost, effort, and time (Lindenberg and Steg 2007). Steg (2008) mentioned that people are less likely to reduce their energy use when saving energy involves high behavioral costs in terms of money, efforts, and convenience. In this research, perceived behavioral control is measured in terms of economic benefits (ECOB) and price, that is, the benefit from people's using of energy-saving appliances (such as the monthly electricity fees by saving electricity) and their high price due to the high cost of advanced technology can influence people's purchase intention on energy-efficient appliances.

Subjective norms. These refer to the perceived social pressure to perform or refrain from a behavior (Abrahamse and Steg 2009). From this point of view, the individuals' perceptions of their external environment affect their behavior significantly (Wang et al. 2011). In this research, subjective norms are measured by the policy and propaganda (POLICY), and social interactions (SOCI). As it is mentioned, a region or country has a good incentive or subsidy policy, no matter whether upstream, midstream, or downstream of the supply chain, policy, and propaganda can promote and stimulate consumer purchases of energy-efficient appliances (Yao et al. 2014; Yang and Zhao 2015; de la Rue Du Can et al. 2014). Social interaction means people who are important to consumers, such as the family, friends, and neighbors around consumers who influence their purchase behavior. As Gaspar and Antunes (2011) confirmed, it is important to consider whether or not the customer is accompanied by other people at the time of purchase. Consequently, when respondents go to a store to buy an appliance with others, they might be more predisposed to consider an appliance's energy efficiency class.

Residual effect. This refers to the past behavior or experience of individuals, which may affect their intentions of performing a particular behavior (Ajzen 1991). People usually have a kind of "accepting the old" or "recognizing the familiar" psychology, for goods which they have purchased. If they do not have any adverse reactions or perceive its superiority in the consumption process, next time they will purchase it again, when required. We measure residual effect by past purchase experience (PPE), and find that the people who are used to purchase energy-efficient appliances are efficient in work.

In addition to several factors in the TPB framework, demographic variables are often treated into behavioral studies as special variables. Regarding gender, women take environmental concerns into consideration more significantly when making their choice. In contrast, men have more concerns on functions, innovative technologies and accessories by the alliances (Gaspar and Antunes 2011). Regarding age, Gaspar and Antunes (2011) concluded that as compared to younger customers, the choices of respondents over 48 years of age were found to have significantly higher frequencies of considering environmental and social issues, country of

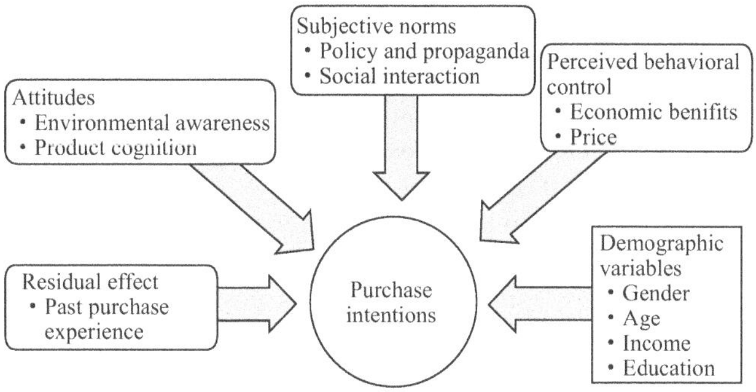

Fig. 4.4 The theoretical model

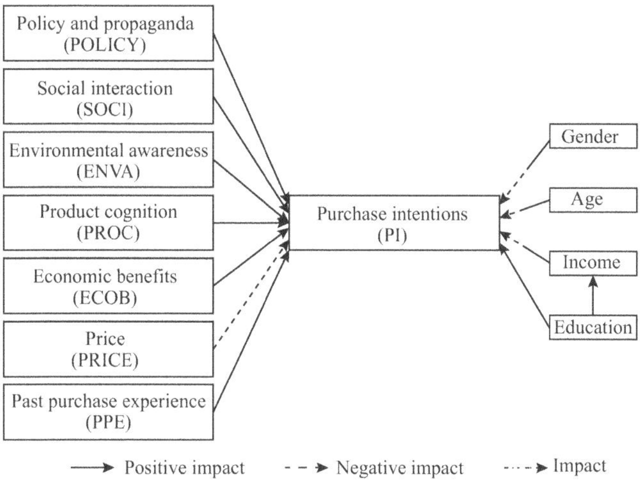

Fig. 4.5 Hypothesized framework

origin, and long-term savings. When debating on income, a high-income family and better familiarization with subsidy incentives can promote the transformation of green attitudes into EERE equipment purchase intention (Yang and Zhao 2015). In addition, the level of education determines their sense of cognition so it is believed that education can influence people's purchase intentions toward energy-efficient appliances. Yao et al. (2014) also proved that the educational level of urban and rural residents has a significant impact on their electricity consumption. Moreover, many studies on population or labor economics have been conducted to prove that educational level also has a significant positive impact on income.

By the expansion of TPB, coupled with the above analysis and discussions, we propose the base model for this research (Fig. 4.4), following diagram (Fig. 4.5) is a hypothesized framework for this book.

Based on this model, we proposed several hypotheses:

Hypothesis 4.6. Policy and propaganda are positively associated with residents' purchase intention on energy-efficient appliances.
Hypothesis 4.7. Social interaction is positively associated with residents' purchase intention on energy-efficient appliances.
Hypothesis 4.8. Residents' environmental awareness is positively associated with their purchase intention on energy-efficient appliances.
Hypothesis 4.9. Residents' product cognition is positively associated with their purchase intention on energy-efficient appliances.
Hypothesis 4.10. Economic benefits of purchase behavior to consumers are positively associated with residents' purchase intention on energy-efficient appliances.
Hypothesis 4.11. Price is negatively associated with residents' purchase intention on energy-efficient appliances.
Hypothesis 4.12. Past purchase experience is positively associated with their purchase intention on energy-efficient appliances.
Hypothesis 4.13–4.15. Gender, age, and income influence the residents' purchase intention on energy-efficient appliances. But the direction is uncertain.
Hypothesis 4.16. Residents' educational level is positively associated with their purchase intention on energy-efficient appliances.
Hypothesis 4.17. Residents' educational level is positively associated with their income.

4.2.2 *Research Method*

For the empirical results of this theoretical model and the hypotheses mentioned in Sect. 4.2.1, we used the structural equation model (SEM): the premise of the construction of the SEM is the development of a reliable and effective scale from which the designed questionnaire would provide realistic, high-quality data.

4.2.2.1 Sample and Data Collection

For each item (except the demographic variables), a five-point Likert scale was used to enable respondents to indicate the extent to which they agree with these items: "strongly disagree" = 1, "disagree" = 2, "neutral" = 3, "agree" = 4, "strongly agree" = 5. Based on the theoretical model and the hypothesis framework, the initial scale was developed after critical discussions within our research group and was finally approved by experts of the related area. On these basis and justifications, we designed a questionnaire in which we divided the order of the items in the initial scale to prevent the situation arising in which respondents answer with the same response to similar items: we also set "trap questions"[2] to help to screen the veracity of the responses.

---

[2]The mean of "trap questions" is that the same problem is asked in the forward and reverse two directions.

In July 2015, after the questionnaires were designed, full-time investigators and research scholars led by our own research team members carried out a country-wide survey, which took more than two months. A total of 689 questionnaires were received where 436 questionnaires were completed by urban households and 253 by rural families. However, there are still very large differences in many aspects of the rural and urban residents in China. In terms of income level, per capita income of rural residents is still far lower than urban residents (according to *Chinese Statistical Yearbook 2015*, per capita disposable income of urban residents was ¥29381, per capita disposable income of rural residents was ¥9892). In terms of environment, the rural population density is far lower than the urban and the rural economy is mainly based on the agricultural industry (natural economy and the first industry). Thus, generally speaking, in the same area, the air quality in the countryside is better than that of the city, and the temperature is also lower than that of the city. In terms of the policy, besides the project to promote energy-efficient products for the benefit of the people, there was "Home appliances to the countryside" subsidy policy specialized for a rural household. In addition to the aspects mentioned above, urban and rural residents are also different in many aspects, like living standards and energy usage habits and so on. Therefore, factors that the urban and rural residents have to consider when buying home appliances will be very different no matter in quantity and type. Our unit of analysis for this research are urban residents. Therefore, the 436 questionnaires representing urban residents were temporarily retained.

After screening, we found 253 valid questionnaires (a 58% effective rate). The survey scope included the seven major geographic regions in China: Eastern China (Shandong, Jiangsu, Zhejiang, Shanghai), North China (Beijing, Tianjin, Hebei, Shanxi), Central China (Hubei, Hunan, Henan), South China (Guangdong, Hainan), Southwestern China (Sichuan, Yunnan, Chongqing, Xizang) and Northwestern China (Shanxi, Inner Mongolia), Northeastern China (Liaoning, Jilin, Heilongjiang), with a total of 22 provinces (autonomous regions, municipalities) which represent almost the entire nation. The descriptive statistics of the demographic variables are shown in Table 4.11. The description of the respondent profile in Table 4.11 also reflects the fact that the sample of this study covers a wide range, with better representation, which could thus be used for further analysis.

4.2.2.2 Measurement Model

This research used the method of exploratory factor analysis (EFA) and confirmatory factor analysis (CFA) to select an appropriate number of questions or indicators. These methods are based on the weight setting of each item and analysis and control of correlation between indicators, so that the items can be simplified and the data purified.

Table 4.11 Respondent profile

Demographic variables		*N*	Percentage (%)
Gender	Female	123	48.6
	Male	130	51.4
Age	Under 20	74	29.2
	21–30	80	31.6
	31–45	71	28.1
	46–60	24	9.5
	Above 60	4	1.6
Education level	Lower secondary or below	19	7.5
	Higher secondary or equivalents	55	21.7
	Bachelor's degree or equivalents	159	62.8
	Master's degree or above	20	7.9
Monthly income (¥)	Below 2500	37	14.6
	2501–4000	53	20.9
	4001–6000	64	25.3
	6001–8000	36	14.2
	8001–10000	27	10.7
	10001–15000	24	9.5
	15001–20000	3	1.2
	Above 20000	9	3.6

First of all, we determined whether original items were suitable for the method of factor analysis or not? Statistical software SPSS offers various statistical test methods: Bartlett's test of sphericity and a KMO test were applied. The original items were tested in SPSS 20.0: the value of KMO was 0.803 which was greater than the critical value (0.7). The *p*-value in the Bartlett sphericity test was 0 which denoted significance. This shows that the original items are suitable for factor analysis.

Then we extracted the common factors and determined the number thereof. Again using SPSS 20.0, we firstly used a principal component analysis to analyze all items, and then used the maximum variance orthogonal rotation method to make linear combinations of the initial common factors in order to find some common factors with more explicit and more obvious meanings. Finally, we eliminated the factors having factor loadings less than 0.5, and then repeated the above steps to find the retained factor loadings of each group of indicators (Table 4.12).

It can be seen from Table 4.12 that retained factors are almost the same as the dimensions from the initial scale, except the initial constructs *Price* (PRICE) and

Table 4.12 Measurement model

Constructs	Indicators	Factor loadings	Average variance extracted (AVE)	Composite reliability (CR)	Cronbach's alpha (α)
Purchase intention (PI)	PI_1	0.696	0.556	0.789	0.721
	PI_2	0.743			
	PI_3	0.795			
Policy and propaganda (POLICY)	POLICY_1	0.776	0.530	0.770	0.734
	POLICY_2	0.786			
	POLICY_3	0.610			
Social interaction (SOCI)	SOCI_2	0.713	0.524	0.688	0.796
	SOCI_3	0.735			
Past purchase experience (PPE)	PPE_1	0.652	0.507	0.754	0.636
	PPE_3	0.795			
	PPE_4	0.681			
Economic effect (ECOE)	PRICE_3	0.703	0.510	0.756	0.671
	ECOB_1	0.650			
	ECOB_2	0.783			
Product cognition (PROC)	PROC_2	0.903	0.742	0.895	0.864
	PROC_3	0.917			
	PROC_4	0.754			
Environmental awareness (ENVA)	ENVA_1	0.774	0.513	0.806	0.764
	ENVA_2	0.825			
	ENVA_3	0.634			
	ENVA_6	0.609			

Economic benefits (ECOB). We merged these two dimensions into one dimension designated *Economic effect* (ECOE) to denote the factor of perceived behavioral control.

To ensure the effectiveness of the study, we tested the convergent validity, composite reliability, and content validity. Convergent validity is used to express the relationship of items under the same structure, the composite reliability denotes the consistency of items under the same structure, and content validity shows the adequacy and compatibility of the questionnaire contents. As shown in Table 4.12, in this study, all of the factor loadings were greater than 0.5 and the average variance extracted more than 0.5, indicating good convergent validity for the questionnaire (Fornell and Larcker 1981). Moreover, the composite reliabilities of all constructs were greater than 0.6, illustrating the ideal intrinsic quality of the questionnaire (Wu 2010). Then the designers of the questionnaire consulted previous research settings, coupled with the determination of the relevant experts in the field, for ensuring its content validity. Finally, to test the reliability of the indicators in the resulting scale, we analyzed the value of Cronbach's Alpha for the seven variables using SPSS-20.0: their α values were: 0.721, 0.734, 0.796, 0.636, 0.671, 0.864, and 0.764, respectively (Table 4.12), which were all greater than 0.6, again indicating that the design of this scale was reliable.

4.2.2.3 Structural Equation Model

The principle behind the structural equation model (SEM) involves the use of some observable variables to measure one unobservable variable, and it can use a variety of indicators to measure the model fitness (Swami et al. 2011). Thus, we used the SEM to analyze the data in this current research. Based on the study of theoretical models and hypotheses proposed in the previous section, we set the structural equation model (Fig. 4.7) through the AMOS17.0 platform.

We applied the model by putting data of survey for identification, fitting, and evaluation, we found that most fitting indicators failed to meet the criterion and the model fit was not very high, so the model required modifications. Then the model was modified several times, according to the directions by AMOS, until the model fitting results reached an ideal level. The results are shown in Table 4.13. We can see that almost all of the evaluation indicators, such as: RMSEA, GFI, CFI, IFI, PGFI, and AIC meet the criteria. The results indicate that the theoretical framework assumed in this study fitting the actual survey data which represents that our model has a good external quality.

Results of the analysis are presented in Fig. 4.6. We can see that, except that the *p*-values of *policy and propaganda* (POLICY) and *income* did not reach a significant level and pass the hypothesis test. Overall, in our study, there were five

Table 4.13 The evaluation results of the modification model

	Indicators	Criterion	Results	Judgment
Absolute fit measures	χ^2	$P > 0.05$	0.147	Yes
	GFI	>0.9	0.897	Almost
	RMR	<0.5	0.071	Yes
	RMSEA	<0.1	0.023	Yes
Incremental fit measures	NFI	>0.9	0.842	No
	CFI	>0.9	0.983	Yes
	IFI	>0.9	0.984	Yes
Parsimonious fit measures	PGFI	>0.5	0.665	Yes
	PNFI	>0.5	0.677	Yes
	PCFI	>0.5	0.790	Yes
	CMIN/DF	<2.00	1.096	Yes

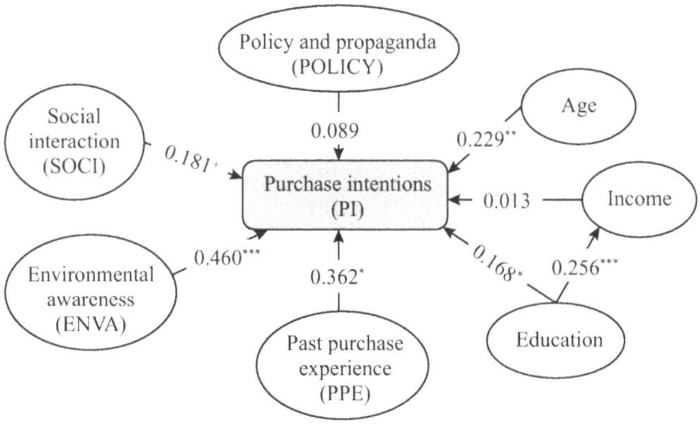

Fig. 4.6 Results of SEM analysis. Notes: + means $p < 0.1$; * means $p < 0.05$; ** means $p < 0.01$; *** means $p < 0.001$

factors: *environmental awareness* (ENVA), *past purchase experience* (PPE), *social interaction* (SOCI), *age*, and *education*, influencing the residents' *purchase intention* (PI) on energy-efficient appliances.

4.2.3 Analysis of Residents' Willingness and Determinants

4.2.3.1 Analysis of the Factors Affecting Energy-Efficient Appliance Purchase Intentions

The estimated results from the SEM are presented in Table 4.14. Firstly, according to the degree of the influence exerted thereon, the significant factors influencing Chinese residents' willingness to pay for energy-efficient appliances were explained as follows:

Environmental awareness (ENVA) is the primary factor affecting Chinese residents' purchase intentions towards energy-efficient appliances. The behavior around purchasing energy-efficient appliances is, in fact, one kind of pro-environmental behavior. With increasing environmental awareness, people will take the initiative and purchase energy-efficient appliances. The impact coefficient (0.460) is the largest, and the p-value is almost zero, indicating that Chinese urban residents may have a strong sense of environmental awareness, and have learned to transform environmental awareness into action. This trend is worthy for China on her way to energy conservation and emission reductions.

The influence coefficient (0.362) of past purchase experience (PPE) is relatively large and positive and its p-value (0.018) is much less than 0.05, which indicates that people's past experience of purchasing energy-efficient appliances has a significant positive effect on people's purchase intentions towards energy-efficient appliances. The three items expressing this variable in our questionnaire were:

Table 4.14 Results of structural equation model and hypothesis testing

Path			β	P	Test results
PI	←	POLICY	0.089	0.235	No
PI	←	ENVA	0.460***	0.000	Yes
PI	←	PPE	0.362**	0.018	Yes
PI	←	SOCI	0.181*	0.062	Yes
PI	←	Age	0.229***	0.003	Yes
PI	←	Education	0.168**	0.035	Yes
Income	←	Education	0.256***	0.000	Yes
PI	←	PROC	−0.005	0.947	No
PI	←	ECOE	−0.080	0.570	No
PI	←	Gender	0.022	0.774	No
PI	←	Income	0.013	0.860	No

Notes * means $p < 0.1$; ** means $p < 0.05$; *** means $p < 0.01$

"There have been energy-efficient appliances in my home," "Most appliances in my home have an energy-saving effect," and "My experience was good when I used energy-efficient appliances in my home." That is to say, my family owns or used to purchase energy-efficient appliances, and their use was beneficial, so I am likely to buy an energy-efficient appliance again. This shows that if people have a purchase experience involving one item, they will have a special preference for such goods. Then they will prefer to choose such an item again. Thus, for China, it is necessary to continue to expand the market share of energy-efficient appliances and make more residents choose energy-efficient appliances to let them develop the habit of purchasing energy-efficient appliances: this may have a long-lasting effect on the popularization of energy-efficient appliances in China.

The influence coefficient of *social interaction* (SOCI) is 0.181 which is not very large, but its *p*-value reaches a significant level which is indicating that Chinese urban residents' social relationships also have a positive impact on their purchase intentions towards energy-efficient appliances. That is, our peer groups can affect our willingness to buy energy-efficient appliances. It is easy to explain the relationship, especially in China: because of the influence of Chinese traditional culture and values, most people have the mean ideology and they advocate the Confucian doctrine. People do not like mavericks and conformity is relatively common. Therefore, if everyone around chooses to buy energy-efficient appliances, people will also join ranks to buy energy-efficient appliances due to face-saving effects and a certain traditional adherence to societal norms.

From demographic variables, age ($\beta = 0.229$, $p = 0.003$) and education level ($\beta = 0.168$, $p = 0.035$) exert a significant influence on people's purchase intentions towards energy-efficient appliances and both have positive influences. As people grow older, and with the improvement of their levels of educational attainment, their social experience and the level of knowledge are constantly improving, so that peoples' awareness of energy saving and environmental protection will be gradually increased. At the same time, people pay more attention to issues such as quality of life and value: they will, therefore, consider purchasing an energy-efficient appliance.

However, *policy and propaganda* (POLICY), *economic effect* (ECOE), p*roduct cognition* (PROC), and two demographic variables (*gender* and *income*) are not significant (Table 4.14) which reflects the fact that these factors cannot exert any influence on Chinese residents to buy energy-efficient appliances. Specific theoretical explanations are described as follows.

The most important thing which we analyze in our study is POLICY variable: its coefficient of influence (0.089) is very small and its p-value (0.235) is over 0.1 which does not reach a significant level. So the policy and propaganda in China exert no effect on Chinese residents' willingness to pay for energy-efficient appliances which are inconsistent with our hypothesis because the research objectives and directions are different from each other. For example, Yao et al. (2014) studied the impact of policy on household electricity consumption, and Wang et al. (2011) studied the energy-saving behavior of residents, while the focus of our study is related to purchasing energy-efficient appliances. According to Zeng et al. (2014), it was mentioned that only 13% of consumers claimed that the subsidy was the primary reason for them to purchase energy-efficient appliances. Besides, they found that consumers lacked an in-depth understanding of the subsidy policy which may be the reason why the variable "POLICY" is not significant. The other specific reasons we analyzed will be presented in Sect. 4.2.3.2.

Economic effect (ECOE) does not have a significant effect as expected ($\beta = -0.080$, $p = 0.570$). In our opinion, there are two reasons for this: first, according to the findings in our interviews, we investigated that people may consider some objective factors like price to a certain extent, but they didn't directly affect their purchase intention. On the contrary, some subjective factors like the willingness of energy conservation and environmental protection are considered much more. The second reason is that due to the financial subsidy for energy-efficient appliances, people did not perceive their actual high prices.

We tried to find the reason why *product cognition* (PROC) is not significant ($\beta = -0.005$, $p = 0.947$). The reason might be that Chinese residents lack the cognition of energy-saving products. However, through our statistics on the results of the survey, we found that people's cognition of energy-saving products is high. For the item "PROC_2: I can recognize the energy-efficient product identification and the energy efficiency grade identification stuck on the energy-efficient appliances," 13% of the respondents choose "1" or "2" (strongly disagree or disagree), 27% of the respondents choose "3" (neutral), and 60% of the respondents choose "4" or "5" (strongly agree or agree). For the item "PROC_3: I understand the exact meaning of grade numbers on the energy efficiency grade identification," 12% of the respondents choose "1" or "2" (strongly disagree or disagree), 30% of the respondents choose "3" (neutral), and 58% of the respondents choose "4" or "5" (strongly agree or agree). The reason for not significant results is not that Chinese residents lack cognition of energy-efficient appliances. Then we made a statistic on the result of the item "PROC_4: I know clearly if the energy-efficient appliances really reduce the daily electricity consumption." 20% of the respondents choose "1" or "2" (strongly disagree or disagree), 43% of the respondents choose "3" (neutral), and 37% of the respondents choose "4" or "5" (strongly agree or agree). Energy-efficient appliances that can save electricity is not very recognized by most of the urban residents. Therefore, we believe that the most important reason why the impact of product cognition on the purchase intentions is not significant is that the high cognition of one product is not equal to the high recognition of it. If people do not recognize the energy-saving appliances, there is no sense even if the degree of the product cognition is very high.

Finally, gender and income are indicating that they both have no significant effects on people's willingness to pay for energy-efficient appliances. There is a big difference in per capita income in different regions of china which may be the most probable cause that income is not significant. Therefore, in this study, we cannot directly deny the impact of income. This may be one of the deficiencies of our research. We need to use a sample of a specific city to test this variable in our following study.

4.2.3.2 Does Policy Work and Why?

As it is mentioned before, that the evaluation of an energy-saving incentive policy has three points, energy saving, cost-effectiveness (de la Rue Du Can et al. 2014), and whether the policy has guided people's intention and behavior or not.

After the end of the implementation of China's subsidy policy for energy-efficient appliances, five categories of energy-efficient appliances' retail market share were greatly enhanced, the growth of the residents' electricity consumption fell, and in the critical time of the global economic crisis, domestic demand was boosted. The government basically achieved its desired policy objectives. Therefore, we think that the subsidy policy, when evaluated in terms of the first two points (energy saving and cost-effectiveness) was effective. However, from our point of view, the more important point is the last one. We want to test whether the policy has any effect on Chinese residents' willingness to pay for energy-saving appliances since the implementation of the subsidy policy, or if it changes Chinese residents at the micro or behavioral level or not: the answer from this study is no.

Next, we analyzed the reasons why China's subsidy policy for energy-efficient appliances has no significant effect on Chinese residents' purchase behavior from three aspects: the time of subsidy, the object of the subsidy, and the method of subsidy.

Firstly, for the time of subsidy, China's subsidy policy for energy-efficient appliances ended on May 31, 2013. We did our research survey during the time period of July 2015–September 2015. There are more than two years in the interval between these two time periods. However, almost all policies were timely. With the passage of time, we believe that the heat of the policy effects is in the past, and maybe people have forgotten the subsidy policy. Then seeing the objective of subsidy: we have mentioned before that China's subsidy policy on energy-efficient appliances is an upstream incentive policy. That is, the government allocates financial subsidy funds directly to manufacturers upstream of the supply chain which can reduce their cost of production to encourage them and to increase the production of energy-saving products (de la Rue Du Can et al. 2014). So the object of subsidy in China was manufacturers, and not consumers: the biggest advantage of this kind of incentive policy is that consumers cannot see and monitor the process by which the financial subsidy funds are offered to manufacturers. So if there is no robust monitoring and verification, it is not easy to ensure that the subsidy is passed on to consumers (Friedmann 2011).

Finally, we analyzed the reasons from the perspective of the subsidy method. Although the end result was that the government allocated financial subsidy funds directly to manufacturers, the process was not as simple. There are four procedures: the application of consumers, the advances made by retailers, retailers doing a deal with manufacturers, and finally manufacturers applying to the financial sector to receive advances in accordance with current provisions. For consumers, the application process is too cumbersome. In addition to providing proof of identity, they also need to fill out the confirmation of energy-saving appliances subsidy and other related materials, which to a large extent affects consumers' buying enthusiasm. More seriously, many residents did not know the subsidy policy because the public was not in place and was insufficient. So some consumers never enjoyed the subsidy policy provided. Moreover, there are some problems when retailers sold energy-efficient appliances: because there is only the information about the amount of subsidy available on the label pasted on energy-efficient appliances, retailers may raise the original price of the appliance and then let consumers get the subsidies. Thus retailers can make more profit from consumers, which will harm the interests of consumers to a certain extent and reduce the consumers' actual perception of subsidies for energy-efficient appliances. Furthermore, because the amount of subsidy was reported to superiors, and the communication chain was poor, combined with a lax audit and supervision regime, the problem of fraud is serious. The phenomenon that sees retailers and manufacturers misrepresenting sales and defrauding the subsidy policy appeared constantly. Thus, subsidies did not fully pass from manufacturers to consumers and consumers did not enjoy the full benefits of the subsidy policy.

In summary, from a macro point of view, after the energy-efficient appliances subsidy policy was enacted in China, the government achieved the policy targets almost perfectly, reached the effect of energy saving and emission reduction targeted and boosted consumption and demand. At a micro-level, people's purchase behavior were not changed substantially, and the policy did not give people the willingness to pay for energy-efficient appliances. It is related to the time of subsidy, the object of subsidy and the vulnerability of the subsidy method when the subsidy policy was implemented in practice.

4.2.4 Policy Implications

Based on the above results and the analysis of the reasons why China's subsidy policy for energy-efficient appliances has no significant effect on Chinese residents' purchase behavior, we propose the following policy recommendations.

Firstly, the government needs to implement a subsidy policy for energy-efficient appliances once again. As mentioned above, it has been almost three years since the end of the last subsidy program: the effect of this policy has vanished. To encourage consumers to purchase energy-efficient appliances, a new subsidy program needs to be enacted. Thereby consumers could have permanent "past purchasing experience" and be accustomed to purchasing energy-efficient appliances; this could be a long-term benefit in that it may make energy-efficient appliances universal. Actually, as we were writing this research, the government republished its second round of subsidy programs for energy-efficient appliances. Two products, range hoods and air purifiers, have been added and it boosts the energy efficiency grade of energy-efficient appliances. It is no doubt a positive step.

Moreover, different types of incentive programs can be implemented simultaneously. A combination of upstream, midstream, and downstream incentives could help the growth and maturity of the market, addressing different barriers and problems faced by different sorts of stakeholders (manufacturers, distributors, retailers, and consumers). Therefore, the direct beneficiaries are not only the manufacturers at the upstream of the supply chain but also the other protagonists in the market. Particularly, the implementation of downstream incentive programs (including reducing tax, and rewarding purchases) could directly encourage and motivate consumers to buy energy-efficient appliances.

Furthermore, improved subsidy methods, including perfecting of the subsidy process, reinforcement of the monitoring and management of the subsidy programs should be considered as an essential part. As for the complicated processes for its application, retailers should conduct training programs to facilitate shopkeepers and consumers for completing the process of buying energy-efficient appliances. As for the price being driven up by retailers in the name of profit, the "subsidy label" should be improved. Except for marking the amount of subsidy on each product, the price ceiling after every announcement and the price ceiling after application of its subsidy should also be mentioned, which could prevent retailers from changing the marked price. As for sales quantities being misrepresented by retailers or manufacturers, the process of subsidization should be standardized and the government should publish its monitoring system thereof. To avoid fraud, a supervision department should establish a perfect system, enhancing supervision and explaining punishment mechanisms explicitly. Those companies, who defraud the scheme, would see goods be confiscated, be banned from entering the market, and be deprived of their qualifications to benefit from the energy-saving subsidy program. Only in this way subsidies can pass from companies to consumers by ensuring that consumers reap the benefits of the subsidy program. This will encourage and motivate people to purchase more energy-efficient appliances.

Finally, based on the results that environmental awareness, social interaction and residents' educational level have a significant effect on purchase intentions, we put forward some policy suggestions. The government should enhance publicity efforts on energy conservation and environmental protection to improve people's environmental awareness. With the help of full network coverage and the popularity of smartphones in China, the specific methods can be started from the social network. Because people's energy-saving behavior is easy to be affected by friends, relatives, and people who are important for them, people's purchase intentions on energy-saving appliances can be promoted by people's sharing behavior in the micro-blog or the circle of friends of WeChat. Most importantly, enhancing the country's technical strength and the ability to transform to real productive forces, and improving the scientific and cultural quality of the whole nation are still an urgent task for China.

4.3 Summary

New energy vehicles and energy-efficient appliances are important low-carbon products in the residential sector. We extend the theory of planned behavior and explore the determinants of residents' purchase intentions for the two low-carbon products.

For new energy vehicle purchase intentions, this research analyzes the effects of perceived usefulness, perceived ease of use, subjective norms, and perceived behavioral control on the purchase intentions for new energy vehicles, and also examines the moderating effect of perceived behavioral control on the relationship between subjective norms and perceived purchase intentions, and establishes a discrete choice model for the purchase intentions. In addition, the research delves further into those factors influencing NEV purchase intentions of urban residents with cars and their corresponding influence on people holding different purchase willingness. Results show that, for the urban residents, perceived ease of use positively affects the NEV purchase intentions of those unwilling to buy NEVs, subjective norms have a significant positive effect on purchase intentions of residents unsure about whether or not to buy, and purchase intentions of urban residents are influenced by the relative usefulness; for urban residents with cars, subjective norms can positively influence purchase intentions of residents unsure about whether or not to buy, and their purchase intentions are also influenced by the relative usefulness. These findings can provide a reference for national policies designed for NEV industry development. The government can propose policies regarding subjective norms and perceived ease of use, creating a NEV-use trend, constructing relevant facilities, and stressing the relative benefits of NEVs. The sales of new energy vehicles can be promoted by public praise and recommendations, especially for residents with uncertain intentions.

For energy-efficient purchase intentions, our research explores the determinants of residents' purchase intentions and conducting the analysis of whether or not subsidy policies have guided people's intentions and behaviors. We investigate 436 urban residents from 22 provinces (autonomous regions, municipalities) in China, covering the seven major geographic regions, extend the theory of planned behavior and conduct an empirical analysis of the factors influencing Chinese urban residents' purchase intention towards energy-efficient appliances. Our results show that policy is insignificant, which indicates that policy and propaganda in China do not have a significant effect on Chinese residents' willingness to purchase energy-efficient appliances. However, the residents' environmental awareness, past purchase experience, social interaction, age, and education all exert a significant influence on Chinese residents' purchase intention. Finally, based on the above research findings, the corresponding policy suggestions which mainly focus on the time of subsidy, the object of subsidy and the method of subsidy are offered for policy-makers.

References

Abrahamse, W., and L. Steg. 2009. How do socio-demographic and psychological factors relate to households' direct and indirect energy use and savings? *Journal of Economic Psychology* 30 (5): 711–720.

Ajzen, I. 1991. The theory of planned behavior. *Organizational Behavior and Human Decision Processes* 50 (2): 179–211.

Arar, J.I. 2010. New directions: the electric car and carbon emissions in the US. *Atmospheric Environment* 44 (5): 733–734.

Baird, J.C., and J.M. Brier. 1981. Perceptual awareness of energy requirements of familiar objects. *Journal of Applied Psychology* 66 (1): 90–96.

Bandura, A. 1977. Self-efficacy: toward a unifying theory of behavioral change. *Psychological Review* 84 (2): 191–215.

Bandura, A. 1982. The assessment and predictive generality of self-percepts of efficacy. *Journal of Behavior Therapy and Experimental Psychiatry* 13 (3): 195–199.

Beck, L., and I. Ajzen. 1991. Predicting dishonest actions using the theory of planned behavior. *Journal of Research in Personality* 25 (3): 285–301.

Berry, L.L., A. Parasuraman, and V.A. Zeithaml. 1988. The service-quality puzzle. *Business Horizons* 31 (5): 35–43.

Bockarjova, M., and L. Steg. 2014. Can protection motivation theory predict pro-environmental behavior? Explaining the adoption of electric vehicles in the Netherlands. *Global Environmental Change* 28: 276–288.

Carley, S., R.M. Krause, B.W. Lane, et al. 2013. Intent to purchase a plug-in electric vehicle: a survey of early impressions in large US cites. *Transportation Research Part D: Transport and Environment* 18: 39–45.

Che, W., Z. Zhang, H. Zhang, et al. 2007. Compositions and oxidative damage of condensate, particulate and semivolatile organic compounds from gasoline exhausts. *Environmental Toxicology and Pharmacology* 24 (1): 11–18.

Coleman, L.J., N. Bahnan, M. Kelkar, et al. 2011. Walking the walk: how the theory of reasoned action explains adult and student intentions to go green. *Journal of Applied Business Research* 27 (3): 107–116.

Cong, R. 2013. An optimization model for renewable energy generation and its application in China: a perspective of maximum utilization. *Renewable and Sustainable Energy Reviews* 17: 94–103.

Cong, R.G., and S.C. Shen. 2014. How to develop renewable power in China? A cost-effective perspective. *The Scientific World Journal* (1): 946932.
Cong, R.G., and Y.M. Wei. 2010a. Potential impact of (CET) carbon emissions trading on China's power sector: a perspective from different allowance allocation options. *Energy* 35 (9): 3921–3931.
Cong, R.G., and Y.M. Wei. 2010b. Auction design for the allocation of carbon emission allowances: uniform or discriminatory price. *International Journal of Energy and Environment* 1 (3): 533–546.
Courter, L.A., C. Pereira, and W.M. Baird. 2007. Diesel exhaust influences carcinogenic PAH-induced genotoxicity and gene expression in human breast epithelial cells in culture. *Mutation Research/Fundamental and Molecular Mechanisms of Mutagenesis* 625 (1–2): 72–82.
Davis, F. 1989. Perceived usefulness perceived ease of use and user acceptance of information technology. *MIS Quarterly* 13 (3): 319–340.
de la Rue Du Can, S., G. Leventis, A. Phadke, et al. 2014. Design of incentive programs for accelerating penetration of energy-efficient appliances. *Energy Policy* 72: 56–66.
Egbue, O., and S. Long. 2012. Barriers to widespread adoption of electric vehicles: an analysis of consumer attitudes and perceptions. *Energy Policy* 48: 717–729.
Ek, K., and P. Söderholm. 2010. The devil is in the details: Household electricity saving behavior and the role of information. *Energy Policy* 38 (3): 1578–1587.
Fornell, C., and D.F. Larcker. 1981. Evaluating structural equation models with unobservable variables and measurement error. *Journal of Marketing Research* 18 (1): 39–50.
Friedmann, R. 2011. A fresh look at evaluation to support energy efficiency in the 21st century. In *Proceedings of the IEPEC 2011 conference. International energy program evaluation conference.* Boston, Mass.
Galarraga, I., L.M. Abadie, and S. Kallbekken. 2016. Designing incentive schemes for promoting energy-efficient appliances: a new methodology and a case study for Spain. *Energy Policy* 90: 24–36.
Gaspar, R., and D. Antunes. 2011. Energy efficiency and appliance purchases in Europe: consumer profiles and choice determinants. *Energy Policy* 39 (11): 7335–7346.
Howell, J.M., and C.M. Shea. 2005. Champions of product innovations: defining, developing, and validating a measure of champion behavior. *Journal of Business Venturing* 20: 641–661.
IEA. 2012. *World energy outlook*. Paris: France, International Energy Agency (IEA).
Korcaj, L., U.J.J. Hahnel, and H. Spada. 2015. Intentions to adopt photovoltaic systems depend on homeowners' expected personal gains and behavior of peers. *Renewable Energy* 75: 407–415.
Lane, B., and S. Potter. 2007. The adoption of cleaner vehicles in the UK: exploring the consumer attitude–action gap. *Journal of Cleaner Production* 15 (11–12): 1085–1092.
Li, X., C.D. Clark, K.L. Jensen, et al. 2013. Consumer purchase intentions for flexible-fuel and hybrid-electric vehicles. *Transportation Research Part D: Transport and Environment* 18: 9–15.
Lindenberg, S., and L. Steg. 2007. Normative, gain and hedonic goal frames guiding environmental behavior. *Journal of Social Issues* 63: 117–137.
Marell, A., P. Davidsson, and T. Garling. 1995. Environmentally friendly replacement of automobiles. *Journal of Economic Psychology* 16 (3): 513–529.
Mau, P., J. Eyzaguirre, M. Jaccard, et al. 2008. The "neighbor effect": simulating dynamics in consumer preferences for new vehicle technologies. *Ecological Economics* 68 (1/2): 504–516.
Møller, P., J. K. Folkmann, L. Forchhammer, et al. 2008. Air pollution, oxidative damage to DNA, and carcinogenesis. Cancer Letters, 266 (1): 84–97.
Mourato, S., B. Saynor, D. Hart, et al. 2004. Greening London's black cabs: a study of driver's preferences for fuel cell taxis. *Energy Policy* 32 (5): 685–695.
Nayum, A., and C.A. Klöckner. 2014. A comprehensive socio-psychological approach to car type choice. *Journal of Environmental Psychology* 40: 401–411.

Nayum, A., C.A. Klöckner, and S. Prugsamatz. 2013. Influences of car type class and carbon dioxide emission levels on purchases of new cars: a retrospective analysis of car purchases in Norway. *Transportation Research Part A: Policy and Practice Psychology of Sustainable Travel Behavior* 48: 96–108.

Nunnally, J. 1978. *Psychometric Theory*, 2nd ed. New York: McGraw-Hill.

Ozaki, R., and K. Sevastyanova. 2011. Going hybrid: an analysis of consumer purchase motivations. *Energy Policy* 39 (5): 2217–2227.

Pfeffer, J. 1982. *Organization and organization theory*. Boston MA: Pitman.

Radner, R., and M. Rothschild. 1975. On the allocation of effort. *Journal of Economic Theory* 10: 358–376.

Rezvani, Z., J. Jansson, and J. Bodin. 2015. Advances in consumer electric vehicle adoption research: a review and research agenda. *Transportation Research Part D: Transport and Environment* 34: 122–136.

Samuelson, C.D. 1990. Energy conservation: a social dilemma approach. *Social Behavior* 5: 207–230.

Sang, Y., and H.A. Bekhet. 2015. Modelling electric vehicle usage intentions: an empirical study in Malaysia. *Journal of Cleaner Production* 92: 75–83.

Schniederjans, D.G., and C.M. Starkey. 2014. Intention and willingness to pay for green freight transportation: an empirical examination. *Transportation Research Part D: Transport and Environment* 31: 116–125.

Schuitema, G., J. Anable, S. Skippon, et al. 2013. The role of instrumental, hedonic and symbolic attributes in the intention to adopt electric vehicles. *Transportation Research Part A: Policy and Practice* 48: 39–49.

Smith, W.J. 2010. Can EV (electric vehicles) address Ireland's CO_2 emissions from transport? *Energy* 35: 4514–4521.

Sovacool, B.K., and R.F. Hirsh. 2009. Beyond batteries: an examination of the benefits and barriers to plug-in hybrid electric vehicles (PHEVs) and a vehicle-to-grid (V2G) transition. *Energy Policy* 37 (3): 1095–1103.

Steg, L. 2008. Promoting household energy conservation. *Energy Policy* 36 (12): 4449–4453.

Swami, V., T. Chamorro-Premuzic, R. Snelgar, et al. 2011. Personality, individual differences, and demographic antecedents of self-reported household waste management behaviours. *Journal of Environmental Psychology* 31: 21–26.

Taylor, S., and P.A. Todd. 1995. Understanding information technology usage: a test of competing models. *Information Systems Research* 6: 144–176.

Tran, M., D. Banister, J.D.K. Bishop, et al. 2013. Simulating early adoption of alternative fuel vehicles for sustainability. *Technological Forecasting and Social Change* 80 (5): 865–875.

Wang, Y.P. 2011. Relationship between road freight transport demand and national economy development based on co-integration analysis. *Journal of Jilin University (Engineering and Technology Edition)* 41: 56–61.

Wang, H., P. Zhou, and D.Q. Zhou. 2013. Scenario-based energy efficiency and productivity in China: a non-radial directional distance function analysis. *Energy Economics* 40: 795–803.

Wang, Z.H., B. Zhang, J.H. Yin, et al. 2011. Determinants and policy implications for household electricity-saving behavior: evidence from Beijing, China. *Energy Policy* 39(6): 3550–3557.

Wen, Z., K. Hou, et al. 2005. A comparison of moderator and mediator and their applications. *Acta Psychologica Sinica* 37 (2): 268–274.

Wu, M. 2010. *Structural equation model-use and application of AMOS*. Chongqing: Chongqing University Press.

Yang, S., and D. Zhao. 2015. Do subsidies work better in low-income than in high-income families? Survey on domestic energy-efficient and renewable energy equipment purchase in China. *Journal of Cleaner Production* 108: 841–851.

Yao, X., Y. Liu, and X. Yan. 2014. A quantile approach to assess the effectiveness of the subsidy policy for energy-efficient home appliances: evidence from Rizhao, China. *Energy Policy* 73: 512–518.

Yin, Z., and F. Wang. 2013. Difference comparison on consumer purchase intention of new energy vehicle. *Shanghai Management Science* 35 (4): 15–19.
Yoon, C. 2011. Theory of planned behavior and ethics theory in digital piracy: an integrated model. *Journal of Business Ethics* 100 (3): 405–417.
Yuan, X., X. Liu, and J. Zuo. 2015. The development of new energy vehicles for a sustainable future: a review. *Renewable and Sustainable Energy Reviews* 42: 298–305.
Zeng, L., Y. Yu, and J. Li. 2014. China's promoting energy-efficient products for the benefit of the people program in 2012: results and analysis of the consumer impact study. *Applied Energy* 133 (15): 22–32.
Zhang, X., K. Wang, Y. Hao, et al. 2013. The impact of government policy on preference for NEVs: the evidence from China. *Energy Policy* 61: 382–393.
Zografakis, N., E. Sifaki, M. Pagalou, et al. 2010. Assessment of public acceptance and willingness to pay for renewable energy sources in Crete. *Renewable and Sustainable Energy Reviews* 14 (3): 1088–1095.

Chapter 5
E-Waste Recycling Behavior in China

5.1 Current Situation of E-Waste Recycling

Electronic waste refers to the loss of electronic and electrical products that should be used, including the unqualified products and their spare parts, leftover parts, spare parts produced in the process of maintenance, discarded products and equipment, etc., which are produced in the process of production.

5.1.1 Characteristics of E-Waste Recycling

There are two main characteristics of E-waste recovery.

(1) Environmental pressure caused by E-waste

Improper recycling and disposal of E-waste will cause secondary environmental problems which are relative to the natural environmental problems caused by natural causes, which means that human activities are not conducive to human changes due to human activities, so that the production and life of human beings are affected and the disasters or hazards are brought to human beings. Materials often contain harmful components to the environment (Table 5.1), which are particularly evident in E-waste.

This chapter takes the following literature for reference:

Wang Z, Zhang B, Guan D. 2016. Take responsibility for electronic-waste disposal. Nature, 536 (7614): 23–25.

Wang Z, Zhang B, Yin J, et al. 2011. Willingness and behavior towards E-waste recycling for residents in Beijing city, China. Journal of Cleaner Production, 19(9): 977–984.

Wang Z, Guo D, Wang X. 2016. Determinants of residents' E-waste recycling behaviour intentions: evidence from China. Journal of Cleaner Production, 137: 850–860.

Z. Wang and B. Zhang, *Low-Carbon Consumption in China: Residential Behavior, Corporate Practices and Policy Implication*,
https://doi.org/10.1007/978-981-15-2792-0_5

Table 5.1 Pollution components in E-waste

Contaminants	Source
Chlorofluorocarbon	Refrigerator
Halogen flame retardant	The outer shell of the circuit board, cable, and electronic equipment
Mercury	Monitor
Selenium	Optoelectronic device
Nickel and cadmium	Batteries and some computer displays
Barium	Cathode-ray tube and PCB
Lead	Cathode-ray tubes, tin soldering, capacitors and displays
Chromium	Metal coating

Source De Winter (2001)

For example, the manufacture of a computer requires more than 700 kinds of chemical raw materials and more than half of these chemicals are harmful to the human body. A 15 inch (1 in = 2.54 cm) computer contains more than 300 types of teratogenic, mutagenic, and carcinogenic harmful substances such as lead, mercury, cadmium, PVC plastic, and so on. If this issue does not deal effectively, under the influence of natural conditions, these harmful substances may turn to the atmosphere, water, and soil, participate in the material cycle of the ecosystem and lead to serious environmental pollution which will result in potential and long-term harms to human beings.

The workshop-style disposal of E-waste with high energy consumption will cause serious environmental pollution. It is easy to produce a large amount of volatile organic compounds and particles through simple dismantling, heating, acidification, incineration, resulting in air and groundwater pollution. Lead, mercury, Fu, six valent chromium, polybrominated biphenyl and polyolated diphenyl ethers in electronic waste are important factors leading to deformity, mutation and cancer. The resultant pollutants of workshop-type simple treatment of E-waste will infiltrate into the underground water, soil and air. Thus bring the burden of environmental pollution and even serious threats to people's life and health. Taking Guiyu Town, Guangdong as an example, as it is one of the largest electronic waste distribution sites in China, the blood sample test of children of 1–6 years has lead as high as 15.3 μg/ml exceeding the expert's recommended vigilance standard by 50% (Ogunseitan et al. 2009). Locally surface water is unavailable; contents of sulfur and nitrogen oxide in the air are exceeding the standards.

(2) The resource value of E-waste

Resources are not only obtained directly from nature but also can be put into production as new raw materials through processing and refining, so as to realize the conservation and recycling of resources. It provides new ideas for solving the problem of resource shortage.

E-waste consists of precious metals such as gold, copper, aluminum, silver, plastic, and others which have a high potential for reuse and recycling (Table 5.2). Improper recycling will result in a huge waste of resources. At present, the total value of recycling resources of major developed countries has reached the US $250 billion a year and is growing at 15–20% annually. 45% of the world's steel production, 62% of the world's copper production, 22% of the world's aluminum production, 40% of the world's lead production, 30% of the world's zinc production, and 35% of the world's paper products come from the recycling of regenerated resources. China's electronic waste disposal industry has an annual output value of 100 billion yuan of which precious metals are the most common. The gold output of Guiyu town in Guangdong accounts for 1/20 of China's total gold output and even affects the international gold price trend.

5.1.2 The Generation and Flow of E-Waste in Major Countries of the World

The world is producing ever more E-waste. The quantity of dumped computers, telephones, televisions, and appliances doubled between 2009 and 2014, reached 42 million tons per year globally (Robinson 2009). Developed countries, especially in North America and Europe produce the most E-waste. The United States generates the largest amount and China the second most. Much of this E-waste ends up in a developing world where regulations are lax. China processed about 70% of the

Table 5.2 The recoverable resources and value of waste household appliances

Material	Proportion/%	Quality/kg	Unit price/(dollar/1b[a])	Total value/dollar
Copper	20	400	0.980	392.0
Iron	8	160	0.045	7.2
Nickel	2	40	2.230	89.2
Tin	4	80	2.350	188.0
Lead	2	40	0.21	8.40
Aluminum	2	40	0.71	28.40
Zinc	1	20	0.48	9.60
Gold	0.1	2	3885.57	7771.14
Silver	0.2	4	34.40	137.60
Palladium	0.005	0.1	5019.16	501.92
Plastic	30	600	0.1	60

[a]1b = 0.45 kg

Source Xu and Ma (2004)

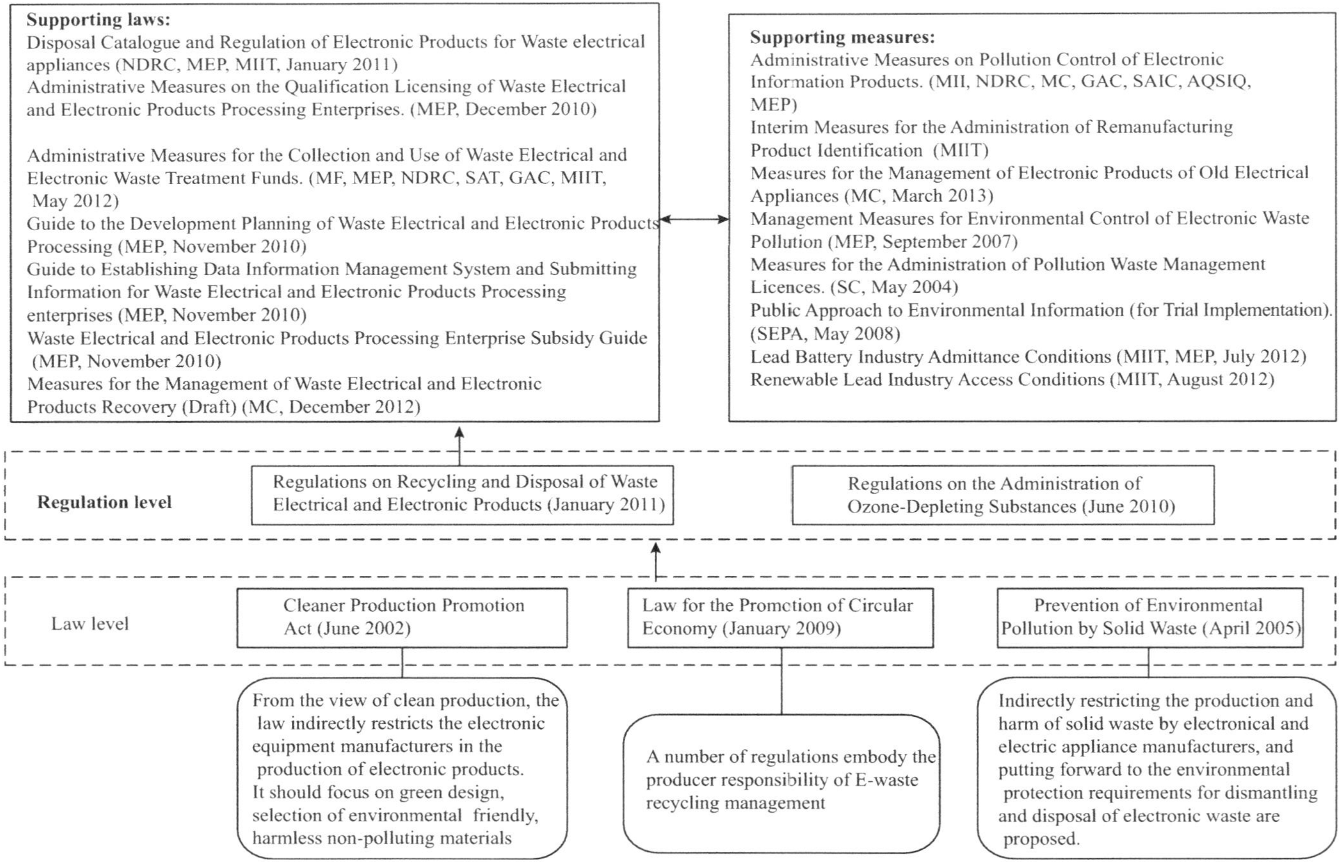

Fig. 5.1 Legal system of E-waste management in China. *Notes* NDRC: National Development and Reform Commission; MEP: Ministry of Environmental Protection; MIIT: Ministry of Industry and Information Technology; MF: Ministry of Finance; SAT: State Administration of Taxation; GAC: General Administration of Customs; MC: Ministry of Commerce; SEPA: State Environmental Protection Administration; SC: The State Council

world's E-waste in 2012 (Zhang et al. 2012); the rest goes to India and other countries in eastern Asia and Africa, including Nigeria (Breivik et al. 2014). Most developed countries have strict regulations governing the disposal of E-waste. European countries, the United States and others have official "take-back" systems, which recover and dispose of E-waste in an environmental-friendly way. The European Union has one comprehensive directive: the Restriction of Hazardous Substances, Waste Electrical and Electronic Equipment. Yet the EU, and the United States dispose of only 40% and 12%, respectively, of E-waste they generate domestically. These rich nations with strict legislation send most of their E-waste to developing countries because many poor nations especially in Africa have a few or no laws on E-waste.

The Basel Convention of the United Nations, which concerns the movement of hazardous waste across borders, is meant to prevent developed countries from illegally dumping hazardous waste in developing countries. But only 87 parties—and not the United States—have ratified it. Few developing countries control imports of toxic E-waste: for example, India's law fails to ban it. This resulted in 50,000 tons of such waste from developed countries being dumped in India in 2012 only (Agoramoorthy and Chakraborty 2012). The shady trading of trash as "used electronics" bypasses such laws entirely.

In China, according to statistics, the electronic waste generated in 2014 amounted to 6 million tons. The scrap value of household appliances and computers accounted for 1.6 million tons (Zeng et al. 2013). This has caused many cities in China to face the problem of environmental pollution instigated by "garbage siege" and improper disposal of E-waste.

In addition, China's E-waste legislation is inefficient and irregularly enforced (Fig. 5.1). China's system is poorly coordinated; it involves more than ten departments publishing regulations, imposing disposal fees, providing subsidies, and monitoring pollution, and illegal imports with little cross talk. This makes nearly 23 million tons of E-waste into China every year from all over the world (Zhou and Xu 2012), and China has already become the "absorption plant" of the world's electronic waste.

5.1.3 E-Waste Recycling Management Policy System and Problems in China

The E-waste management legal system of E-waste management in China is shown in Fig. 5.1. In general, China's E-waste legal system is mainly consisting of three parts: legal layer, regulatory layer, and policy management method. The legal system about e-waste was developed gradually from the restraint that enterprises must produce clean production. However, with the increasing of E-waste and resource shortage, the government realizes the importance of resource recycling and strengthens the legal system. Lack of relevant laws and regulations on E-waste management causing a lot of environmental pollution problems. For this reason, the Ministry of Environmental Protection in 2011 issued the "regulations on recycling and disposal of waste electrical and electronic products".

Unlike the European and American E-waste administration system, China's E-waste management involves many government regulators (Zeng et al. 2013). Figure 5.2 shows that China's E-waste management involved government administrative departments and their respective functional labor division. The law promulgated by the Standing Committee of the National People's Congress is at the top of the legal system and it is a fundamental guiding principle. Particularly, Circular Economy Promotion Law clearly puts forward the principles and methods of E-waste recycling and disposal. The State Council puts forward the Regulation on Recycling and Disposal of Waste Electrical and Electronic Products, which is a special regulation for the recycling and disposal of E-waste. The regulation stipulates disposal norms, enterprise qualification standards, and punishment procedures on all aspects including E-waste recovery, dismantling, processing, and refining. The specific supervision and administration of E-waste recovery and disposal is coordinated and managed by the State Council under the overall planning of various ministries and commissions within their respective functions. So far, China has formed a three-level E-waste management system, from top to bottom, overall planning, and layer by layer, to promote the E-waste management comprehensively.

5.1.4 Status and Problems of China's E-Waste Recycling Industry

Because of its economic and recycling value, E-waste has been paid attention by many investors and has gradually formed a set of industrial systems of "recycling-dismantling-disposal and reuse". Unlike developed countries such as Europe, America, and Japan, China's E-waste industrial chain is more complex and involves many players. Due to lack of responsibility and supervision loopholes illegal recycling channels coexists with regular recycling channels. The overall schematic diagram of China's E-waste industrial chain is shown in Fig. 5.3.

China's E-waste recycling channels are generally divided into three categories: producer's self-managed recycling through their distribution networks, personal consumer recovery through small traders, and recycling through professional recycling agencies. Among them, individual retail recovery occupies the dominant

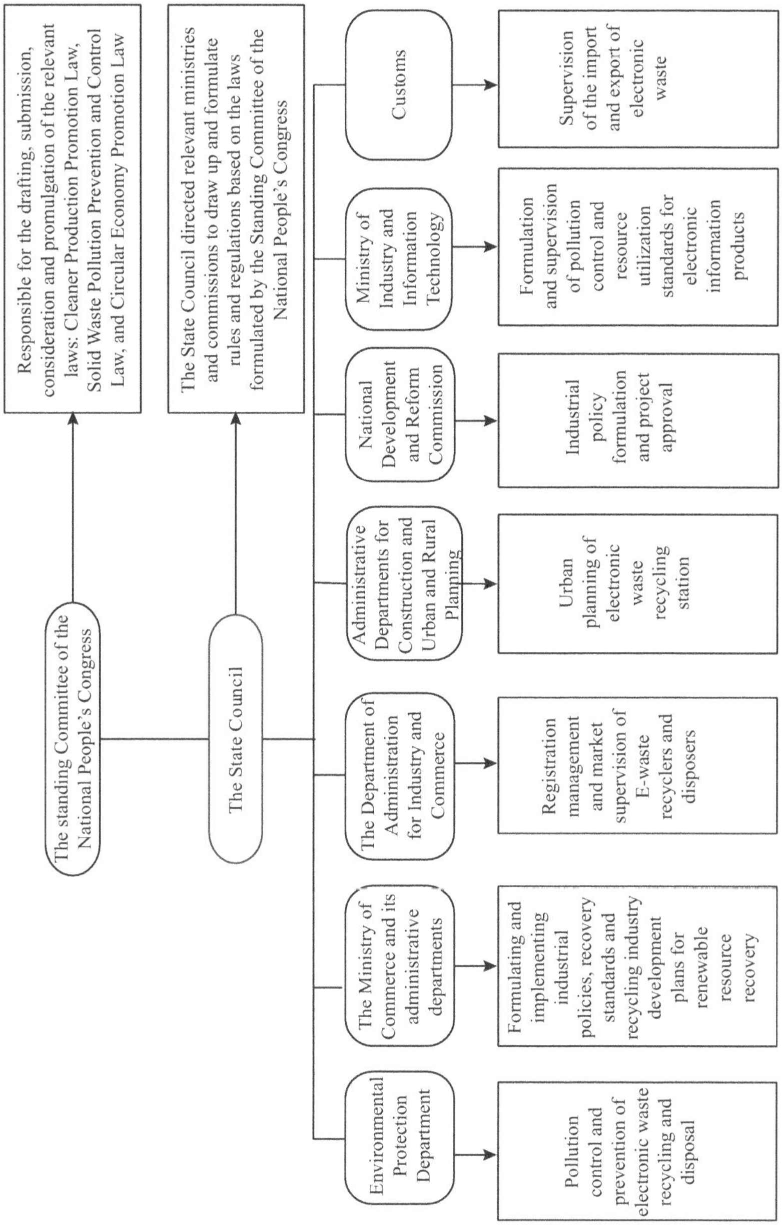

Fig. 5.2 Administrative system of recycling and disposal of E-waste

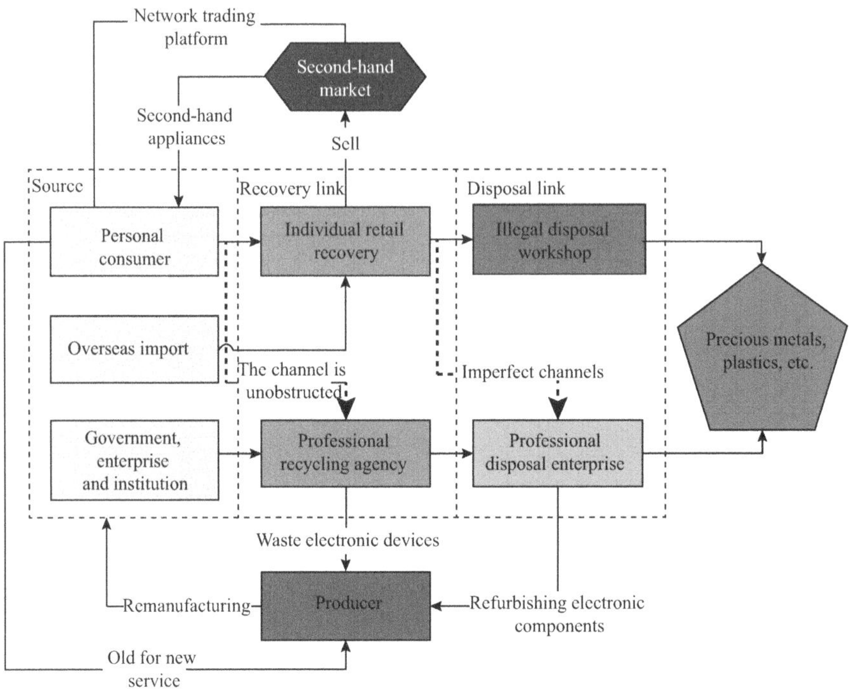

Fig. 5.3 The overall schematic diagram of China's E-waste industrial chain

position; nearly 60% of E-waste enters the recycling market through individual retail recycling channels (Liu et al. 2006). Producer's direct participation in recycling is not very common in China. The Chinese government's policy of "old for new service" from June 2009 to May 2010 has shared the cost of recycling waste of household appliances in the form of financial subsidies and greatly promoted the enthusiasm of production enterprises to participate in recycling. It promotes the construction of the producer's owned recycling network but with the expiration of the policy its progress slows down. The professional recycling network is dominated by the qualification of electronic waste disposal.

After recycling E-waste mostly flowed in two ways: First, reflowing into the hands of the consumers through the second-hand market; Second, refining and reusing the resources with economic value. The E-waste flow through the second-hand market is mostly electronic appliances that have a long service life and better usage value (utility value). After simple maintenance and refurbishment, they are reflowed into the market. As for dismantling and disposal of E-waste are concerned, there are two types of industrial systems coexisting in China: the professional disposal enterprise and the illegal disposal workshop.

The disposal of professional electronic waste pays attention to both resources and environmental benefits, and emphasizes the maximum recovery of resources and economic benefits under the environmental standards stipulated by the state. Geographically the professional E-waste recycling enterprises are mostly distributed in the eastern and in central coastal areas, while in the western part E-waste recycling and disposal industry is relatively missing. Although the eastern coastal areas are economically developed, densely populated, have concentrated electronic waste production but many professional E-waste disposal enterprises are still facing difficulties to survive. Investigation shows that the lack of sufficient supply of electronic waste is the most important dilemma faced by professional E-waste disposal enterprises.

By category, disposal abandoned electronic products are dominated by TV in current professional E-waste disposal business, accounting for 64% of the total processing scale; the second is the microcomputer accounting for 12%; and the volume of the air conditioner is only 5% as shown in Fig. 5.4a. This reflects to some extent that the disposal of professional E-waste disposal in China is relatively simple, most of the disposers only have the ability to dismantle and dispose of TV sets, and the development of other categories of E-waste dismantling and disposal industry is still lagging behind, resulting in a great deal of E-waste flowing into illegal disposal workshops.

More than 50% of professional E-waste dismantling dealers in China have more than 1 million annual dismantling capacity and only 6% of enterprises have more than 3 million disposal capacity as shown in Fig. 5.4b. However, as a result of the

Fig. 5.4 Disposal types and capabilities of China's E-waste professional dismantling and disposal enterprises, *Source* Wang F et al. (2013)

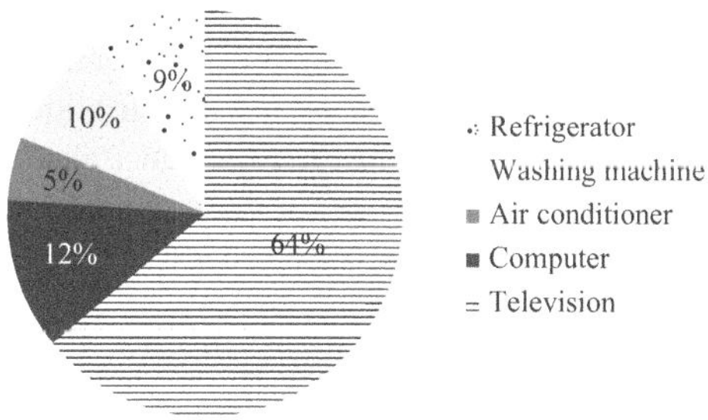

(a) The constitution of E-waste disposed by formal companies

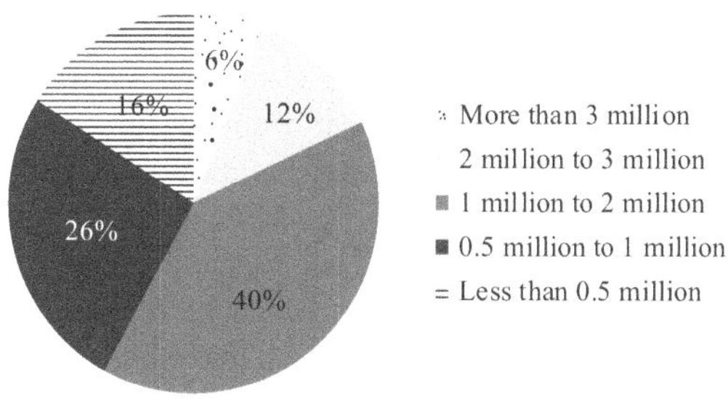

(b) The disposal capacity of formal companies

relatively insufficient construction and docking capacity of the recycling channels, the supply of electronic disposal products is often insufficient, propelling not to utilize production capacity up to scale, resulting in wastage of resources and difficulties for enterprises management.

The disposal of illegal workshop-type E-waste only pays attention to the economic value of electronic waste, while neglects or even sacrifices the environmental benefits, which often brings serious damage to the local ecological environment, threatens the health and life safety of the residents. In the process of E-waste treatment, workers often lack the necessary protective measures to direct exposure and contact with all kinds of toxic substances, which poses a huge threat to their life and health.

The disposal of illegal E-waste has developed rapidly in China and has formed a gray industrial chain which includes the recycling, transportation, classification, disassembling, and processing of electronic waste. Beijing, Shanghai, Guangzhou, Shenzhen and other major cities are the most important recycling hubs for electronic waste with huge local E-waste, relatively centralized recycling sites, convenient transportation and other expedient factors. Only in Beijing, there are 300 thousand persons engaged in the recovery and disposal of E-waste. Shijiazhuang in Hebei is a distributed and dismantling center of E-waste in the north of China. It has presumed the responsibility of E-waste recycling in major cities of the north, especially Beijing. Shijiazhuang in Hebei is primarily used for dismantling and rough processing of E-waste, refining precious metal materials of copper and aluminum. The leftover materials and electronic components are transported to the South China E-waste centers, Guiyu Guangdong, to be processed in depth, to extract gold, silver, and other precious metals, or to refurbish used components. Taizhou in Zhejiang Province is the distribution center of E-waste disposal along the eastern coast of China. It also undertakes the dismantling and primary processing of many imported E-waste. Guiyu in Guangdong is one of the most important areas of E-waste disposal in China. In addition to simple dismantling and processing, many disposal facilities are engaged in the dissolution and refining of precious metals while environmental problems are very prominent in the area.

5.2 Willingness Behavior Toward E-Waste Recycling in China

With the rapid development of the electrical industry, the problem of electrical and electronic waste (E-waste) generation and management has become more and more serious in China. Researchers have conducted many studies about E-waste disposal based on Extended Producer Responsibility (EPR) (Milanez and Bührs 2009; Lin et al. 2002; Fleckinger and Glachant 2010). However, most electrical companies in China are relatively small and have intense competition with each other. They cannot have enough money to bring recycling activities into effect as large foreign companies. Therefore, to solve the E-waste problem in China, researchers should not only focus on the EPR but also consider the performance of government and the behavior of residents (Nnorom et al. 2009). In developing countries, government and municipalities play an important role in product take-back and provision of recycling infrastructure (Hicks et al. 2005). Many recent studies on E-waste disposal in China have been focusing on how to implement EPR and the function of government in E-waste recycling (Fleckinger and Glachant 2010; Nnorom et al. 2009; Zhao et al. 2010; Mo et al. 2009; Tucker et al. 1998), while the research on the willingness or behavior in E-waste recycling has not attracted enough attention. Nevertheless, without the willingness of residents to participate in the recycling activities, the policies by government and arrangements by producers cannot be well put into practice. It is necessary to conduct an in-depth study of the residents' willingness and behavior in E-waste recycling in China.

5.2.1 E-Waste in Beijing, China

Beijing is facing large quantities of E-waste generation and various channels of E-waste sources. Beijing E-waste sources include both domestic generation and imports. E-waste imports are divided into legal and illegal imports. The former is managed and controlled by China Customs and the Ministry of Ecology and Environment of the People's Republic of China. Domestic generation mainly comes from post-consumer E-products. Only small quantities come from industrial processes (Yang et al. 2008).

E-waste recycling and disposal in Beijing are not well put into effect. According to Yang Jianjin, 70% of end-of-life E-products are stored in homes or offices, which make these reusable resources unable to enter the recycling system. The storage time varies from several months to several years (Yang et al. 2008). Only 12% of E-waste is disassembled or dismantled for material recovery collection, which will be sorted and treated for reuse or recovery of base materials. Most of the reclaimed E-waste by various backward channels are flowed to the second-hand market after simple repairing, upgrading, or refurbishing. Although these second-hand products are still technically in working order, many of them do not follow the latest technical standards, which may cause serious energy dissipation. Some of them are even beyond the durable years, which exists a potential safety hazard for consumers. Besides, at the end-of-life of these second-hand products, they will reenter in E-waste flow. There are four kinds of reclaiming channels for E-waste recycling in Beijing, which are peddlers, dealers/retailers, specialized collectors, and second-hand market.

The disposal technology and process of E-waste in Beijing are relatively backward compared with Western countries. Small and medium-sized plants are the main hubs of E-waste disposal. About 80% of recycled E-waste is simply repaired or refurbished before flowing to the second-hand market. 4.2% of E-waste is dismantled into small parts which are often reused as input materials especially in the

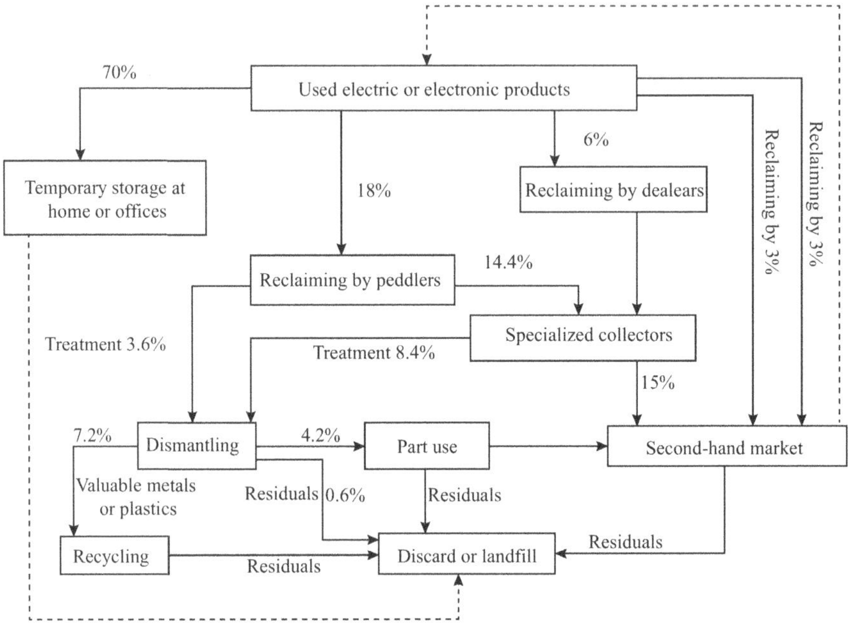

Fig. 5.5 E-waste flow in Beijing. *Sources* Yang et al. (2008)

electronic toys. 7.2% of E-waste is disassembled for the recovery materials, majority of which are plastics and precious metals such as gold, silver, which are used in chemical, metallurgy, and manufacturing industries. Still, disposal processing methods mainly comprise of burning and acidolysis, which usually cause serious environmental pollution. The rest of uneasily disposed E-waste is often discarded as normal waste, which results in a vicious waste of resources.

The integral profile of E-waste flow in Beijing can be seen in Fig. 5.5.

5.2.2 *Resident's Behavior in E-Waste Recycling*

As different countries have different circumstances, scholars from various countries hold diverse attitudes toward the responsibility the residents should be taken in E-waste recycling. Jofre and Morioka compared the E-waste management strategies of Japan, United States, and European Union. It is found that although all these countries follow the principle of extended product responsibility, there are different requirements for the responsibilities and obligations of residents involved in E-waste recycling (Jofre and Morioka 2005). Many EU countries (e.g., Germany, Spain) have emphasized the process of E-waste recycling fees should be mainly imposed on the producers, while the residents would only fulfill the obligations of sending the E-waste to collection points. The cost of recycling electronic waste is borne by consumers in Japan. EPR related laws and regulations allow producers to pass on costs to consumers by embedding costs in product prices. While in the United States the responsibility of E-waste recycling is shared by producers, government, and the residents. Different from the developed countries, the income level of residents and environmental awareness of developing countries are relatively low, and there is still a certain gap in the enthusiasm of the residents involved in the recovery. According to Saphores et al., the willingness to pay for waste management and green electronics may not be as high as to pay the water and electricity fees in most of developing countries (Saphores et al. 2006).

According to Nnorom, less knowledge of the toxicity of E-waste and the dangers of inappropriate recycling techniques are considered the primary barriers in E-waste recycling (Nnorom et al. 2009). The willingness and behavior of residents in E-waste recycling bring some uncertainties (Ravi et al. 2005): the diversity of consumers' recycling habits directly affects the time and place of E-waste recycling; the willingness of residents to afford the E-waste recycling fee could also affect the quantity of reclaimed E-waste. Therefore, even where recycling infrastructure is available, an understanding of the household's willingness to participate in the recycling program is still necessary (Saphores et al. 2006).

Hornik et al. (1995) had conducted detailed literature research on determinants of recycling behavior. In their study, they reviewed more than 400 articles in more than 20 years and did a meta-analysis on 67 empirical studies. The factors affecting the recycling behavior were divided into five categories: Extrinsic Incentives; Intrinsic Incentives; External Facilitators; Internal Facilitators; Demographic Variables (Hornik et al. 1995).

Based on Jacob's study, many scholars have explored and verified a number of more specific influencing factors toward the recycling behavior. Darby and Obara (2005) did a case study on small electronics waste recycling of Cardiff and explored the relationship between income and recycling behavior. The study also showed that the recovery behavior of other resources (waste paper, plastic, etc.) has important spillover effects on recycling behavior (Darby and Obara 2005). Van Beukering et al. (2006) shared the view that consumers with high-income levels are more likely to participate voluntarily in recycling. However, they also pointed out that this willingness would decline if such recycling was time intensive.

Demographic variables are given special attention by scholars in E-waste recycling. Some surveys indicate that age is a factor in recycling with more older persons participating to a large extent in recycling schemes than younger persons while others indicate no such significant relationship between age and recycling (Tonglet et al. 2004; Ajzen 1985). Some studies note that gender is not a vital determinant of recycling behavior (Guagnano et al. 1995; Ajzen 1985). However, the demographic factors explain only part of differences in recovery behavior (Oskamp et al. 1991).

According to Hansmann, the individual's behavior is largely determined by the social and cultural background. Educating consumers about the importance of recycling, as well as how and where to recycle can enhance public participation in recycling (Guagnano et al. 1995). Individual recycling behavior can be easily influenced by the family and friends around. The family typology and the socioeconomic stratum where the family belongs can also result in differences in the E-waste recycling of residents. As Ojeda-Benitez said, "Family exerts a profound and lasting influence on the perception and behavior of its members, both in the consumption and in waste generation and handling" (Ojeda-Benitez et al. 2008).

The common research methods toward recycling behavior are usually based on empirical surveys behavior and the research models can be divided into three categories: econometric model (Guerin et al. 2001; Hansmann et al. 2006), system simulation model (Guagnano et al. 1995), and psychology model (Jones 1990; Boldero 1995; McCarty and Shrum 1994; Tonglet et al. 2004; Davis et al. 2006).

5.2.3 Data Sources: A Questionnaire from Beijing, China

A questionnaire was designed to elicit residents' willingness in E-waste recycling and preferences of recycling patterns in 2009. In the questionnaire introduction, the objective of research and the importance of truthiness of answers were clearly emphasized in order to reduce hypothetical mark bias. An information session about E-waste was also stated to avoid the confusion of the scope of E-waste. Moreover, the formulation of questions included in the questionnaire was based on the literature review in Sect. 5.2.2, with the content covered by three main areas: demographic information; E-waste recovery behavior; preference patterns of E-waste recycling.

The survey in Beijing urban area includes six administrative divisions: Dongcheng, Xicheng, Xuanwu, Chongwen, Haidian, and Chaoyang. The specific distribution of the questionnaire is shown in Table 5.3.

5.2.4 Methodology: Logistic Regression Model

When the dependent variable is in 0–1 style, researchers have a choice between logistic regression and probit regression. According to Borsch–Supan, the Logit model is a better choice if the response decision is made based on maximization of utility (Borsch-Supan 1990). Considering that residents' willingness in E-waste recycling is mainly depending on the expected utility from the recycling behavior, the Logit model was selected in this work. The following specification was used:

$$\mathrm{Logit}(R) = \frac{1}{1+\mathrm{e}^{-Z}} Z = \beta_0 + \sum_{i=1}^{n} \beta_i x_i + \varepsilon_i \tag{5.1}$$

Table 5.3 The first distribution of samples and response rate

Location	Information of administrative divisions	Quantities of issued questionnaire	Quantities of responded questionnaire	Response Rate/%
Xicheng	In the center of the northwest side of Beijing with a population of 112 thousand	195	90	46.15
Dongcheng	On the east of Xicheng with a population of 101 thousand	120	111	92.50
Haidian	On the northwest of Xicheng with a population of 906 thousand	285	285	100
Chaoyang	On the northeast of Dongcheng with a population of 998 thousand	360	294	81.67
Xuanwu and Chongwen	In the south of Beijing with a population of 168 thousand	213	198	92.96
Total		1173	978	83.38

Note In 2010, Chongwen District merged with Dongcheng District, and Xuanwu District merged with Xicheng District

where Z = latent and continuous measure of residents' willingness in E-waste recycling, x_i = a vector of observations of explanatory variables, β = a vector of parameters to be estimated, ε = a random error term (assumed to follow a standard normal distribution).

The observed and coded discrete willingness variable, R, is determined from the model as follows: 0—willing to participate in E-waste recycling; 1—not willing to participate in E-waste recycling. In the questionnaires, the designed question for R is "How do you deal with the end-of-life electrical products?", and five answers are given as follows: A—disposal as normal waste; B—sell to the collectors as scrap metal; C—sell to the collectors as second-hand product; D—store at home; E—other choices. This work considers the selection of choice B or C as 0—willing to participate in E-waste recycling.

5.2.5 *Specimen Description of Residents' E-Waste Recycling Behavior in Beijing*

The effective sample size of this book is 957 after the questionnaires whose response rates are lower than 60% are deleted. The criteria, mean and standard deviation of all variables are shown in Table 5.5. It is shown that some independent variables are not strictly following the normal distribution due to the special characteristics of the population. However, there are no special requirements for the independent variables in logistic regression model to be normal (Wang 2001). All the items in Table 5.4 are directly from the corresponding questions in the questionnaire. The values of independent variables x_1–x_5 come from the part of demographic information in the questionnaire. Residential condition is measured by the question "the house you living now is rented or your own". Family population, income and education are ordinary variables according to the actual value that the respondents fill out. Engel's coefficient is measured by the ratio of income divided by consumption. The rest independent variables x_6–x_{10} come from the part of E-waste recovery behavior in the questionnaire. Recycling habits are measured by past experience of whether participating in recycling or not. The convenience of recycling facilities and service is measured by the recycling condition of the recycling spot around their residential areas x_8 and x_9 are, respectively, evaluated by the awareness about environmental protection and laws. The significant level that residents perceived toward economic benefits is used to measure x_{10}.

In our questionnaire, respondents were given the option of filling the disposal status of their E-waste. It is shown that each family had 1.93 end-of-life electrical products on average need to be disposed of in the recent three years, 41.72% of which were collected by various recycling channels, 36.71% of which were discarded as normal municipal waste. The other 21.57% were stored at home.

It is the residents' responsibility to afford a part of the charge for the disposal of their E-waste products in some developed countries, such as Japan (Jofre and Morioka 2005). However, there are no such systematic strategies for household payment on E-waste recycling in China. It is necessary for China to increase the emphasis on the probability of implementing such a payment system. Accordingly, we conducted a survey about household willingness to pay for E-waste recycling. It is shown that only 22.57% of respondents said they were willing to afford the E-waste recycling payment; most respondents (54.23%) said that they would accept only if there were mandatory laws or regularities. As shown in Fig. 5.6, there are no significant differences in the willingness to afford the payment between rich residents and poor residents (Pearson Chi-square = 10.544, Sig. = 0.229).

As for the preference payment style of the E-waste recycling fee, two critical factors have to be considered: time of payment and amount of payment (Forslind 2009). There are three common payment styles in terms of payment time: (1) Paying in advance: paying when purchasing the product with the payment embedded in the product price; (2) Paying afterward: paying when the product is asked for disposal; (3) Deposal system: defraying the deposit when purchasing the product and returning a certain proportion of money when handing over E-waste product in accordance with the provisions. Most respondents (50.62%) prefer the first style since it is convenient for consumers to carry out E-waste recycling. The third style is not widely accepted as expected, which might be largely accounted for the complicated process of the system. Furthermore, according to the amount of payment is given four kinds of recovery styles are implemented: selling to second-hand markets (least costly); selling to peddlers; delivering to the professional recycling spots; asking the specialized collectors to reclaim on-site with certain payment (most costly). We conduct a cross-tab analysis between the

Table 5.4 Variables included in the analysis: definitions and statistics

Variables	Scale	N	Mean	S.D.
Dependent variable Recycling behavior (R)	0 = participate in recycling, 1 = not participate in recycling	951	0.42	0.495
Independent variables Residential condition (x_1)	1 = rent for temporarily living, 2 = rent for long living, 3 = own a house, 4 = own more than one house	951	1.77	0.971
Family population (x_2)	1 = one, 2 = two, 3 = three, 4 = more than three	954	2.97	0.778
Income level (x_3)	1 = less than 1000 yuan per month, 2 = 1000 to less than 2000 yuan per month, 3 = 2000 to less than 3000 yuan per month, 4 = 3000 to less than 4000 yuan per month, 5 = no less than 4000 yuan per month	951	2.86	1.198
Education level (x_4)	0 = no college education member in the family, 1 = the ratio of college education member above 0 but less than 0.5, 2 = the ratio of college education member between 0.5 and 1	942	0.85	0.744
Engel's coefficient (x_5)	Actual value	843	0.60	0.171
Recycling habit (x_6)	0 = has recycling experience, 1 = not participate in recycling before	957	0.18	0.386
Convenience of recycling facilities and service (x_7)	0 = inconvenient, 1 = a little convenient, 2 = very convenient	957	1.60	0.601
Laws and regularities (x_8)	0 = clearly known, 1 = unknown	957	0.48	0.500
Environmental awareness (x_9)	1 = low, 2 = media, 3 = high	957	1.48	0.730
Economic benefits (x_{10})	0 = not important, 1 = no matter, 2 = important	948	1.31	0.641

payment style and recovery style, and find that there are some relationships for residents' attitudes toward these two styles (Pearson Chi-square = 324.377, Sig. = 0.000). The respondents who would like to pay in advance largely prefer to dispose of the E-waste in their comfortable ways. As they had taken the disposal charge, they believed that they had the right to deal with the E-waste as they like. Most of them would like to sell their E-waste to the second-hand market (58.59%) or the peddlers (11.36%) to get some economic benefits. Only 12.6% of respondents select to deliver the E-waste to recycling spots. The respondents who preferred to pay afterward held more responsibilities to deliver the E-wastes to recycling spots. The respondents who selected the deposit system might be more willing to ask the specialized collectors to reclaim on-site with certain payment (Fig. 5.7).

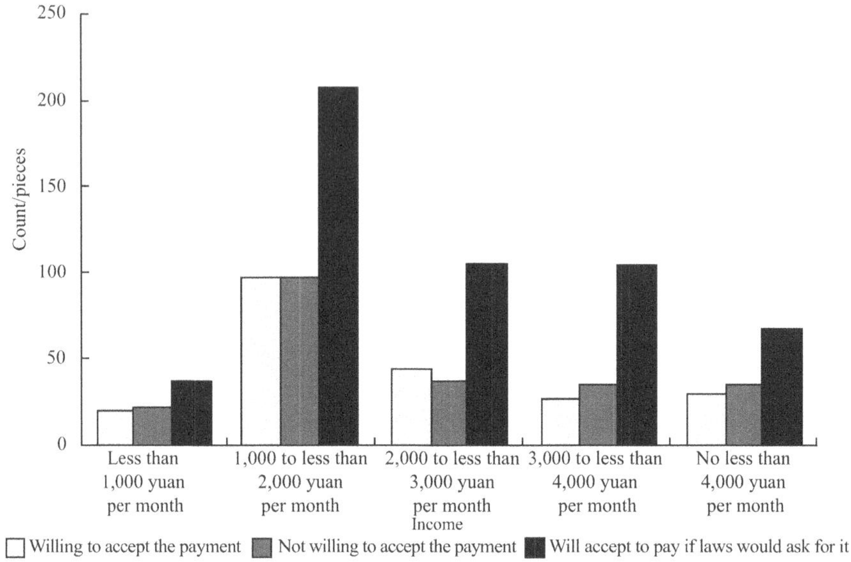

Fig. 5.6 Cross-tab analysis between income and willingness to pay for recycling (pieces)

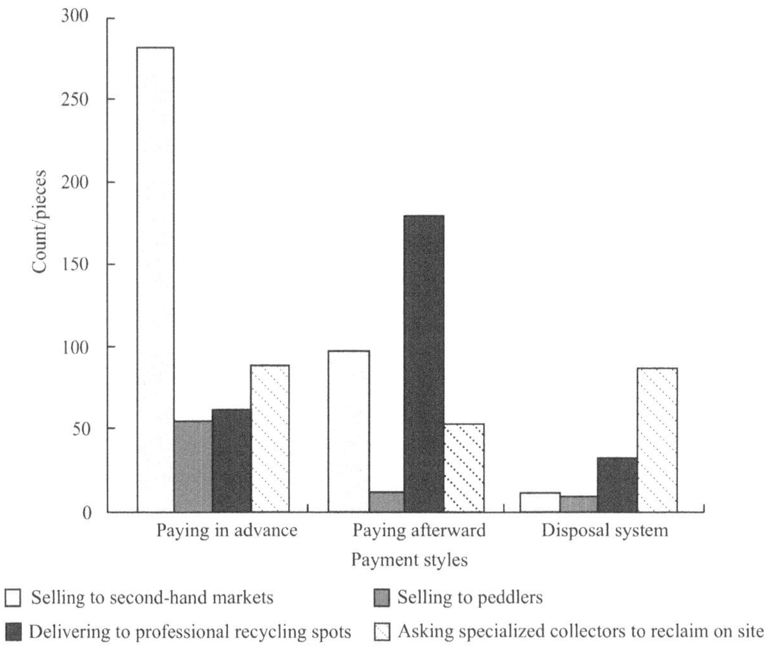

Fig. 5.7 The crosstab analysis between the payment style and the recovery style

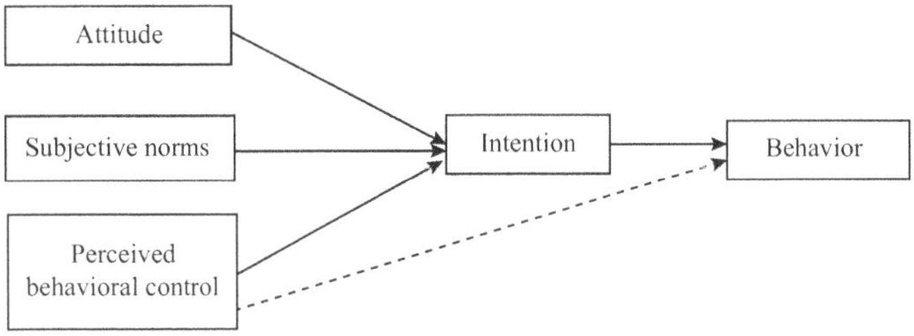

Fig. 5.8 The framework of TPB

Above all, two factors have a vital impact on the preference of recycling style: economic benefits and convenience. As shown in Fig. 5.8, only a few respondents (7.63%) would like to select peddlers as their recovery style, despite the fact that inappropriate disposal by peddlers may cause a heavy environmental burden. But reclaiming by peddlers is still the main channel for E-waste recycling in Beijing because the price they offer is higher and their on-site services are convenient. It is necessary to implement the license system to restrain the environmental threat of E-waste on reclaiming channels. Based on this study, a centralized recycling pattern by professional recycling spots would have a broad prospect as its low economic benefits and environmental-friendly operations. Moreover, the convenient auxiliary convenient services and various reclaiming channels are inevitable.

5.2.6 Influencing Factors on Residents' Behavior in E-Waste Recycling

Logistic regression as shown in Sect. 5.2.4 is introduced to explore the influencing factors toward residents' willingness and behavior in E-waste recycling in Beijing. For the questionnaire investigation during the choice of the survey, location did not meet the conditions of random sampling. The "location" was introduced in the regression as a control variable which was defined as follows: 1—Xicheng, 2—Dongcheng, 3—Chaoyang, 4—Haidian, 5—Xuanwu and 6—Chongwen. The instrument software for logistic regression was SPSS 12.0. According to the Hosmer and Lemeshow test, Sig. = 0.521 > 0.05, which means the regression model is well fitted. The total sound judgment of the model was 74.3%, and the correct rate of judging the residents participated in recycling was 84.6% while the correct rate of judging the residents that did not participate in recycling was 63.1%.

The logistic regression results after the iterated operation were shown in Table 5.5. This indicated that *Residential condition*, *Recycling habits*, and *Economic benefits* are statistically significant at a 5% significant level, and *Convenience of recycling facilities and service* are significant as well at a 10% significant level. Moreover, the coefficient of *Residential properties* and *Recycling*

Table 5.5 Logistic regression results in SPSS ($N = 942$, $N' = 957$)

Variables	β	S.E.	Wald	df	Sig.
Residential condition (x_1)	0.469	0.152	9.496	1	0.002[b]
Family population (x_2)	0.126	0.183	0.479	1	0.489
Income level (x_3)	0.169	0.151	1.240	1	0.265
Education level (x_4)	0.282	0.212	1.769	1	0.184
Recycling habit (x_6)	1.551	0.382	16.481	1	0.000[b]
Convenience of recycling facilities and service (x_7)	−0.427	0.250	2.922	1	0.087[a]
Laws and regularities (x_8)	0.161	0.281	0.325	1	0.568
Environmental awareness (x_9)	−0.187	0.196	0.906	1	0.341
Economic benefits (x_{10})	−0.480	0.231	4.332	1	0.037[b]
Location = 1 (Xicheng)	1.904	0.526	13.114	1	0.000[b]
Location = 2 (Dongcheng)	−0.153	0.506	0.091	1	0.763
Location = 3 (Chaoyang)	0.419	0.510	0.677	1	0.411
Location = 4 (Haidian)	1.896	0.393	23.229	1	0.000[b]
Constant	−1.823	0.930	3.841	1	0.050[b]

Notes [a] denotes significant at the 10% level; [b] denotes significant at the 5% level

habits are positive, while the coefficient of *Convenience of recycling facilities and service* and *Economic benefits* are negative. Other variables such as *Law*, *Environment awareness*, *Education*, and *Income* are not statistically significant. After *Engel's coefficient* was introduced into the regression, the total correct judgment decreased to 70.1% and *Engel's coefficient* is also not statistical significant.

From the perspective of influence power, *Recycling habit* plays the main role in residents' willingness in E-waste recycling compared with the other three factors ($\beta = 1.551$). Once there are 1% changes in the *Recycling habit*, proportional odds of the Logit model would change 4.72%. The results also show that the effects of *Residential condition*, *Economic benefits*, *Convenience of recycling facilities and service* have few disparities with each other. Because the absolute estimates of those three parameters are more or less the same ($\beta = 0.469$, -0.480, -0.427 respectively).

It's also indicated that there are regional disparities in the residents' E-waste recycling behavior in Beijing as the variable of *Location* is also statistically significant at the 5% level in logistic regression. Specifically, residents living in Haidian and Xicheng are more willing to participate in E-waste recycling compared with those living in other districts, as *Location* = Xicheng and *Location* = Haidian are statistically significant at the 5% level when *Location* = Xuanwu and Chongwen is considered as a reference item. This might be largely attributed to different E-waste disposal situations in each district. Haidian and Xicheng are both located in the northwest of Beijing, with the largest electrical and electronic product markets (e.g., Zhongguancun electronic city) in Beijing. As a result, there are more E-product demand and E-waste generation than other districts.

5.3 Determinants of Residents' E-Waste Recycling Behavioral Intentions in China

The amount of E-waste generated by Chinese enterprises is huge, but the ratio of recycling through normal channels is very small, most of the E-waste flows to small traders. Why have small traders become the main force in Chinese E-waste recycling? There is little research into this issue; Chi et al. (2014) used Taizhou, China as a case study and pointed out that the informal workers had an advantage with respect to the recycling range, service convenience, flexibility, and availability. In addition, Orlins and Guan (2016) have found that informal sectors' lack E-waste recycling environmental awareness, and their investigation revealed that they would not like to accept unified management from the government due to fear of losing their jobs or profits. Thus, the following questions arise: do residents think that informal sectors lack environmental awareness? How does resident cognition of E-waste recycling by informal sectors influence resident behavior? To solve these problems, this section investigated seven geographic regions of China and residents in 22 provinces: (autonomous regions, municipalities) based on the theory of planned behavior (TPB), and by using a structural equation model to analyze the factors affecting residents' E-waste recycling behaviors, it focused on the influence of residents' perceptions of informal recycling on their recycling behaviors.

5.3.1 Theory of Planned Behavior (TPB)

To study the factors affecting E-waste recycling behavior and intentions, many scholars have used TPB as a framework: (Tonglet et al. 2004; Hicks et al. 2005; Nixon and Saphores 2007; Wang et al. 2011; Yu et al. 2014; Yla-Mella et al. 2015; Zhang et al. 2016). The TPB postulates three conceptually independent determinants of the purchase intention. The first is the attitude toward the behavior and refers to the degree to which a person has a favorable or unfavorable evaluation or appraisal of the behavior in question. The second predictor is a social factor termed as a subjective norms; it refers to the perceived social pressure to perform or not to perform the behavior. The third antecedent of the purchase intention is the degree of perceived behavioral control, which refers to the perceived ease or difficulty of performing the behavior and it is assumed to reflect past experience as well as anticipated impediments and obstacles (Ajzen 1991).

Besides the influencing factors of the TPB, demographic variables are often referred to as influencing factors in behavioral studies: in the research about E-waste recycling behavior, many scholars have also proved that the demographic variables exert a significant influence. In the literature review about the factors influencing E-waste recycling, income and education have been found to be key, and have significant influence on residents' recycling behavior (Nixon and Saphores 2007; Song et al. 2012; Yin et al. 2014). Saphores et al. (2012) think that

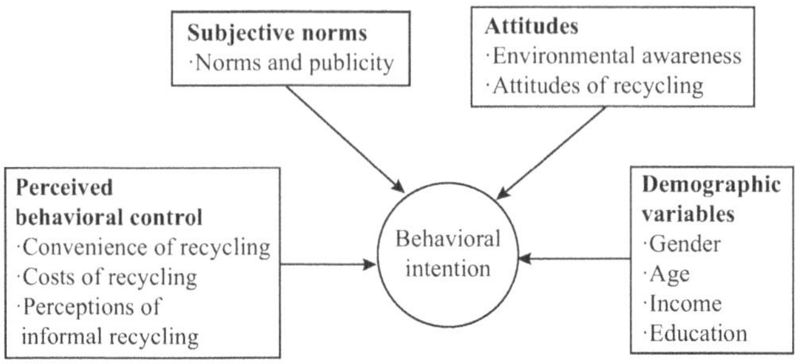

Fig. 5.9 The conceptual framework

age has a negative effect on residents' willingness to recycle because usually young people have a higher education level, are more knowledgeable about environmental protection and want to take some actions to improve their living environment. Saphores et al. (2006) investigated the willingness of people toward E-waste recycling and found that the number of people in the "Very willing" category decreased by 0.215 for people younger than 36 or older than 65. Many scholars believe that gender also affects recycling behavior, and they think that women are more inclined toward recycling (Darby and Obara 2005; Saphores et al. 2006, 2012; Sidique et al. 2010). We also took gender, age, income, and education (as four demographic variables) into consideration in our model. The conceptual framework is shown in Fig. 5.9.

5.3.2 Research Hypothesis

Tonglet et al. (2004) and Nixon and Saphores (2007) confirmed that the consciousness of environmental protection and attitude toward recycling would actively promote the behavioral intention of residents' E-waste recycling. In addition, many scholars believed that the E-waste recycling infrastructure or recycling management systems would positively affect the desire and behavior of residents' toward recycling (An et al. 2015; Bouvier and Wagner 2011; Tonglet et al. 2004; Wang et al. 2011; Yla-Mella et al. 2015; Zhang et al. 2016). We, therefore, proposed the following three research hypotheses:

Hypothesis 5.1. Residents' *Environmental awareness* positively affects E-waste recycling behavioral intention.
Hypothesis 5.2. *Convenience of recycling* affects the E-waste recycling behavioral intention.
Hypothesis 5.3. *Attitude toward recycling* of residents positively affects E-waste recycling behavioral intention.

Yu et al. (2010, 2014) proved that the laws and regulations, related government propaganda positively influenced the willingness of residents to recycle E-waste.

Orlins and Guan (2016) researched China's informal sectors and found their lack of environmental protection consciousness during E-waste recycling. Thus, the law should help residents to realize that it is potentially harmful to the environment to recycle via informal vendors. However, combined with the actual situation of China's E-waste recycling, implementation and publicity of the laws and regulations will enhance resident environmental awareness and make them willing to recycle E-waste. Due to a lack of formal channels, residents can only choose to sell their E-waste to vendors and erroneously believe that vendors play the role of environmental protectors in E-waste recycling. On this basis, we proposed the following three research hypotheses:

Hypothesis 5.4: *Norms and publicity* positively influence the behavioral intention toward E-waste recycling.
Hypothesis 5.5: *Perceptions of informal recycling* negatively influence the behavioral intention toward E-waste recycling.
Hypothesis 5.6: *Norms and publicity* negatively influence Perceptions of informal recycling.

Nixon and Saphores (2007), and Wang et al. (2011) pointed out that income affected resident desire and behavior toward E-waste recycling. Wang et al. (2011) investigated residents in Beijing, China, and proved that income positively affected the behavioral intention toward E-waste recycling. However, we covered 22 provinces (autonomous regions, municipalities), across seven geographical areas in China and found that residential income gaps were significant and large: the higher the income of a person is, the less they care about the benefit of E-waste recycling, so they do not tend to recycle. In addition, the higher the cost of E-waste recycling is, the weaker the intention is. We, therefore, proposed another three research hypotheses:

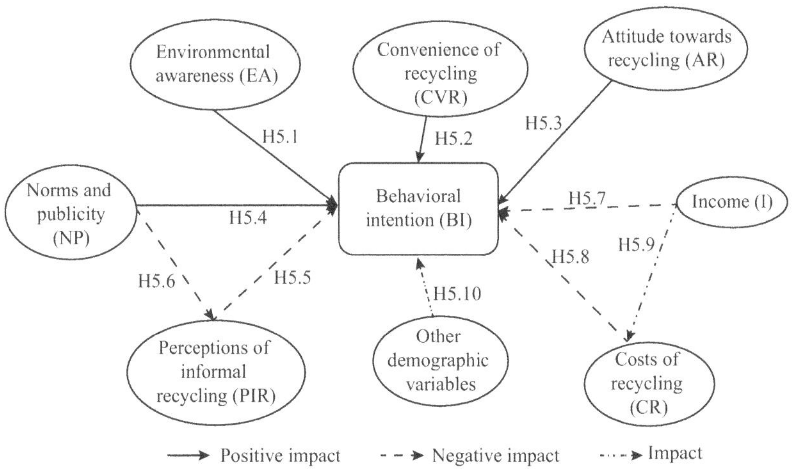

Fig. 5.10 Hypothesis framework

Hypothesis 5.7: *Income* of residents negatively influences recycling behavioral intentions toward E-waste.
Hypothesis 5.8: *Costs of recycling* negatively influence recycling behavioral intentions toward E-waste.
Hypothesis 5.9: *Residential income* has an impact on *Costs of recycling.*
In addition, we include *Gender*, *Age*, and *Education* as three demographic variables in the model and then proposed the following hypothesis:
Hypothesis 5.10: *Gender*, *Age,* and *Education* have an impact on residents' E-waste recycling behavioral intentions.

To summarize the above hypotheses, the overall research framework is shown in Fig. 5.10.

5.3.3 Data Sources: A Questionnaire from China in 2015

At first, we designed a questionnaire to collect accurate data. Six kinds of major E-waste are selected as the research objects in our survey: refrigerators, air conditioners, washing machines, computers, TV sets, and mobile telephones. In the introduction of our questionnaire, we clearly emphasize the importance of the truth worthy data by reducing an invalid questionnaire. We also ensured that the collected data were only used for scientific research and keeping the personal information secret. The detail information is shown in Table 5.6.

The questionnaires were divided into two groups where one was from the urban area and the other was from a rural area. Then the independent sample t-test was used for the two groups, respectively. The results showed that there was no significant difference between the two groups which means that they can be combined for further analysis. In addition, we analyzed the 525 samples deeply and made a descriptive statistics for the demographic variables. We found gender, education and income were consistent with China's actual conditions. About age, we hoped the person who responded the questionnaire was the head of the house and clearly knew the handling information of E-waste in the house. So responders between the ages of 21–45 years were the most followed by 46–60 years, we also survey other age groups such as the age under 21 years or above 60 years. The results of the descriptive statistics are shown in Table 5.7.

5.3.4 Methodology: Structural Equation Model (SEM)

The methodology mainly contains the exploratory factor analysis (EFA) and structural equation model (SEM). At first, we used the method of EFA to test the initial scale. Then, we used the method of SEM to carry out this research and find out the determinants of resident's E-waste recycling behavioral intentions.

Table 5.6 The second distribution of samples and response rate of questionnaires

Location	Issued provinces (autonomous regions, municipalities)	Quantities of issued questionnaire	Quantities of responded questionnaire	Response rate/%
East China regions	Shandong, Jiangsu, Zhejiang, Shanghai	125	93	74
North China regions	Beijing, Tianjin, Hebei, Shanxi	120	91	76
Central China regions	Hubei, Hunan, Henan	105	76	72
South China regions	Guangdong, Hainan	44	39	89
Southwest regions	Sichuan, Yunnan, Chongqing, Xizang	95	84	88
Northwest regions	Shaanxi, Inner Mongolia	75	67	89
Northeast regions	Liaoning, Jilin, Heilongjiang	103	75	73
Total		667	525	79

Step one: EFA is used to find out the essence of the multivariate observation variable structure and deal with dimension reduction. We mainly focused on testing the validity and analyzing items, and the specific contents were shown in Table 5.8.

Step two: The principle behind the SEM involves the use of some observable variables to measure one unobservable variable and it can use a variety of indicators to measure the degree of fitness of the model (Swami et al. 2011). Based on the theoretical model and hypotheses proposed in the previous section, and combined with the questionnaire data mentioned in Sect. 5.3.3, we used the method of SEM to carry out this research and used the software of AMOS 17.0 to establish our model.

Firstly, we tested the convergent validity, composite reliability, and content validity of each variable. As shown in Table 5.9, all the factor loadings and the average variances extracted (AVE) were greater than 0.5, indicating good convergent validity for the questionnaire (Fornell and Larcker 1981). The composite reliabilities of all constructs were greater than 0.6, illustrating the ideal intrinsic quality of the questionnaire (Wu 2010). Then the designers of the questionnaire consulted previous research settings, coupled with the advice of other experts in the field, and thus ensuring content validity.

Then overall fitness of the model was tested. We put the survey data into the model. After identification, fitting, and evaluation, we found that most fitting indicators failed to meet the criterion and the model fitness was poor, so the model required modification. Then the model was modified, according to suggestions from

Table 5.7 Respondent profile: gender, age, education and income

Demographic variables		N	Percentage/%
Gender	Female	233	44.38
	Male	292	55.62
Age	Under 21	68	12.95
	21–30	178	33.90
	31–45	181	34.48
	46–60	84	16.00
	61 or above	14	2.67
Education level	Lower secondary or below	211	40.19
	Upper secondary	126	24.00
	Bachelor's degree or sub-degree	148	28.19
	Master's degree or above	40	7.62
Monthly income	Below 2500 yuan	77	14.67
	2501–4000 yuan	101	19.24
	4001–6000 yuan	164	31.24
	6001–8000 yuan	63	12.00
	8001–10,000 yuan	34	6.48
	10,001–15,000 yuan	45	8.57
	15,001–20,000 yuan	26	4.95
	Above 20,000 yuan	15	2.86

Table 5.8 The results of factor analysis and the reliability test of constructs

Constructs	Items	Cronbach's alpha
Environmental awareness (EA)	3	0.751
Costs of recycling (CR)	3	0.772
Attitude towards recycling (AR)	3	0.695
Norms and publicity (NP)	2	0.754
Convenience of recycling (CVR)	3	0.696
Perceptions of informal recycling (PIR)	3	0.754
Behavioral intention (BI)	3	0.731

AMOS, until the fitness reached an ideal level. The results are shown in Table 5.10. However, the significant probability value p of χ^2 was 0.022 which reached a significant level indicating that the fitness between the hypothesis model and sample data was not high. We think that this may have been caused by large sample size. Rigdon (1995) proposed that χ^2 value was affected by the estimated parameters and the number of samples. When using real data to evaluate the theoretical model, the χ^2 statistic is usually of a little help. Therefore, when judging the fitness of the model, other suitable indicators are also needed to consider to form a

Table 5.9 Composite reliability and convergent validity of constructs and factor loading of indicators

Constructs	Indicators	Factor loading	Average variance extracted (AVE)	Composite reliability (CR)
Environmental awareness (EA)	EA1	0.627	0.518	0.762
	EA2	0.773		
	EA3	0.751		
Costs of recycling (CR)	CR1	0.540	0.528	0.765
	CR2	0.840		
	CR3	0.765		
Attitude towards recycling (AR)	AR1	0.604	0.531	0.770
	AR2	0.809		
	AR3	0.758		
Norms and publicity (NP)	NP1	0.976	0.617	0.748
	NP2	0.530		
Convenience of recycling (CVR)	CVR1	0.618	0.507	0.754
	CVR2	0.763		
	CVR3	0.746		
Perceptions of informal recycling (PIR)	IR1	0.561	0.504	0.745
	IR2	0.898		
	IR3	0.626		
Behavioral intention (BI)	BI1	0.634	0.506	0.753
	BI2	0.702		
	BI3	0.790		

comprehensive judgment. The ratio of χ^2 divided by the degree of freedom (CMIN/DF) was 1.218 which showed that the model had good adaptation to the actual sample data. Therefore, the theoretical framework assumed in this study fits the actual survey data. So it can be concluded that our model has good external quality (Wu 2010).

Next, we tested the research hypotheses. The factors that significantly affect the behavioral intention are *Environmental awareness*, *Attitude to recycling*, *Perceptions of informal recycling*, *Income,* and *Costs of recycling*. The effect of the five factors (positive or negative) is also consistent with the hypothesis that the assumptions of H 5.1, H 5.3, H 5.5, H 5.7, and H 5.8 are correct. The *Perceptions of informal recycling* are affected by the *Norms and publicity* significantly, but their effect is opposite to that of hypothesis H 5.6. In addition, hypothesis H 5.4 is not verified. This shows that the intermediate variable (*Perceptions of informal recycling*) fully mediates the impact of *Norms and publicity* on the behavioral intention of residents. Hypotheses H 5.2, H 5.9, and H 5.10 did not pass the test, so it could be inferred, the convenience of recycling and the three demographic variables

Table 5.10 The evaluation results of the modified SEM model

	Indicators	Criterion	Results	Judgment
Absolute fit measures	χ^2	$p > 0.05$	0.022	No
	GFI	>0.9	0.966	Yes
	AGFI	>0.9	0.945	Yes
	RMR	<0.5	0.041	Yes
	RMSEA	<0.1	0.02	Yes
Incremental fit measures	NFI	>0.9	0.938	Yes
	CFI	>0.9	0.988	Yes
	IFI	>0.9	0.988	Yes
	RFI	>0.9	0.909	Yes
Parsimonious fit measures	PGFI	>0.5	0.605	Yes
	PNFI	>0.5	0.639	Yes
	PCFI	>0.5	0.673	Yes
	CMIN/DF	<3.00	1.218	Yes

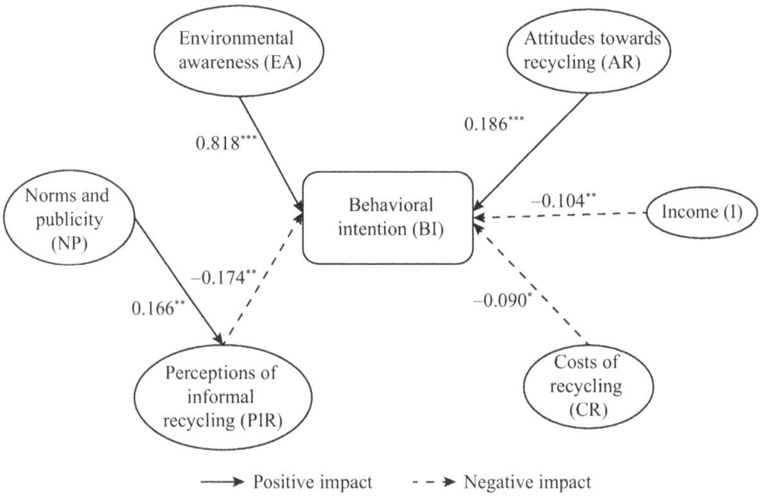

Fig. 5.11 The effects of the model

(gender, age, and education) had no significant effect on behavioral intention and income had no effect on the cost of recovery.

Moreover, we analyzed the effect of the model. The causal effects between latent variables in structural equation models include direct effects and indirect effects. Among them, the direct effect is to test the hypotheses. The results of the six paths (BI ← EA, BI ← PIR, BI ← AR, BI ← I, PIR ← NP, BI ← CR) are significant and their standardized regression coefficients are the direct influence coefficients. Perceptions of informal recycling as the intermediate variable fully mediate the impact of *Norms and publicity* on the behavioral intentions. So, the effect of *Norms and publicity* is an indirect effect. Finally, the effects of the model are shown in Fig. 5.11.

5.3.5 Determinants of Direct Influence on E-Waste Recycling Behavioral Intentions

According to the testing of hypotheses (above), we know that *Environmental awareness*, *Attitude toward recycling*, *Perceptions of informal recycling*, *Income* and *Costs of recycling* pass the test and they significantly affect E-waste recycling behavioral intentions.

The influence coefficient of *Environmental awareness* is 0.818 which is very high and its p-value reaches a significant level ($p < 0.001$) indicating that *Environmental awareness* has a positive impact on E-waste recycling behavioral intentions. According to relevant information, E-waste contains a variety of toxic and hazardous substances, if handled improperly may heavily pollute the environment and cause serious harm to health. E-waste recycling is a meaningful act of environmental protection: if resident awareness about environmental protection grows, environmental protection behavior will enhance and participation in E-waste recycling activities will also improve. Moreover, residents' environmental awareness is a major construct that requires slow training and long concept-retention. This project not only promotes the recycling of E-waste but also helps residents to participate in other environmentally protective acts.

The influence coefficient of *Attitude toward recycling* is 0.186, which is not very high but its p-value reached a significant level ($p < 0.001$) indicating that *Attitude toward recycling* also has a positive impact on E-waste recycling behavioral intentions. If residents' participation in E-waste recycling can engender satisfaction, and residents consider recycling makes a contribution to society, then the E-waste recycling behavior will improve. As *Environmental awareness* and *Attitude toward recycling* requires cultivation in a continuous propaganda campaign for environmental protection, only those residents that believe that everyone has a responsibility and obligation to contribute to E-waste recycling, will see that their attitudes become relatively upbeat and more active.

Many scholars believe that income positively affects E-waste recycling behavioral intentions (Hansmann et al. 2006; Hornik et al. 1995; Tonglet et al. 2004), but Wang et al. (2011) investigated this in Beijing and showed that income had no impact on the E-waste recycling behavioral intentions. However, our survey covers China's 22 provinces (autonomous regions, municipalities) and the income gap among residents is large, so residential income plays an important role in E-waste recycling. Our results show that residents' income ($\beta = 0.104$, $p < 0.05$) has a negative impact on E-waste recycling behavioral intentions. Current participants in E-waste recycling have a corresponding interest compared to lower income residents, higher income residents do not care about the benefits arising from recycling, and they are more willing to part with the benefits to avoid the trouble and inconvenience of recycling. Similarly, *Costs of recycling* ($\beta = 0.090$, $p < 0.1$) has a negative impact on E-waste recycling behavioral intentions. If residents pay more through formal channels, they may feel negative toward E-waste recycling, even though they do not participate in recycling through formal channels and hold a bias toward informal recycling, discarding, or stockpiling at home.

5.3.6 Determinants of Indirect Influence on E-Waste Recycling Behavioral Intentions

The results show that *Perceptions of informal recycling* plays only an intermediate role in the relationship between *Norms and publicity* and *Behavioral intention*. The influence coefficient of *Norms and publicity* is 0.166 which is not very large but its *p*-value reaches a significant level ($p < 0.05$) indicating that *Norms and publicity* positively affect *Perceptions of informal recycling*. But the influence coefficient of *Perceptions of informal recycling* is 0.174 and its *p*-value also reaches a significant level ($p < 0.05$), so *Perceptions of informal recycling* negatively affect *Behavioral intention*.

Many scholars believe that *Norms and publicity* have positive impacts on residents' behavioral intention of E-waste recycling, for example, Hicks et al. (2005), Fleckinger and Glachant (2010), Mo et al. (2009) and Nnorom et al. (2009). However, we took *Perceptions of informal recycling* as an intermediate factor between *Norms and publicity* and *Behavioral intention*. We found *Perceptions of informal recycling* have negative impacts on *Behavioral intention* and common sense also agrees with this result. But *Norms and publicity* have positive impacts on *Perceptions of informal recycling* so we need to combine China's actual conditions to analyze this result.

With the implementation of laws and regulations relevant to E-waste recycling, Chinese residents have come to realize that E-waste should be recycled. Whether stockpiling at home or discarding, it goes against the benefits of recycling E-waste, and it also causes environmental pollution and even injurious to human health. Shantou University Medical College had made a health spot check in Guiyu Town, Shantou City, Guangdong Province where E-waste recycling and decomposition was relatively concentrated and 165 children aged 1–6 years were tested for physical examination. The results show that all the children's blood lead levels exceeded and poisonous. Considering the environmental protection and human health, residents begin to recycle E-waste. How to recycle more E-waste becomes another problem facing residents. China has 105 enterprises dismantling E-waste at fixed-point places by 2014, but the Chinese's perception about E-waste recycling by dismantling enterprises is poor. Thus, residents who are willing to recycle E-waste have a certain distance with these regular dismantling enterprises and they cannot readily recycle E-waste through formal channels. The amount of dismantling E-waste products was 2.0085 million units by May 19, 2014 relative to 200 million units per year of major waste appliances, the amount of dismantling E-waste accounted for only 1% of the total. In sharp contrast to this, small-scale dealers engaged in E-waste recycling are very advantageous. Door-to-door recovery not only is very convenient but also gives the residents higher economic compensation, so selling the E-waste to small-scale dealers has become the main mode of recycling for most residents.

However, small-scale dealers engaged in E-waste recycling, lack environmental awareness, and are unwilling to accept government management because of the fear of losing jobs and income (Orlins and Guan 2016). Small-scale dealers regard profit as their main purpose instead of properly dealing with E-waste, and they sell E-waste to "cottage industry" dismantling enterprises. These dismantling enterprises just dismantle useful components thereof and due to lack of advanced technology, most dismantling enterprises use manual disassembly, landfill, incineration, acid extraction, dumping and so on, which produces a large amount of heavy metals and toxic substances, cause serious environmental pollution and a waste of precious resources. The economic value of E-waste is comparable to a mine, for example, a ton of ore contains about 5 g of gold that can be called bonanza. As in a ton of TV set, the content of gold is at least 80 g, some have as high as 150 g, as well as copper, plastic, and other renewable resources. In the case of a market dominated by individual dismantling enterprises, some formal dismantling enterprises have been embarrassed by a situation in which they have no E-waste to dismantle. In other words, China's current E-waste recycling industry lacks formal and effective recycling channels, so that residents have to choose small-scale dealers first, although they are clearly unable to deal with E-waste properly. In short, the lack of formal and effective recycling channels puts a formidable obstacle in front of residents seeking to recycle E-waste safely and to some extent, this also hinders the healthy development of improved E-waste recycling industry.

Finally, among the four demographic variables of *Gender*, *Age*, *Education*, and *Income*, only *Income* has passed the hypothesis test, while *Gender*, *Age*, and *Education* have no influence on E-waste recycling behavior. In the field of E-waste recycling, residents are not stratified by *Gender* or *Age*. Thus, *Gender* and *Age* have no influence on their behavior. Many scholars believe that education has a positive impact on the behavioral intention to E-waste recycling (Hansmann et al. 2006; Hornik et al. 1995; Tonglet et al. 2004). Education is not the main factor affecting residential environmental awareness, so education levels have no significant impact on the behavioral intentions toward E-waste recycling. In addition, the *Convenience of recycling* also did not pass the hypothesis test. Our research took "no time to send E-waste to recycling points," "no ability to send E-waste to recycling point," and "the traffic inconvenience to the recovery point" to measure *Convenience of recycling*.

5.4 Summary

This chapter mainly discusses the recycling behavior of E-waste in China. First of all, it introduces the characteristics, the generation, and the flow of E-waste. It also analyzes the reasons for the dilemma of E-waste in China from the aspects of laws, policies, and industrial systems. We analyze the willingness and behavior toward E-waste recycling in China. The results show that a significant proportion of residents in Beijing are still not very willing to participate in E-waste recycling. At last, we discuss the determinants of residents' E-waste recycling behavioral intentions in China. The results show that the main factors which affect Chinese residents' E-waste recycling behavioral intentions are: *Environmental awareness*, *Attitude toward recycling*, *Perceptions of informal recycling*, *Income* and *Costs of recycling*. However, *Norms and publicity*, *Convenience of recycling*, *Gender*, *Age*, and *Educational background* have no significant impact on residents' E-waste recycling behavioral intentions. *Norms and publicity* indirectly affect resident behavioral intentions to E-waste recycling via the intervening variable *Perceptions of informal recycling*: *Norms and publicity* also has a positive effect on *Perceptions of informal recycling*.

References

Agoramoorthy, G., and C. Chakraborty. 2012. Control electronic waste in India. *Nature* 485: 309.

Ajzen, I. 1985. From intentions to actions: a theory of planned behavior. In *Action control: from cognition to behavior*, ed. J. Kuhl, and J.F Beckmann, 11–39. New York: Springer-Verlag.

Ajzen, I. 1991. The theory of planned behavior. *Organizational Behavior and Human Decision Processes* 50 (2): 179–211.

An, D., Y. Yang, X. Chai, et al. 2015. Mitigating pollution of hazardous materials from WEEE of China: Portfolio selection for a sustainable future based on multi-criteria decision making. *Resources, Conservation and Recycling* 105: 198–210.

Boldero, J. 1995. The prediction of household recycling of newspapers: The role of attitudes, intentions, and situational factors. *Journal of Applied Social Psychology* 25: 440–462.

Borsch-Supan, A. 1990. On the compatibility of nested logit models with utility maximization. *Journal of Econometrics* 43 (3): 373–388.

Bouvier, R., and T. Wagner. 2011. The influence of collection facility attributes on household collection rates of electronic waste: The case of televisions and computer monitors. *Resources, Conservation and Recycling* 55 (1): 1051–1059.

Breivik, K., J.M. Armitage, F. Wania, et al. 2014. Tracking the global generation and exports of e-waste. Do existing estimates add up? *Environmental Science & Technology* 48 (15): 8735–8743.

Chi, X., M.Y.L. Wang, and M.A. Reuter. 2014. E-waste collection channels and household recycling behaviors in Taizhou of China. *Journal of Cleaner Production* 80: 87–95.

Darby, L., and L. Obara. 2005. Household recycling behaviour and attitudes towards the disposal of small electrical and electronic equipment. *Resources, Conservation and Recycling* 44: 17–35.

Davis, G., P.S. Phillips, A.D. Read, et al. 2006. Demonstrating the need for the development of internal research capacity: Understanding recycling participation using the theory of planned behaviour in West Oxfordshire, UK. *Resources, Conservation and Recycling* 46: 115–127.

De Winter, K.C. 2001. From here to eternity: Recycling hi-tech junk. *Waste Age* 32 (3): 186–190.

Fleckinger, P., and M. Glachant. 2010. The organization of extended producer responsibility in waste policy with product differentiation. *Journal of Environmental Economics and Management* 59 (1): 57–66.

Fornell, C., and D.F. Larcker. 1981. Evaluating structural equation models with unobservable variables and measurement error. *Journal of Marketing Research* 18 (1): 39–50.

Forslind, K.H. 2009. Does the financing of extended producer responsibility influence economic growth? *Journal of Cleaner Production* 17: 297–302.

Guagnano, G.A., P.C. Stern, and T. Dietz. 1995. Influences on attitude-behavior relationships: A natural experiment with curbside recycling. *Environment and Behavior* 27: 699–718.

Guerin, D., J. Crete, and J. Mercier. 2001. A multilevel analysis of the determinants of recycling behavior in the European countries. *Social Science Research* 30: 195–218.

Hansmann, R., P. Bernasconi, T. Smieszek, et al. 2006. Justifications and self-organization as determinants of recycling behavior: The case of used batteries. *Resources, Conservation and Recycling* 47: 133–159.

Hicks, C., R. Dietmar, and M. Eugster. 2005. The recycling and disposal of electrical and electronic waste in China—Legislative and market responses. *Environmental Impact Assessment Review* 25: 459–471.

Hornik, J., J. Cherian, M. Madansky, et al. 1995. Determinants of recycling behavior: A synthesis of research results. *The Journal of Socio-Economics* 24: 105–127.

Jofre, S., and T. Morioka. 2005. Waste management of electric and electronic equipment: Comparative analysis of end-of-life strategies. *Journal of Material Cycles and Waste Management* 7: 24–32.

Jones, R.E. 1990. Understanding paper recycling in an institutionally supportive setting—An application of the theory of reasoned action. *Journal of Environmental Systems* 19: 307–321.

Lin, C.K., L. Yan, and A.N. Davis. 2002. Globalization, extended producer responsibility and the problem of discarded computers in China: an exploratory proposal for environmental protection. *Georgetown International Environmental Law Review* 14: 525–579.

Liu, X.B., M. Tanaka, and Y. Matsui. 2006. Electrical and electronic waste management in China: Progress and the barriers to overcome. *Waste Management and Research* 24: 92–101.

McCarty, J.A., and L.J. Shrum. 1994. The recycling of solid-wastes: Personal values, value orientations, and attitudes about recycling as antecedents of recycling behavior. *Journal of Business Research* 30: 53–62.

Milanez, B., and T. Buehrs. 2009. Extended producer responsibility in Brazil: The case of tyre waste. *Journal of Cleaner Production* 17: 608–615.

Mo, H., Z. Wen, and J. Chen. 2009. China's recyclable resources recycling system and policy: A case study in Suzhou. *Resources, Conservation and Recycling* 53: 409–419.

Nixon, H., and J.D.M. Saphores. 2007. Financing electronic waste recycling Californian households' willingness to pay advanced recycling fees. *Journal of Environmental Management* 84: 547–559.

Nnorom, I.C., J. Ohakwe, and O. Osibanjo. 2009. Survey of willingness of residents to participate in electronic waste recycling in Nigeria—A case study of mobile phone recycling. *Journal of Cleaner Production* 17 (18): 1629–1637.

Ogunseitan, O.A., J.M. Schoenung, J.D.M. Saphores, et al. 2009. The electronics revolution: From e-wonderland to e-wasteland. *Science* 326: 670–671.

Ojeda-Benitez, S., C. Armijo-de Vega, and M.Y. Marquez-Montenegro. 2008. Household solid waste characterization by family socioeconomic profile as unit of analysis. *Resources, Conservation and Recycling* 52 (7): 992–999.

Orlins, S., and D. Guan. 2016. China's toxic informal e-waste recycling: Local approaches to a global environmental problem. *Journal of Cleaner Production* 114: 71–80.

Oskamp, S., M.J. Harrington, T.C. Edwards, et al. 1991. Factors influencing household recycling behavior. *Environment and Behavior* 23: 494–519.

Ravi, V., R. Shankar, and M.K. Tiwari. 2005. Analyzing alternatives in reverse logistics for end-of-life computers: ANP and balanced scorecard approach. *Computers & Industial Engineering* 48: 327–356.

Rigdon, E.E. 1995. A necessary and sufficient identification rule for structural models estimated in practice. *Multivariate Behavioral Research* 30: 359–383.

Robinson, B.H. 2009. E-waste: An assessment of global production and environmental impacts. *Science of the Total Environment* 408: 183–191.

Saphores, J.D.M., H. Nixon, O.A. Ogunseitan, et al. 2006. Household willingness to recycle electronic waste: An application to California. *Environment and Behavior* 38: 183–208.

Saphores, J.D.M., O.A. Ogunseitan, and A.A. Shapiro. 2012. Willingness to engage in a pro-environmental behavior: An analysis of e-waste recycling based on a national survey of U. S. households. *Resources, Conservation and Recycling* 60: 49–63.

Sidique, S.F., F. Lupi, and S.V. Joshi. 2010. The effects of behavior and attitudes on drop-off recycling activities. *Resources, Conservation and Recycling* 54: 163–170.

Song, Q., Z. Wang, and J. Li. 2012. Residents' behaviors, attitudes, and willingness to pay for recycling e-waste in Macau. *Journal of Environmental Management* 106: 8–16.

Swami, V., T. Chamorro-Premuzic, R. Snelgar, et al. 2011. Personality, individual differences, and demographic antecedents of self-reported household waste management behaviours. *Journal of Environmental Psychology* 31: 21–26.

Tonglet, M., P.S. Phillips, and A.D. Read. 2004. Using the theory of planned behaviour to investigate the determinants of recycling behaviour: A case study from Brixworth, UK. *Resources, Conservation and Recycling* 41: 191–214.

Tucker, P., G. Murney, and J. Lamont. 1998. Predicting recycling scheme performance: A process simulation approach. *Journal of Environmental Management* 53: 31–48.

van Beukering, P.J.H., C.J.M. Jeroen, and J. van den Bergh. 2006. Modelling and analysis of international recycling between developed and developing countries. *Resources Conservation and Recycling* 46: 1–26.

Wang, F., R. Kuehr, D. Ahlquist. et al. 2013. China e-waste research report. https://www.docin.com/p-683730446.html [2015-10-13]. (In Chinese)

Wang, J.C. 2001. Logistic Regression Models: Methods and Application. Beijing: Higher Education Press.

Wang, W.W., M. Zhang, and M. Zhou. 2011. Using LMDI method to analyze transport sector CO_2 emissions in China. *Energy* 36 (10): 5909–5915.

Wu, M. 2010. *Structural Equation Model-use and Application of AMOS*. Chongqing: Chongqing University Press.

Xu Z, Ma S. 2004. How to fullfill the value of reverse supply chain tactics. Logistics Technology, 45-47.

Yang, J., B. Lu, and C. Xu. 2008. WEEE flow and mitigating measures in China. *Waste Management* 28: 1589–1597.

Yin, J., Y. Gao, and H. Xu. 2014. Survey and analysis of consumers' behaviour of waste mobile phone recycling in China. *Journal of Cleaner Production* 65: 517–525.

Yla-Mella, J., R.L. Keiski, and E. Pongracz. 2015. Electronic waste recovery in Finland: Consumers' perceptions towards recycling and re-use of mobile phones. *Waste Management* 45: 374–384.

Yu, J., E. Williams, M. Ju, et al. 2010. Managing e-waste in China: Policies, pilot projects and alternative approaches. *Resources, Conservation and Recycling* 54: 991–999.

Yu, L., W. He, G. Li, et al. 2014. The development of WEEE management and effects of the fund policy for subsidizing WEEE treating in China. *Waste Management* 34: 1705–1714.

Zeng, X., J. Li, A.L.N. Stevels, et al. 2013. Perspective of electronic waste management in China based on a legislation comparison between China and the EU. *Journal of Cleaner Production* 51: 80–87.

Zhang, B., Z.H. Wang , J.H. Yin, et al. 2012. CO_2 emission reduction within Chinese iron & steel industry: practices, determinants and performance. *Journal of Cleaner Production* 33: 167–178.

Zhang, S., M. Zhang, X. Yu, et al. 2016. What keeps Chinese from recycling: Accessibility of recycling facilities and the behavior. *Resources, Conservation and Recycling* 109: 176–186.

Zhao, W., R.B. Leeftink, and V.S. Rotter. 2010. Evaluation of the economic feasibility for the recycling of construction and demolition waste in China—The case of Chongqing. *Resources, Conservation and Recycling* 54: 377–389.

Zhou, L., and Z. Xu. 2012. Response to waste electrical and electronic equipments in China: legislation, recycling system, and advanced integrated process. *Environmental Science and Technology* 46: 4713–4724.

Chapter 6
Motivation of Energy-Intensive Industries on CO_2 Reduction: Learning From the Iron and Steel Industry

The international communities have recognized the severity of climate change, and the pressure for the reduction of CO_2 emissions has become more prominent among the whole globe. China is one of the world's biggest CO_2 emitters, accounting for one-quarter of global CO_2 emissions. The pressure on China to reduce CO_2 emissions has been increased at a greater level. In many global climate change negotiations in recent years, China has been pressured by developed and developing countries which are affected directly by climate change, to promise a concrete goal for reductions. In addition, there are also great pressures resulting from domestic economic development. According to the estimation by Li et al. (2012), China would need to cut 1651 Mt of CO_2 emission in 2020 in order to achieve the target of cutting CO_2 emissions per unit of GDP by 40–45% since 2005, even in a slow economic growth scenario.

Steel, cement, and other high-energy-consuming industries are the main sources of CO_2 emissions in China. According to statistics, carbon dioxide emissions from China's high-energy-consuming industries account for more than 70% of the total carbon dioxide emissions. Considering the great challenge of CO_2 abatement, a sector-based approach has been suggested to deal with the CO_2 emission problem. The iron and steel industry (ISI) ranks as the third biggest CO_2 emitter in China (after the power and construction material sectors), due to its coal-dominated energy structure and high consumption of limestone. It accounts for 10% of total domestic CO_2 emissions and 35–40% of CO_2 emissions in major cities (Zeng et al. 2009). There is a great potential for CO_2 reduction in ISI of China. It is imperative to mitigate the incremental pressures of CO_2 emissions by improving energy efficiency in ISI.

This chapter takes the following literature for reference: Zhang B, Wang Z H, Yin J H, et al. 2012. CO_2 reduction within Chinese iron and steel industry: practices, determinants and performance. Journal of Cleaner Production, 33: 167–178.

Z. Wang and B. Zhang, *Low-Carbon Consumption in China: Residential Behavior, Corporate Practices and Policy Implication*,
https://doi.org/10.1007/978-981-15-2792-0_6

However, it is not clear which pressures predominate as drivers or barriers for Chinese ISI enterprises in implementing the practice of CO_2 reduction. Even though there has been some research on how to reduce CO_2 emission, e.g., energy efficiency innovation, clean energy substitution, and production reduction, from the total industrial perspective (e.g., Rynikiewicz 2008; Guo and Fu 2009), few scholars have investigated whether these practices influence the performance of ISI, especially from individual enterprises' perspective. What is the relationship between the CO_2 reduction practices, determinants, and the performance of the enterprises in ISI?

Therefore, this chapter explores a conceptual model of the relationship between CO_2 abatement determinants and practices. This model is being tested by using the data collected from 85 IS enterprises of China with the help of a questionnaire survey, which is designed to address the following three questions:

What CO_2 reduction practices are being employed or introduced by Chinese ISI?

Which determinants act as drivers or hindrance for the implementations of these practices?

Do these practices have a positive effect on the environmental and economic performance of ISI?

6.1 CO_2 Emission Structure and Reduction Practice of Chinese Iron and Steel Industry

Carbon dioxide emissions and emission reduction in the iron and steel industry have always been a hot topic for scholars and policy-makers. Many scholars have recognized the important role of ISI in CO_2 reduction, and conducted corresponding research on CO_2 abatement in ISI. Some of them focused on the trend of CO_2 emissions and the characteristics of energy consumption of this sector in specific countries or regions. For instance, Hu et al. (2006) explored the current status and future scenarios of CO_2 emission in Chinese ISI. The results showed that the future improvement on emission mitigation for Chinese ISI should be replied on reducing production and by increasing the use of scrap and promoting carbon capture and storage. Sheinbaum et al. (2010) analyzed energy and CO_2 emission trends of iron and steel sector in Mexico from 1970 to 2006. The findings of their study revealed that Mexican steel production led to a 2.4% annual increase in primary energy consumption and 2.3% in its energy and process-related CO_2 emissions. Some studies have focused on the ways or methods of reducing CO_2 emission for ISI. For instance, Zeng et al. (2009) proposed mitigation paths for the Chinese IS industry to tackle global climate change. They found that carbon reduction should be supported by the encouragement of the clean development mechanism (CDM), stimulation of social responsibility, the undertaking of strict energy auditing and promotion of emission reducing-oriented investment. Hidalgo et al. (2005) focused on the carbon reduction in ISI from a technological perspective. Market mechanisms were also considered as an effective way for carbon reduction in ISI, as proposed by several studies (e.g., Demailly and Quirion 2008; Lund 2007). All these studies enrich the understanding of CO_2 abatement in ISI, and provide a solid foundation for further analysis. As the world's largest steel producer and consumer, CO_2 emission structure and reduction practice of China's steel industry are particularly noteworthy.

6.1.1 Energy Consumption and CO_2 Emission Structure of Iron and Steel Industry

The iron and steel sector is one of the largest CO_2 emitters in China. There are three main emission sources in the processes of iron and steel production: (1) the direct, onsite burning of fossil fuels; (2) indirect emission from electricity consumed during the production process; and (3) directly, non-energy-related emissions (Wang et al. 2007). It is widely accepted that the first and second sources are the main cause of CO_2 emission. About 395.69 Mt related CO_2 was discharged in 2007, and this amount was expected to grow due to the increasing production and carbon-intensive energy structure. Coal is the dominant primary energy source, and the major source of CO_2 emissions. Take the year 2007, for instance, coal provided about 74.5% of total energy needs, in which coke coal accounted for 85.6% and power coal accounted for 14.4%. Electricity and petroleum are the second and third sources of primary energy. In addition, the energy efficiency of Chinese ISI lags behind the world's advanced level. The average consumption per ton of steel in the key Chinese IS enterprises is 15% higher than the world's advanced enterprises (Guo and Yin 2007). It is estimated that there are 705 kgce (kg coal equivalent) per ton of steel consumed by the key IS enterprises in 2004, which is 8.3% higher than the world average (Zeng et al. 2009). Therefore, the corresponding CO_2 emissions should be even higher.

Iron production is the most energy-intensive process, and accounts for nearly 40% of China's total CO_2 emissions emanating from the iron and steel sector. Each process of iron making (e.g., coking and sintering,) consumes more energy per unit of production in comparison to the international advanced level (Fig. 6.1). There are two main technology paths for the transformation of iron into steel: the oxygen blown converter and the electric arc furnace (EAF). The first path is the dominant technical path in China, and accounts for 81.6% of steel production. The electric arc furnace is based on the use of scrap and electricity, while the basic blast furnace (BOF) uses iron ore and coal. It is estimated that the CO_2 emission of BOF is about 1700 kg/t (steel), while the scrap—EAF is only about 400 kg/t (Ren and Wang 2011).

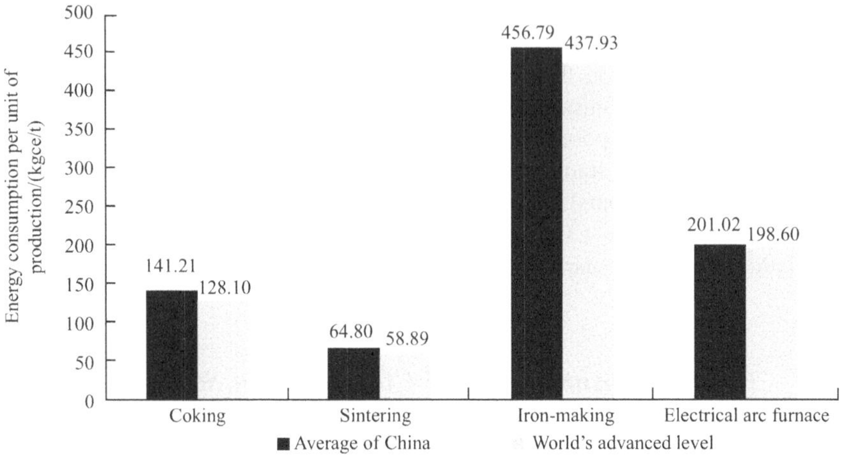

Fig. 6.1 Energy efficiency of several main processes of ISI. *Source* Zeng et al. 2009

Energy consumption is not the only source of CO_2 emission. CO_2 is discharged during several processes of iron and steel production due to the decarbonization of limestone ($CaCO_3$) and dolomite ($MgCO_3$). There was 42.5 Mt of limestone consumed by ISI in 2007 (Li et al. 2010). This is, as a result, an important source of 18.7 Mt of CO_2 emission.

6.1.2 The Practices of CO_2 Reduction in Iron and Steel Companies

CO_2 reduction channels can be split into two ways: consumption side and production side (Gielen and Moriguchi 2002). CO_2 reduction from consumption depends on reducing the steel demand, but it is not directly controlled by producers. CO_2 reduction from production can be influenced by a number of practices on energy efficiency improvement, fuel substitution, material recycling, etc. (Rynikiewicz 2008).

Increasing energy efficiency is largely dependent on producers' innovation practices. These innovation practices include "technological improvement, application of technology, application of new equipment or materials, and structural adjustment of products, techniques or energy consumption" (Zeng et al. 2009). R&D investment plays a vital role in the application of technology on energy conservation or carbon abatement (e.g., Carbon capture and storage). It strengthens producers' potential abilities in response to future pressures from climate change. In addition, production process innovation deals with radical or incremental innovations that would decrease the CO_2 emissions per ton of material (Rynikiewicz

2008). It can help to adjust the structure of products and energy consumption to the cleaner ones. Moreover, replacing the equipment with energy-efficient ones is often necessary to introduce new materials and technology in order to increase energy efficiency. Here we define all these kinds of enterprises' technological improvement and innovative activities on energy conservation or carbon reduction as "innovation" on energy conservation and carbon reduction. All these innovation activities require a large amount of financial and human resource investment.

It was proposed by some scholars that the control of materials, by-products and residues is an effective way to reduce the discharge of CO_2, (e.g., Gielen and Moriguchi 2002; Rynikiewicz 2008). There are significant amounts of energy by-products or residual energy generated during the production process in ISI, for example, coke oven gas, blast furnace gas, and blast furnace slag. Coke oven gas and blast furnace gas are good substitutes for fossil energy. If these gas energy carriers are sold and being used, the IS enterprises can not only get economic benefits, but also allocate their carbon content to the user of the gas. The blast furnace slag can be used as a cement substitute. It is estimated that the use of blast furnace slag as a substitution would result in 0.8 t CO_2 saved from the production of 1 t of cement production (Gielen and Moriguchi 2002). Some IS companies in China have highlighted the optimization of the flow of material and energy between different processes or with other industries. Zero waste and 4R (reuse, remanufacture, recycling, and recovery) have become common concepts in their production. For instance, more than 90% of blast furnaces that are above 1000 m^3 have been installed with the top-pressure recovery turbine (TRT) which is expected to recover 10 billion kW·h every year in China (Huang 2010).

In addition, the strategy of energy conservation and carbon abatement offers the direction of the company to conduct CO_2 abatement in the short-term or long-term future. Several IS companies in China start to draw up their schemes of energy saving and CO_2 reduction. Baosteel,[1] for example, has formulated "Management Process of Energy Saving Goal", which describes the future development of high energy efficiency and resource efficiency of manufacturing processes. Moreover, several projects of CO_2 abatement are planned to be implemented in the short term. An industry-scale plant for recovering flue gas from the lime kiln will be constructed, which is expected to reduce 10000 t CO_2 emissions annually (Zou 2008).

The implementation of the carbon credit mechanism is another practice of IS companies in China related with CO_2 reduction. Although it has no direct effect on producers' CO_2 abatement, it can motivate producers to indulge in emission reduction practices. Since carbon trading is becoming an inevitable trend in the world market, credit mechanism can help the company to get the opportunity to change the amount of CO_2 reduction into economic benefits by carbon trading. At present, China has no regulated cap-and-trade carbon market in the iron and steel industry. The IS companies of Chinese are not required to participate in carbon

[1]Baoshan Iron and Steel Co., Ltd (Baosteel) is the largest and most advanced integrated steel company in China, which is located in Shanghai.

trading in the domestic market. However, some companies have participated in selling their carbon credit to international buyers by the CDM project. Moreover, some pressures of carbon trading are coming from the international business through exporting products or being supplied by foreign iron ore companies. Far-sighted companies have realized the opportunities in the carbon trading, and plan to work on the practices related to the implementation of the carbon credit mechanism.

6.2 Determinants of Corporate Practices of CO_2 Reduction: Empirical Analysis from Iron and Steel Industry

A conceptual model is developed in this section to identify the determinants that drive or hinder the companies' practices for CO_2 reduction in ISI, and examine the effect of these determinants on these practices. Besides, the relationship between companies' CO_2 reduction practice and performance (including both economic performance and environmental performance) is also explored as well. Firm size is considered as the control variable in this model. Figure 6.2 is the assumed framework and structure in this chapter. It reflects the relationship between CO_2 reduction practice and some specific determinants, and the assumed effects of CO_2 reduction practice on enterprises' performance.

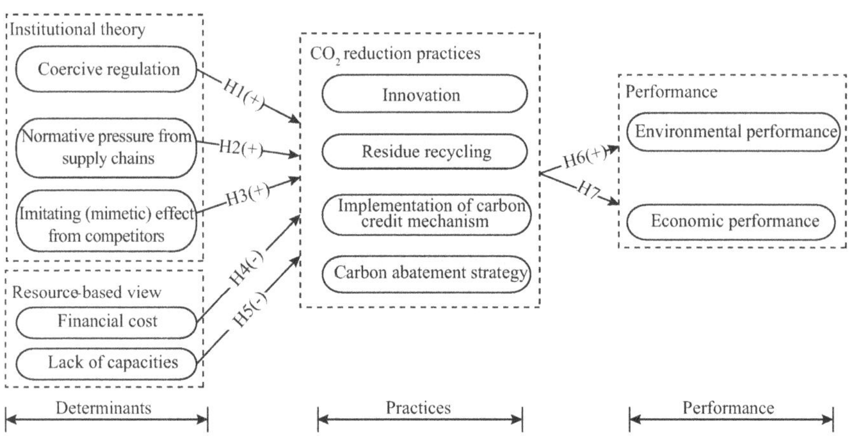

Fig. 6.2 Conceptual model for the determinants of corporate practices of CO_2 reduction notes: (+) indicates positive effect; (−) indicates a negative effect

6.2.1 *Hypotheses for the Determinants of CO_2 Reduction Practices in Iron and Steel Companies*

Institutional theory and resource-based view are introduced as the theoretical basis for identifying the determinants of CO_2 reduction practices in IS companies. It is posited by institutional theory that the drivers of environmental management practices can be motivated by three kinds of determinants: coercive, normative, and mimetic (DiMaggio and Powell 1983). Thus, barriers to proactive environmental management practices can be attributed to the lack of resource and capabilities from the resource-based view (Ebinger et al. 2006).

Coercive driver refers to the compulsive pressure coming from power groups such as governmental agencies (Rivera 2004). Environmental regulations can be considered as an important factor for effective practices of energy saving and CO_2 reduction (Jones 2010). Chinese government has released many environmental regulations to decrease resource consumption and mitigate the emission problems. Many of these regulations evolve to coercive pressures driving iron and steel companies to implement the practices of energy saving and CO_2 reduction. For instance, it is announced by an administrative decree of the State Council that backward productivity of 31.22 Mt iron and 27.94 Mt steel must be obsoleted in 2011, involving 154 IS companies in China. Moreover, there are some laws in China enacted to guide environmental practices. Energy Conservation Law (first effective in 1997 and amended in 2007) especially sets the responsibilities of energy-intensive industries (e.g., ISI), and provides criteria for the maximum quota of certain products' energy consumption and the energy efficiency level of facilities during production. Cleaner Production Promotion Law (effective from January 1, 2003, and amended in 2012) and Circular Economy Promotion Law (effective from January 1, 2009, and amended in 2018) have initiated regulations on the efficient use of materials, and drive the IS companies to reduce, reuse, and recycle the residue energy and material scraps. With such situations, the Chinese have experienced legal pressures as they have to comply with these regulations. López-Gamero et al. (2010) believed that regulatory pressures had positive impacts on proactive environmental practices. Based on the discussions above, we proposed the following hypothesis:

Hypothesis 6.1 Regulatory pressures motivate the practices of CO_2 reduction in Chinese ISI.

Normative drivers derive "from external stakeholders who have interests in the organization" (Zhu and Geng 2010). From this perspective, pressures from supply chains, customers, and communities could have a positive effect on iron and steel companies and practices of CO_2 reduction. Similar results have been found by Zhang et al. (2008) who believed that environmental management practices of Chinese manufacturers are influenced by their suppliers and customers. It is shown that there are almost 60% of iron ores imported from foreign countries for Chinese ISI, 75.4% of which are monopolized by several suppliers (e.g., BHP Billiton, Vale of Brazil, Rio Tinto) in Australia, Brazil, and India (Liao 2011).

Environmental preference from these suppliers can motivate IS companies to care about their environmental practices, such as energy conservation and emission reduction. In addition, customer requirement is a key normative pressure to implement environmental practices as well (Jørgensen et al. 2010). Many companies like Ford, GM, and Toyota encouraged their Chinese suppliers to be certified with the ISO 14001 standard. And some of these customers bring the cooperation on carbon trading or CDM project, which can provide motivation as well as capital and technology for ISI to promote its own CO_2 reduction practices. Based on the above discussion, the second hypothesis is formulated:

Hypothesis 6.2 The pressures from the supply chain motivate the practices of CO_2 reduction in Chinese ISI.

The imitating effect refers to the tendency of individuals that inclined to imitate the successful practices of others around them. It is common for the companies to learn from the experiences in the industry so as not to lag behind their competitors or substitutions, including the environmental practices. With the increasing globalization, Chinese IS companies have gained more opportunities to learn from their leading international competitors in developed countries. Some environmental practices, such as the implementation of the environmental management system, are introduced in some IS companies in China. These environmental practices of competitors can be evolved as drivers for the practices of energy conservation and emission reduction (Zhu and Geng 2010). Therefore, we have the following hypothesis:

Hypothesis 6.3 The environmental practices of competitors motivate the practices of CO_2 reduction in Chinese ISI.

The successful implementation of CO_2 reduction practices is based on the expenditure of various resources. The financial cost can be considered as a key barrier to the practices of CO_2 abatement. R&D practices of new technology for energy saving and carbon removal or storage often cost a huge amount of investments, and cannot get clear benefit before application. And replacement of energy-intensive equipment and retrofit of industrial processes increase the short-term cost significantly. It is demonstrated by Presley et al. (2007) that Chinese companies may hesitate to conduct the practices of energy saving and CO_2 reduction since values may exist in nonfinancial ways such as improved image and better relationships with suppliers/customers. Moreover, the lack of capacities is another barrier that hinders the practices of CO_2 abatement from the resource-based view. Human resource capability is an important resource for IS companies to implement CO_2 reduction practice. Lack of knowledge, skill, and professional advice which are all supported by human resources, impose the restrictions on the implementation of environmental practices (del Brio et al. 2008). Training of human resources, as a result, is an important stage for implementing environmental practices. Technology capacity is another resource for IS companies to implement CO_2 reduction practice. Many environmental practices come to an untimely end because of lacking advanced technology (Simonsson 2002). The successful implementation of CO_2 reduction practice needs smooth information collecting capacity as well. Information that are closely related to CO_2 reduction (e.g., information about climate change policies, development of advanced technology, carbon trading, etc.), are a major concern for companies' decision-making on the practices of CO_2 abatement. Lack of corresponding information would be a barrier to the CO_2 reduction practice of IS companies in China. Based on the above analysis, the following two hypotheses are posited:

Hypothesis 6.4 Financial cost impedes the practices of CO_2 reduction in Chinese ISI.

Hypothesis 6.5 Lack of capacities impedes the practices of CO_2 reduction in Chinese ISI.

6.2.2 *Questionnaire Survey for the Determinants and Practices of CO_2 Reduction in Iron and Steel Companies*

The questionnaire is covered by four sections: (1) CO_2 reduction practices; (2) drivers and barriers; (3) Performance; and (4) Basic information (See the Appendix). Measurement items in each section are developed based on the review of the literature, and then are organized to evaluate the theoretical framework as shown in Fig. 4. The four dimensions of CO_2 reduction practice are evaluated by 12 items, which are referred by the previous studies (Zsidisin and Hendrick 1998; Zhu et al. 2007). All the items are measured by using a five-point Likert-type scale (1—not considering it; 2—plan to consider it; 3—just implemented for less than one year; 4—implemented for 1–3 years; 5—implemented for more than 3 years). There are 15 items to evaluate the 5 determinants that drive or hinder the implementation of CO_2 reduction practice in the second section of our questionnaire. These items are developed based on a previous study (Zhu and Geng 2010). And questions for the items are measured by the agreement on the description with the five-point Likert-type scale (1—strongly disagree; 2—tend to disagree; 3—neither agree nor disagree; 4—tend to agree; and 5—strongly agree). Eight items in the third section of the questionnaire are developed with a focus on environmental performance and economic performance. These items are adopted from the experience of Zhu et al. (2010). Questions about the company's performance improvement in the recent 3 years are also answered using a five-point Likert-type scale (1—not at all, 2—some but insignificant, 3—some but slightly significant, 4—significant, and 5—highly significant). To avoid the misunderstanding of the items in the whole questionnaire, we give a brief explanation at the beginning of each section.

In order to validate these measurement items, we first obtained comments and suggestions from three academicians in the field of environmental management. We requested these experts to give their opinions on whether the items in the questionnaire are comprehensible according to the theoretical framework, and how to make further improvement. In the next step, we conducted a pilot test by distributing the manuscript of the questionnaire to seven senior managers from IS companies who were taking charge of environmental management in their companies. They were asked to fill in the questionnaire and provide comments on the appropriateness of the questionnaire items and whether they are readily understandable by our target respondents in the ISI. Minor modifications on the wording of items are conducted during these processes of validating and pilot test.

The unit of analysis for our research work is the middle/senior-level managers who are responsible for the environmental management in IS companies to fill in the questionnaire. The questionnaire distribution was conducted with the help of the China Iron and Steel Association (CISA).[2] It provided a list of 203 iron or steel companies that were registered in CISA. We sent the questionnaires to them by email, and the survey was undertaken over a period of five months from April 2011 to August 2011. Ninety-six responses were finally received, but 11 of them were incomplete with many missing items. Therefore, the usable responses were 85, with an effective rate of 41.87%.

The description of sampling characteristics and distribution are displayed in Table 6.1. All the respondents are evenly distributed among the four kinds of ownerships and five different firm scales in terms of the number of employees. Most of the respondents (58.8%) were from state-owned IS companies. This corresponds to the fact that this sector is dominated by state-owned enterprises (Guo and Fu 2009). Since ISI in China is a labor-intensive industry, 47% of the responses we collected are from the enterprises with employees above 10,000.

To further verify the reliability of the responses, the test–retest method was conducted in this chapter. A random sampling of 10% of companies from the 85 responses was initiated. This was completed using the software SPSS 20.0. Managers who were in charge of environmental management in nine companies were interviewed using the questionnaire as a guide. We asked them to fill in the questionnaire again, and offer examples of CO_2 reduction practices and the difficulties that they faced. The interviews were undertaken for three months from September 2011 to November 2011. Then a one-way analysis of variance (ANOVA) was performed between the previous responses and the interview responses. The results suggested that there were no significant differences between the answers of these two kinds of responses.

Table 6.2 shows the Cronbach's alpha values for each dimension in the questionnaire. The high values of Cronbach's alpha (>0.70) suggest good internal consistency of the questionnaire (Nunnally and Bernstein 1994). It is indicated that the reliability of all the items underlying CO_2 reduction practice, determinants, and performance are reasonably well.

[2]CISA was found in 1999, which is dedicated to providing service to steel producers and functioning its role as bridge, link, adviser, and assistant.

Table 6.1 Profile of the respondents in ISI

Characteristics of respondents		Number	Percentage/%	Cumulative percentage/%
Ownership	Chinese state-owned enterprises	50	58.8	58.8
	Foreign enterprises	3	3.5	62.3
	Joint ventures	5	5.9	68.2
	Chinese private enterprises	27	31.8	100
Total		85	100	–
Number of employees	<1000	4	4.7	4.7
	[1000–5000)	22	25.9	30.6
	[5000–10000)	19	22.4	53.0
	[10000–20000)	15	17.6	70.6
	≥ 20000	25	29.4	100
Total		85	100	–

Table 6.2 Regression results between CO_2 reduction practices and corresponding determinants (standard estimations)

Independent variables	Dependent variables				VIF
	Innovation (*t* value)	Residue recycling (*t* value)	Implementation of carbon crediting system (*t* value)	CO_2 reduction strategy (*t* value)	
Firm size	0.246 (2.413)*	0.076 (0.693)	0.305 (3.050)**	0.197 (1.929)$^{+}$	1.134
Regulation	0.138 (1.218)	0.037 (0.305)	-0.041 (−0.249)	0.310 (2.727)**	1.404
Pressures from supply chains	0.191 (1.895)'	0.205 (1.888)'	0.225 (2.267)*	0.228 (2.311)	1.114
Imitating effect	0.147 (1.343)	−0.183 (1.550)	0.289 (2.707)*	−0.070 (−0.640)	1.310
Financial cost	−0.218 (−2.152)*	−0.009 (−0.084)	0.216 (2.188)*	−0.267 (−2.633)*	1.119
Lack of capacities	−0.139 (−1.322)	0.163 (1.433)	−0.265 (−2.579)*	−0.036 (−0.342)	1.212
Adjusted R^2	0.232	0.107	0.269	0.226	
F value	5.227	2.686	6.144	5.088	

Notes * means $p < 0.1$; ** means $p < 0.05$; *** means $p < 0.01$

6.2.3 Regression Analysis for the Relationships Between Determinants and Practices of CO_2 Reduction

In this section, we perform a regression analysis to examine if the determinants explored in Sect. 6.2.1 are related to CO_2 reduction practices. There are four regression models designed for examining the impact of five determinants (regulation, pressures from supply chains, imitating effect, financial cost, and lack of capacities) on the four CO_2 reduction practices (CO_2 abatement innovation, residue recycling, carbon trading, and CO_2 abatement strategy). We include firm size as a control variable in each regression. This is due to the possibility that the size of the firm may influence the extent of their implementation of environmental practices (Zhu et al. 2007).

$$\text{CO}_2\text{ reduction practice}_{(N)} = F(\text{Determinants, Organizational size, error})\text{ for }(N = 1 - 4) \quad (6.1)$$

The regression results for examining the impact of each determinant on the practices of CO_2 reduction are displayed in Table 6.2. Totally, the F values in the four regression models are all high enough to reject the null hypotheses that the independent variables are not associated with the dependent variable. The hypotheses from one to five, at least, are verified partially. In order to check out the multi-collinearity of these regression models, variance inflation factor (VIF) was calculated checked among independent variables. The largest of the resulting VIF score in all of the regression models in Table 6.2 is 1.404, which is well below the maximum level of 10.0 suggested by Mason and Perreault (1991). This means multi-collinearity should not be a serious concern in our regression.

For the impact on the CO_2 reduction practices of *Innovation*, *Pressures from supply chains* impose a significant positive effect ($\beta = 0.191$, $p = 0.062 < 0.1$), while *Financial cost* plays a negative role ($\beta = -0.218$, $p = 0.035 < 0.05$). Hypotheses 2 and 4 are verified at this point. Innovation on energy conservation and CO_2 reduction has to be costed with high investment. For instance, the retrofit of energy-efficient equipment would increase the constant cost of the company. Moreover, the application of new technology on energy conservation and CO_2 reduction needs high investment in R&D or substantive expenditure on purchasing. However, the benefits from energy conservation or CO_2 reduction are often not clear in the short term (van Hemel and Cramer 2002). As a result, few companies would like to afford the high cost of innovation on energy conservation and CO_2 reduction without direct financial benefits. High cost becomes a barrier for innovation practices. Contrarily, *Pressures from supply chains* give a push to the practice of innovation on energy conservation and CO_2 reduction, although the effect is a bit slightly significant. Many customers begin to care about the environmental ability of their suppliers, and require the product to be environmentally friendly. For instance, Baosteel gets the certification of ISO 140001 in order to satisfy the requirements in the export market. In addition, suppliers (e.g., iron ore producers) raise the price in the name of environmental protection or CO_2 reduction, which forces the IS companies to retrofit their production process with energy conservation and CO_2 reduction. Therefore, more innovation activities would be implemented due to the higher pressures from both the demand and supply sides. Besides, the results also suggest that larger IS companies would be more willing to implement innovation practices on energy conservation and CO_2 reduction. Larger companies have stronger capacities to afford the innovation cost.

Only *Pressures from supply chains* have a certain effect on *Residue recycling* ($\beta = 0.205$, $p = 0.63 < 0.1$). It is indicated that the implementation of *Residue recycling* is largely dependent on coordinating with suppliers and customers. There is a need for some smooth channels for the end-of-life product to be taken back from the customers to reuse. Moreover, the residue energy and by-products need to find proper buyers to get the recycling moving on. For instance, the coke and oven gas are collected from the electricity supplier in some places in China. In addition, the insignificant effect of *Regulation* on *Residue recycling* could be understood. Until then, there seems no direct regulations that require IS companies to recycle their residue in China. Most IS companies recover the residues spontaneously in order to get economic benefits or improve their environmental performance. From this point, it could tentatively promote to implement specific regulations or standards on *Residue recycling* in the IS industry. This would be helpful in motivating IS companies and residue recycling activities. However, the insignificant effect of *Financial cost* on *Residue recycling* is far beyond our expectation. We assumed that the *Financial cost* would hinder the implementation of *Residue recycling* since there might be a high investment in the infrastructure or upgrading the equipment. But we neglected the economic benefits of the residue recycling. It would reduce the consumption of resources or energy, and might bring profits from the selling of recycled waste and by-product. These positive effects might neutralize the negative effects brought from *Financial cost*. Moreover, we do not expect *Lack of capacities* has no significant effect on *Residue recycling*. It indicates that most IS companies in China have the capacities (e.g., technology and talents) of *Residue recycling*. It could not be a hinder for them to reuse and recycle the waste resource and by-products.

From Table 6.2, we can see that the *Implementation of carbon credit mechanisms* is positively related to three determinants (namely *Pressures from supply chains*, *Imitating effect,* and *Financial cost*), and negatively related to lack of capacities. From this point, hypotheses 6.2, 6.3, and 6.5 are verified. It is surprising to see that *Financial cost* becomes a driver for the practice of *Implementation of carbon credit mechanisms*. For one thing, this might be somewhat attributed to the fact that the high investments of CDM projects are afforded by the buyers to reduce CO_2. The financial cost, as a result, has an insignificant negative effect on iron and steel companies' CDM projects. For another thing, it is high costed to conduct carbon reduction for energy-intensive IS companies. The carbon credit mechanism provides an opportunity to buy carbon credits instead of reducing CO_2 emission themselves. In addition, our results also indicate that the *Imitating effect* plays a significant role in the *Implementation of carbon credit mechanisms* for Chinese IS companies. Some companies' attempts on carbon credit mechanism would motivate other companies' practices on it. In recent years, some IS companies in China have successfully conducted CDM projects, and get attractive benefits from them. Take *Wuhan Iron and Steel Group,* for example, it benefits 204 million yuan from CDM annually. This would appeal other IS companies to apply for CDM. Our results show that *Pressures from supply chains* are another traitor of *Implementation of carbon credit mechanisms*. Since a lot of businesses have happened in the global market for many Chinese IS companies, many of them realize the requisite trend of carbon trading in the future. In contrast, *Lack of capacities* hinders the Chinese IS companies conducting carbon trading as shown by our results. At present, few of the companies in China have the talents who are good at the rules and knowledge about carbon credit mechanism or carbon trading. And the information about how to trade and how to benefit from the carbon market is minor. These barriers impose restrictions on the *Implementation of carbon credit mechanisms*. Our results show that larger companies also seem to take advantage of the carbon credit mechanism. This might be due to their stronger capacities to get the information or cultivate the talents.

Regulation has no significant influence on the practices of energy saving and CO_2 reduction except for the CO_2 reduction strategy. This indicates that the regulations of CO_2 emission in China are not sophisticated enough to have a significant effect on detailed practices. However, it does give directions for companies to face up to the CO_2 emission problems. Larger companies still have the advantage of strategy-making on energy conservation and CO_2 reduction. But the superiority is much smaller compared with the implementation of innovation and carbon trading.

6.3 Relationship Between Practices of CO_2 Reduction and Corporate Performance

6.3.1 Hypotheses for the Performance of CO_2 Reduction Practices in Iron and Steel Companies

From the perspective of individual companies, the implementation of CO_2 reduction practices aims finally to improve the performance of themselves. Without significant benefits, companies are unwilling to commit themselves to CO_2 abatement voluntarily. There are two kinds of performance concerned by modern enterprises: environmental performance and economic performance, which are applied in our analysis.

Since environmental problems (e.g., climate change, pollution) have drawn great attention to society, values from environmental performance start to attach more importance to the companies. The practices of reducing CO_2 emission would improve the social image of the company, and then could get more supports from the society and government. The relationship with customers or suppliers might be improved, especially during international business. Less trading barriers related to carbon reduction would exist by virtue of the endeavor on CO_2 reduction practices. From the above analysis, we give the sixth hypothesis:

Hypothesis 6.6 The practices of CO_2 reduction can improve the environmental performance of IS companies.

The effects of CO_2 emission practices on companies' economic performance are a bit more complicated. On the one hand, the implementation of CO_2 reduction practice increases the operational cost of the company. It requires high investment in the technology innovation or the energy-efficient equipment retrofit. On the other hand, there would be great potential benefits from the CO_2 reduction practices. The reduced CO_2 emission could be changed into economic benefits during the trading on the international carbon market in the future. And the environmental fine or the carbon taxes would be reduced during the international business. Based on the above analysis, we frame the seventh hypothesis:

Hypothesis 6.7 The practices of CO_2 reduction have an effect on the economic performance of Chinese IS companies.

6.3.2 *Descriptive Statistic for the Performance of CO_2 Reduction*

The mean values for the four dimensions of CO_2 reduction are shown in Table 6.2. There are relatively high commitments on CO_2 abatement strategy (3.53 for short-term objectives on energy saving or carbon abatement; 3.32 for long-term vision; 3.20 for the clear plan). This means that in recent years, many IS companies have realized the importance of CO_2 reduction and consider it an important assignment for their future development. For instance, *Wuhan Iron and Steel Group* has established a series of strategic plans for energy conservation and emission reduction, such as Plan for *Energy Conservation and Dynamic Facilities, Plan for Construction of Circular Economy Style Company and Plan for Pro-environmental Production*. The objective and performance of energy conservation and emission reduction are published in its annual Social Responsibility Report. The innovations on energy saving and CO_2 abatement get higher scores than average 3 as well. Many iron and steel companies in China implemented new technology on energy conservation in recent years, and replaced some backward equipments with energy-efficient ones. For instance, Baosteel invested 4.02 billion yuan on 346 innovation projects for energy conservation and emission reduction from 2006 to 2010. However, all the standard deviations in this item are a bit high (all values over 1.00). This indicates that although some companies did well on their energy saving and CO_2 abatement innovation, there are still some companies which lag behind on it. From this point, there are still certain potential spaces for Chinese IS companies to continue their efforts to improve the energy efficiency of the backward IS companies.

However, the mean values for *Implementation of carbon credit mechanisms* are much lower, which are all below 2.5. This means that most Chinese IS companies in our questionnaire survey have not yet put the carbon credit mechanisms into practice. This is due to the domestic underdeveloped carbon market[3] in China. There are few smooth channels for carbon trading and making economic benefits from carbon reduction, so that the companies are lacking the enthusiasm to conduct carbon credit mechanisms to calculate their reduced carbon. But many of them have realized the trend of carbon trading and potential opportunity in carbon market, and planned to prepare for the carbon credit mechanisms. And some companies have gotten benefits from the implementation of clean development mechanism (CDM). The relative large deviation value (over 1.00) of *Application of CDM* indicates that there are still several companies did well on the practice. A few Chinese IS companies have implemented CO_2 reduction projects with foreign investment and then sold credits of reduced carbon to the investors under the framework of CDM. For instance, Jigang Group[4] undertook a CDM project to reduce CO_2 emission by captive power generation through a waste heat recovery system. The corresponding amount of reductions reached 167,055 metric tons of CO_2 equivalent per annum. And corresponding to our result, the score of *Construction of carbon reduction accounting system* was much low (mean value = 1.6, deviation value = 0.727). It indicates that Chinese IS companies still lack the administrative capacities for carbon trading. It appears that it will take some time for them to reach the criteria of MRV (measurable, reportable and verifiable) data. As a result, speeding up the process of domestic carbon market construction is a requisite way to raise IS companies' awareness of carbon credit. Only providing a smooth channel for IS companies to exchange their reduced carbon credit with economic benefits, can they actually motivate to conduct carbon credit mechanisms and reduce CO_2 emission actively.

The mean values for the three items of residue recycling are between 2.5 and 3.0. It is indicated that the recycling of residual energy and waste materials is still under consideration or just implemented for a short time for many Chinese IS companies. And all the standard deviations are a bit high (all values over 1.00), which indicates that different clusters of Chinese IS companies may exist due to different expressions of residue recycling. There are several pioneering companies that initiate good practices on residue recycling. For instance, Jigang Group uses the recycled blast furnace gas, coke oven gas, and converter gas to generate electrical power. There

---

[3]Until now, there has been no compliance carbon market in China. Several voluntary carbon exchanges (e.g., Beijing Environmental Exchange, Shanghai Environmental Energy Exchange, and Tianjin Climate Exchange) have been constructed while with a small volume of business.

[4]*Jigang Group* is an iron and steel company located in Shandong Province, China.

was 3.1 billion kW·h electricity produced in 2010, which guaranteed 52.3% of power use in Jigang Group and saved 0.27 Mt standard coal.

Almost all performance items are ranked higher to some degree, with mean values consistently within the 3.00 (3 = to some degree) and 4.00 (4 = relatively significant) ranges. This indicates that many companies think that they have improved certain performance on both environmental and economic perspectives in recent three years.

6.3.3 Regression Analysis for the Relationships Between the Practices of CO_2 Reduction and Corporate Performance

In this section, we perform a regression analysis to examine if these practices of CO_2 reduction have effects on the companies' performance. There is a second set of two separate regression models in which each of the four CO_2 reduction practices is treated as a dependent variable, while the two performance dimensions (environmental performance and economic performance) are defined as independent variables. We include company size as a control variable in each regression. This is due to the possibility that the size of the firm may influence the extent of their implementation of environmental practices (Zhu et al. 2007). As a result, nine total regressions are determined with the following functional forms:

$$\text{Performance } (M) = F(\text{CO}_2 \text{ reduction practice, Organizational size, error}) \text{ for } (M = 1 - 2) \tag{6.2}$$

In the regression models, each factor for CO_2 reduction practice, determinants, and performance are averaged from their underlying measurement items to form a single indicator factor. The summary factors were subsequently used for the regression analyses. The same processing method was also adopted by Zhu et al. 2007 who indicated that such a summary process could "reduce the model complexity and allow us to test the relationships based on small sample size."

Table 6.3 shows the regression results that examine the impact of CO_2 reduction practices on companies' performance. Hypothesis 6 is well supported since F value reaches 10.674 that significantly passes through the verification of variance analysis. However, the F value ($F = 3.063$) for hypothesis 6–7 is a bit small, which indicates that the effect of CO_2 reduction on economic performance is less significant. Similar findings were proposed by Testa and Iraldo (2010) who thought that environmental practices in supply chains bring the significant environmental performance while the financial performance from such practices is still ambiguous. In addition, VIFs for all independent variables in these two regressions range from

Table 6.3 Regression results between CO_2 reduction practices and corresponding performance (standard estimations)

Independent variables	Dependent variables		VIF
	Environmental performance (*t* value)	Economic performance (*t value*)	
CO_2 reduction strategy	0.308 (2.821)***	0.084 (0.649)	1.580
Carbon trading	−0.084 (−0.827)	0.093 (0.771)	1.370
Residue recycling	0.324 (3.322)***	0.254 (2.200)**	1.256
Innovation	0.207 (1.851)*	0.118 (0.893)	1.656
Firm size	0.033(0.343)	−0.040 (−0.344)	1.247
Adjusted R^2	0.365	0.109	
F value	10.674	3.063	

Notes * means $p < 0.1$; ** means $p < 0.05$; *** means $p < 0.01$

1.247 to 1.656, well below the maximum level of 10.0. There are no problems of multi-collinearity in the two regressions as well.

Strategy of energy saving and carbon reduction can be a source of benefit for environmental performance, while it has no clear effect on the improvement of economic performance. It is indicated that high environmental commitment could improve companies' environmental performance (e.g., emission reduction and energy conservation), but hardly bring direct economic benefits at least in the short term. For instance, some strategic plans for energy conservation and emission reduction could improve companies' social image when they are published to the public. However, the corresponding benefits from these strategic plans would not come out until they are successfully implemented for several years.

Innovation on energy conservation and CO_2 emission appears to have a positive effect on the environmental performance of company. Both replacing backward equipment with energy-efficient ones and application of advanced technology for energy conservation could have a direct effect on CO_2 reduction. However, their effect on economic performance is not clear. For one thing, while investment in innovation of energy conservation and CO_2 reduction is high, this increases the economic cost of the company. For another thing, the reduced emission of carbon or waste can bring indirect economic benefits, such as decreasing the waste disposal fee and trading the reduced CO_2 through implementing CDM projects. As a result, the economic performance appears to be dependent on the balance of innovation investment and benefits.

Residue recycling can bring both environmental and economic performance. On the one hand, the residue recycling practices can reduce energy consumption and waste emissions, which will improve the environmental performance of companies. On the other hand, residue recycling can save the cost of material and energy consumption. Furthermore, some recycled by-products or waste materials can be sold to other stakeholders for economic benefits. Take *Baogang Group*[5] for example, there was 1.385 billion yuan expenditures saved on electricity power due to residue energy reuse and recycling from 2002 to 2009. Meanwhile, the discharge rate of blast furnace gas dropped from 14.5 to 2.94% and the recycling rate of converter gas increase from 13.6 to 82.2%.

Our results show that there are no significant relationships between the implementation of carbon credit mechanisms and IS companies' performance (including both economic and environmental performance). This is likely due to the fact that the carbon credit mechanisms has not been taken into practice for many IS companies in China. And carbon credit is a long-term investment, and significant benefits would not appear in a short time.

Finally, firm size does not appear to be an important factor that influences the effect of CO_2 reduction practices of IS companies on their performance. This might be attributed to the fact that many small and backward IS companies have been closed down after several years of eliminating backward productivity by the central government. The remains have certain capacities for improving their performance. This also indicates that the potential space for further eliminating backward productivity in ISI is getting smaller. The government needs to find other ways (e.g., market measures) instead of administrative intervention to reduce national CO_2 emissions in the future.

6.4 Summary

To achieve sustainable development in response to the pressures from climate change, it is important for Chinese IS companies to implement various kinds of CO_2 reduction practices. This chapter is to identify the determinants that drive or impede the CO_2 reduction practices of Chinese IS companies, and find out the relationship between the practices and performance. We offer a conceptual model consisting of 7 hypotheses.

The results show that although some CO_2 reduction practices can bring significant environmental performance, few of them impact on the improvement of economic performance clearly. Hypothesis 6.6 is better verified than Hypothesis 6.7. This indicates that the implementation of CO_2 reduction practices is still in an early stage for Chinese iron and steel companies. Economic benefits should be exploited to motivate iron and steel companies to conduct CO_2 reduction proactively and voluntarily.

Every hypothesis from 6.1 to 6.5 is partially verified. Firstly, *Regulation* has no significant effect on most practical CO_2 reduction activities, but can promote strategy-making for energy conservation and CO_2 reduction. Most regulations related to energy conservation or CO_2 reduction in China are generally flexible or voluntary so far (Zhu and Geng 2010). Stricter regulations and higher enforcement

[5]*Baogang Group* is an iron and steel company located in Inner Mongolia, China.

levels particularly focused on ISI are needed to promote the CO_2 reduction practices. Secondly, *Pressures from supply chains* show a positive influence on most CO_2 reduction practices except CO_2 reduction strategy. Chinese IS companies have experienced market pressure from external relationships such as green purchasing and customer cooperation focusing on environmental concerns, and it's beneficial for their implementation of pro-environmental practices. From this perspective, some governmental purchase programs should be considered to apply green procurement to promote iron and steel companies' practices on energy conservation and CO_2 reduction. Thirdly, *Financial cost* is found to be an important factor which is seriously considered in the implementation of most CO_2 reduction practices. Financial subsidies or tax breaks from the government, as a result, could be applied for promoting CO_2 reduction practices, especially at the beginning stage of the practices. In addition, both *Imitating effect* and *Lack of capacities* play a significant role in *Implementation of carbon credit mechanisms*. Although several companies have got great benefits from the implementation of CDM, many companies still lag behind in carbon credit due to the lack of capacities to get related information and knowledge. However, the practice of carbon credit could be driven by the excellent performance of the leaders on CDM projects. Government, therefore, can provide more platforms for the collection and distribution of the information related to carbon crediting, and strengthen the education system on training the talents of carbon credit mechanisms.

Although some valuable findings have been explored, there is still great potential to enrich this study. Firstly, it would be valuable to further identify the mediation or moderation effects of those determinants we explored in this chapter on the IS companies' CO_2 reduction practices and their performance. Secondly, the questions for the items of *Residue recycling* is designed according to how long the IS companies have implemented corresponding activities. Further investigation would be conducted on the percentage of residue recycling of IS companies to better understand their performance of residue recycling. Moreover, some case studies about CO_2 reduction practices in Chinese IS companies could be made to further verify the relationship among CO_2 reduction practices, determinants, and performance. Finally, questions about how to conquer the barriers of CO_2 reduction practices and improve both environmental and economic performance require further research.

References

del Brio, J.A., B. Junquera, and M. Ordiz. 2008. Human resources in advanced environmental approaches—a case analysis. *International Journal of Production Research* 46 (21): 6029–6053.

Demailly, D., and P. Quirion. 2008. European emission trading scheme and competitiveness: A case study on the iron and steel industry. *Energy Economics* 30 (4): 2009–2027.

DiMaggio, P.J., and W.W. Powell. 1983. The iron cage revisited: Institutional isomorphism and collective rationality in organizational fields. *American Sociological Review* 48 (2): 147–160.

Ebinger, F., M. Goldbach, and U. Schneidewind. 2006. *Green supply chains: A competence-based perspective// Sarkis J*, 251–270. Greening the Supply Chain. Berlin: Springer.

Gielen, D., and Y. Moriguchi. 2002. CO_2 in the iron and steel industry: An analysis of Japanese emission reduction potentials. *Energy Policy* 30 (10): 849–863.

Gielen, D.J. 1997. *Technology characterization for ceramic and inorganic materials. Input data for Western European MARKAL, ECN-C-97-064*. Petten: Energy Research Centre of The Netherlands.

Guo, H., and Z. Yin. 2007. Range of recovery of exhaust heat in metallurgical enterprises. *Iron &Steel* 2: 77–81 (in Chinese).

Guo, Z.C., and Z.X. Fu. 2009. Current situation of energy consumption and measures taken for energy saving in the iron and steel industry in China. *Energy* 35 (11): 4356–4360.

Hidalgo, I., L. Szabo, J.C. Ciscar, et al. 2005. Technological prospects and CO_2 emission trading analyses in the iron and steel industry: A global model. *Energy* 30 (5): 583–610.

Hu, C.Q., L.Y. Chen, C.X. Zhang et al. 2006. Emission mitigation of CO_2 in steel industry: Current status and future scenarios. *Journal of Iron and Steel Research, International*, 13(6): 38-42, 52.

Huang, D. 2010. Energy efficiency in China iron and steel industry. http://220.156.189.23/seminars/document/2010/Energy Efficiency in CHN Steel Huang Dao.pdf. [2010-10-13]

Jones, C. 2010. Exploring new ways of assessing the effect of regulation on environmental management. *Journal of Cleaner Production* 18: 1229–1250.

Jørgensen, M.S., U. Jørgensen, K. Hendriksen, et al. 2010. Environmental management in Danish transnational textile product chains. *Management Research Review* 33 (4): 357–379.

Li, H., L. Wang, L. Shen, et al. 2012. Study of the potential of low carbon energy development and its contribution to realize the reduction target of carbon intensity in China. Energy Policy. In Press, Corrected Proof, Available online 17 November.

Li, J., W. Fan, and J. Guo. 2010. Analysis of producer gas desulphurization for metallurgical lime kiln. *Fire-resistant and Lime* 35 (1): 15–18. (In Chinese).

Liao. 2011. Accelerating the diversification of China's iron ore imports. Shanghai Securities News, http://news.xinhuanet.com/fortune/2011-08/24/c_121902221.htm (In Chinese).

López-Gamero, M.D., J.F. Molina-Azorín, and E. Claver-Cortés. 2010. The potential of environmental regulation to change managerial perception, environmental management, competitiveness and financial performance. *Journal of Cleaner Production* 18 (10–11): 963–974.

Lund, P. 2007. Impacts of EU carbon emission trade directive on energy-intensive industries—Indicative micro-economic analyses. *Ecological Economics* 63 (4): 799–806.

Mason, C., and W.D. Perreault. 1991. Collinearity, power, and interpretation of multi-ple regression analysis. *Journal of Marketing Research* 28 (3): 268–280.

Nunnally, J.C., and I.H. Bernstein. 1994. *Psycometric Theory*, 3rd ed. New York: McGraw-Hill.

Presley, A., L. Meade, and J. Sarkis. 2007. A strategic sustainability justification methodology for organizational decisions: A reverse logistics illustration. *International Journal of Production Research* 45 (18–19): 4595–4620.

Ren, L., and W. Wang. 2011. Analysis of existing problems and carbon emission reduction in Shandong's iron and steel industry. *Energy Procedia* 5: 1636–1641.

Rivera, J. 2004. Institutional pressures and voluntary environmental behavior in developing countries: Evidence from the Costa Rican hotel industry. *Society and Natural Resources* 17 (9): 779–797.

Rynikiewicz, C. 2008. The climate change challenge and transitions for radical changes in the European steel industry. *Journal of Cleaner Production* 16 (7): 781–789.

Sheinbaum, C., L. Ozawa, and D. Castillo. 2010. Using logarithmic mean Divisia index to analyze changes in energy use and carbon dioxide emissions in Mexico's iron and steel industry. *Energy Economics* 32 (6): 1337–1344.

Simonsson, A. 2002. Application of sustainable product development and designda pilot study of the state of the art, barriers and enablers. Master's thesis, LiTH-IFM-Ex-1072. Sweden: Linköping University: Department of Physics and Measurement Technology.

Testa, F., and F. Iraldo. 2010. Shadows and lights of green supply chain management: Determinants and effects of these practices based on a multi-national study. *Journal of Cleaner Production* 18: 953–962.

van Hemel, C., and J. Cramer. 2002. Barriers and stimuli for ecodesign in SMEs. *Journal of Cleaner Production* 10: 439–453.

Wang, K., C. Wang, X. Lu, et al. 2007. Scenario analysis on CO_2 emissions reduction potential in China's iron and steel industry. *Energy Policy* 35 (4): 2320–2335.

Zeng, S., Y. Lan, and J. Huang. 2009. Mitigation paths for Chinese iron and steel industry to tackle global climate change. *International Journal of Greenhouse Gas Control* 3 (6): 675–682.

Zhang, B., J. Bi, Z. Yuan, et al. 2008. Why do firms engage in environmental management? An empirical study in China. *Journal of Cleaner Production* 16: 1036–1045.

Zhu, Q., and Y. Geng. 2010. Drivers and barriers of extended supply chain practices for energy saving. *Journal of Cleaner Production*. In Press, Corrected Proof, Available online 8 October 2010.

Zhu, Q., J. Sarkis, and K. Lai. 2007. Green supply chain management: Pressures, practices and performance within the Chinese automobile industry. *Journal of Cleaner Production* 15: 1041–1052.

Zhu, Q., Y. Geng, and K.H. Lai. 2010. Circular economy practices among Chinese manufacturers varying in environmental-oriented supply chain cooperation and the performance implications. *Journal of Environmental Management* 91 (6): 1324–1331.

Zou, K. 2008. Baosteel energy-saving and emission-reducing practice. http://www.asiapacificpartnership.org/pdf/Steel/5th_meeting/4-3%28Baosteel%29.pdf (In Chinese).

Zsidisin, G.A., and T.E. Hendrick. 1998. Purchasing's involvement in environmental issues: A multi-country perspective. *Industrial Management and Data Systems* 7: 313–320.

Chapter 7
Energy Efficiency and CO_2 Emission Abatement Technology

With the rapid pace of domestic industrialization and urbanization, the energy consumption of China has grown so fast that it has become the second-largest energy consuming and CO_2 emitting country in the world. According to the prediction of International Energy Agency (IEA), China and India will be the main drivers of non-OECD primary energy demand growth until 2030, and China will account for 39% of global increase in primary energy use, whose share in total demand will jump from 16% in 2007 to 23% in 2030. Furthermore, there is no reason to feel optimistic about the energy intensity of China. According to the latest data from World Bank, GDP per unit of energy use of China in 2008 was only $3.6 per kg, while the United States had $5.8 per kg and Japan reached even as high as $8.1 per kg, which means that China consumes too much energy to support its economic growth, and economic efficiency of energy use is rather low.

Energy use is a major source of GHG emission (Herring 1999; Miketa and Mulder 2005; Li and Cheng 2008; Sari and Soytas 2009), causing environmental problems, which is a serious question given that the extreme weather conditions are spreading rapidly across China. Continually rising energy consumption and highly positioned energy intensity (Andrews-Speed 2009) have not only sounded the alarm for Chinese energy security, but also increased the pressure on whether or not China should bear more responsibility in cutting emissions in the -post-Kyoto-protocol era. Hence, cutting energy consumption has become a matter of urgency in China.

This chapter takes the following literature for reference:

Wang Z, Zeng H, Wei Y, et al. 2012. Regional total-factor energy efficiency: an empirical analysis of industrial sector in China. Applied Energy, 97(9): 115–123.

Wang Z, Chao F, Zhang B. 2014. An empirical analysis of China's energy efficiency from both static and dynamic perspectives. Energy, 74(C): 322–330.

Wang Z, Yang Z M, Zhang Y X, et al. 2012. Energy technology patents—CO_2, emissions nexus: an empirical analysis from China. Energy Policy, 42(2): 248–260.

Z. Wang and B. Zhang, *Low-Carbon Consumption in China: Residential Behavior, Corporate Practices and Policy Implication*,
https://doi.org/10.1007/978-981-15-2792-0_7

In recent years, the data envelopment analysis (DEA) method has been widely used to calculate energy efficiency (Zhou et al. 2010). According to Wang et al. (2013), the DEA methods utilized in the existing literature related to energy efficiency can be roughly divided into two groups according to their research perspectives. Herein, through calculation and analysis of China's historical performance regarding energy efficiency and revealing the significant difference in energy efficiency between regions and provincial units, we tried to deepen the understanding of international readers' about the evolution of China's energy utilization at national, regional, and provincial levels. In addition, through analyzing the key factors responsible for a change in energy efficiency and keeping in view the technical progress, productive scale, and management level, we opted to put forward corresponding suggestions for the Chinese government to further improve the energy efficiency from the production technology, management and scale aspects.

And the current study also explores the relationship between energy technology patents and CO_2 emissions in China. Considering the potentially different roles of fossil-fueled and carbon-free technology innovations in reducing the CO_2 emissions, we distinguished the patents for fossil-fueled technologies from patents for carbon-free energy technologies. The former mainly refers to the patents relevant to fossil-fueled (coal, oil, and natural gas, etc.) technologies in energy sectors and energy user sectors; the latter mainly refers to patents relevant to nuclear and renewable energy technologies. Specifically, this research tries to answer the following questions: (1) is there a significant relationship between patents for fossil-fueled technologies and CO_2 emissions? (2) Is there a significant relationship between patents for carbon-free energy technologies and CO_2 emissions? (3) Are eastern, central, and western China regions in the same relationship? To avoid the omitted variable bias, GDP is also included in the current study due to its important role in affecting CO_2 emissions and energy technology patents. To examine such types of relationships, a dynamic panel data approach will be applied.

7.1 Regional Total-Factor Energy Efficiency in China

The rapid growth of the Chinese economy resulted in great pressure on energy consumption, especially the energy-intensive sector and the industrial sector. To achieve sustainable development, China has to develop the modes to promote energy efficiency to meet the demand of Chinese rapid economic growth, as the energy efficiency of China is relatively low. Meanwhile, the appeal of energy saving and emission reduction has been made by the Chinese central government. Therefore, it is important to evaluate the energy efficiency of the industrial sector in China and assess efficiency development probabilities. The framework of the total-factor energy efficiency index is adapted to determine the discrepancy of energy efficiency in the Chinese industrial sector.

7.1.1 Single- and Total-Factor Energy Efficiency

There are two ways to measure the energy efficiency indicators, one is the total-factor energy efficiency, which considers the energy efficiency of each kind of input factor interaction; the other is single-factor energy efficiency, which only compares the energy elements to output without considering the other factors of production. Total-factor energy efficiency is closer to the actuality but is computationally complex. While the single-factor energy efficiency is simple, it exaggerates the energy efficiency and does not consider the substitution between elements.

7.1.1.1 Single-Factor Energy Efficiency

Because of the singleness and intuition of measuring index, the research at domestic and international level on the single element of energy efficiency is relatively mature. At present, there are many aspects in the study of single-factor energy efficiency, which can be roughly grouped into the following four categories according to different research purposes: (1) Studying the changing characteristics or trend of single-factor energy efficiency with economic development; (2) Comparing the single-factor energy efficiency between different countries and industries at the macro level; (3) Analyzing the difference of single-factor energy efficiency between different regions and different administrative units in the same country at the microcosmic level; (4) Studying the factors that influence the change of single-factor energy efficiency.

Single-factor energy efficiency is often defined as the ratio of an economy's useful output to energy input. The most common single-factor energy efficiency index is GDP energy consumption. The value is measured by the ratio of per unit GDP to energy consumption. The energy efficiency index of a single factor cannot be neglected in the analysis of power consumption. GDP energy efficiency indicators do not reflect the basic technological efficiency of energy use (Wilson 1944). Since GDP efficiency indicators reflect the overall state of social production, they do not reflect the differences in technology and energy efficiency between industries, therefore they can't characterize the development of different industries in the national economy in energy use efficiency (Jenne and Cattell 1983). GDP is a comprehensive manifestation of various production factors in an economy. Energy efficiency indicators can't reflect the substitution effect of the labor force and other factors of production on energy input, only if the input of power elements cannot fulfill the social productive activities (Renshaw 1981; Hu and Wang 2006). Because the type of energy used in an economy is difficult to be the only one, many studies have shown that different types of energy can be used depending on existing technologies (Patterson 1996; Patterson et al., 2005). However, energy efficiency indicators cannot describe the differences between economies and different energies, because they are only the monetary performance of overall economic output and the ratio of total energy investment.

All the research on single-factor energy efficiency has been widely accepted by domestic and foreign scholars. New progress has been made, but at the same time there are many deficiencies. The study of single-factor energy efficiency cannot reflect the substitution of the labor force and other factors of production on energy factors.

7.1.1.2 Total-Factor Energy Efficiency

In order to make up the insufficiency of the single-factor energy efficiency research, scholars began to consider adding the input factors or output factors which affect the economic development, such as labor input, capital investment, pollution output, and so on, so as to make a more comprehensive evaluation of energy efficiency, thus forming the research of total-factor energy efficiency. Generally speaking, total-factor energy efficiency includes the configuration efficiency and technical efficiency. Among them, the allocation efficiency is defined as the price structure of all resource elements, and the ability to minimize the total cost by optimizing the proportion of resource elements in the total demand. Technical efficiency refers to the ability to minimize the input of input elements under the precondition of keeping the output unchanged. At present, the research on the total-factor energy efficiency is mainly aimed at the technical efficiency, whose research method originates from the theory of total-factor production in microeconomics. The research methods of total-factor energy efficiency are various, and the most commonly used is the production frontier analysis method. The production frontier is generally measured by a production function, which is used to indicate the maximum set of outputs corresponding to a certain level of technology. The production frontier analysis method is divided into parametric method and nonparametric method. The former is called the Stochastic Frontier Analysis, represented as the SFA, and the latter is called the Data Envelopment Analysis, respect of the DEA.

Because the SFA method originates from foreign countries, the study of total-factor energy efficiency by foreign scholars using the SFA method is relatively mature. The SFA method is still rarely used to study the total-factor energy efficiency in China, and only a few scholars have used this method to study the influence of energy input and pollution emission efficiency on economic growth. The SFA method requires a certain form of the production function to be preset, then the measurement method should be used to fit the sample data. But the form of the production function may be inconsistent with the reality, and the fitting may cause inconsistency. Therefore, the SFA method has not been widely used in the study of total-factor energy efficiency. At present, most of the research on total-factor energy efficiency uses the DEA method. The advantage of the DEA method is that there is no need to construct a production frontier function. The efficiency is evaluated by calculating the distance of the decision unit from the front curve. And it can be evaluated effectively under the condition of multiple inputs and multiple outputs. In addition, some scholars, under the framework of total-factor energy efficiency, rely on the DEA evaluation model to analyze the influencing factors of energy efficiency change in China.

Considering the combination of energy and other factors we can evaluate the energy efficiency more scientifically, and the domestic and foreign scholars have made some achievements in the research of energy efficiency under the total-factor frame. However, whether it is total-factor energy efficiency or single-factor energy efficiency, its numerical size can only reflect the change of energy efficiency in a certain period of time, and we can measure the differences in regional energy efficiency only by introducing the different research methods for energy efficiency.

7.1.2 Models for Measure Total-Factor Energy Efficiency with Nonparametric Approach

In this book, we focused on analyzing the total-factor energy efficacy of the Chinese industrial sector using industrial data. Since China is eager to change its way of economic development, it is important to recommend the relevant policies based on the results of the research.

7.1.2.1 Descriptive Regions in China

For geographical and political reasons, study area is divided into three regions: eastern, central, and western. For decades since the reform and opening-up in 1978, development priorities have been different among regions. As a result of the division and preference policy, the level of development in each region differs very much both economically and culturally. Because of the research purpose of this study herein, the regional analysis will be done, and the fact that readers may lack the knowledge of the regions, it is necessary to introduce the regional divide of China in the first place.

The eastern region, consists of 12 provincial units that stretch from Liaoning to Guangxi, including nine coastal provincial units (Shandong, Hebei, Jiangsu, Zhejiang, Fujian, Guangdong, Hainan, Liaoning and Guangxi) and three municipalities (Beijing, Tianjin and Shanghai). The region is famous for its GDP contribution and economic development level. The central region is constituted by nine inland provincial units: Heilongjiang, Jilin, Inner Mongolia, Henan, Shanxi, Anhui, Hubei, Hunan, and Jiangxi. This region is the base of agriculture and has a large population. The western region, covers almost half of China, but has the lowest population density and is least developed due to geographical reasons. Ten provincial units make up the region, and the members are Gansu, Guizhou, Ningxia, Qinghai, Shaanxi, Xizang, Yunnan, Xinjiang, Sichuan, and Chongqing. However, due to the unavailability of statistical data, Xizang will be excluded from this analysis.

7.1.2.2 Data Envelopment Analysis

Total-factor energy efficiency provides us an effective frame to evaluate energy efficiency. Under this frame, new concepts can be introduced as either input or output factors according to the research purpose. Li and Cheng (2008) evaluated the energy efficiency in multiple scenarios using non-parameter method while taking into account the undesirable environmental impact and non-energy influencing factors faced by China and concluded that the massive use of energy consumption economy has drawbacks in the form of wastage of a large amount of energy. All those concepts including TFEE were developed based on the Data Envelopment Analysis method.

Data envelopment analysis (DEA) was developed in 1978 by an American operation researcher named Charles (Charnes et al. 1978) and is a nonparametric statistical method based on the linear programming technique to assess the efficiency of decision-making units (DMU). In other words, if we consider a DMU is producing outputs with various inputs, DEA can be used to assess the efficiency that inputs produce the outputs. Therefore, it can be effectively used as a multiple input–output framework to directly compute the index of TFEE, which means that the DEA model is well matched with the concept of TFEE. Hu and Wang (2006), who firstly introduced the concept of total-factor energy efficiency, adopted DEA as a methodology for analysis.

Generally speaking, there are five kinds of DEA models that are widely used. The five basic models were named, respectively, as C2R (Charnes et al. 1978), BC2 (Banker et al. 1984), C2GS2 (Charnes et al. 1985), C2W, and C2WH (Charnes et al. 1978). BC2 and C2GS2 were used to verify technical efficiency, C2W extended the basic C2R model to the condition of infinite DMUs, and C2WH introduced the preferences of decision-makers. In this study, as we have planned to comprehensively evaluate the overall efficiency, technical efficiency, and scale efficiency in China, therefore C2R model is applied because it has the ability to analyze these efficiency. More technically, DEA aims to construct a nonparametric envelopment frontier composed of DMUs. The DMUs located on or above the frontier are efficient while those within the frontier are inefficient. C2R model enables the overall efficiency (OE) to be divided into pure technical efficiency (PTE) and scale efficiency (SE), which offers more adequate information to analyze the overall efficiency of inputs producing outputs (Hao and Zong 2007). The relationship among three could be stated as

$$\text{Overall efficiency} = \text{pure technical efficiency} \times \text{scale efficiency} \tag{7.1}$$

To clearly state the method of DEA, a graph is used to help to understand the decomposition of OE and measure the multipliers. As shown in Fig. 7.1, the round in black, represents any random convex set, though a little rough is data envelopment which holds all the samples and line ON is tangent to the round, which represents the constant returns to scale (CRS). For any point E in the data envelopment, under the calculation of C2R, three equations can be defined, which are, respectively, stated as

$$\text{Overall efficiency} = \|BC\|/\|BE\| \tag{7.2}$$

$$\text{Pure technical efficiency} = \|BD\|/\|BE\| \tag{7.3}$$

$$\text{Scale efficiency} = \|BC\|/\|BD\| \tag{7.4}$$

In equations (7.2)–(7.4), the operator $\|\cdot\|$ means the module of the hyper line segment in the space. As can be conducted from the equation, it is obvious that Eq. (7.1) is valid.

It can be concluded from Fig. 7.1 that for any point inside the data envelopment, its pure technical efficiency and scale efficiency is less than 1, and as a result of which, the overall efficiency is also less than 1. However, for any point on the envelopment, D, for example, according to the equation, the pure technical efficiency is 1, but the scale efficiency is less than 1, thus the overall efficiency is less than1. For those points doting on the surface of data envelopment, i.e., points whose pure technical efficiency is 1, they are usually called DEA-effective; otherwise, they are usually called non-DEA-effective. Consider an extreme condition, point A, where C, D, and E have been coincident, according to the equation of decomposition. OE, PTE, and SE at

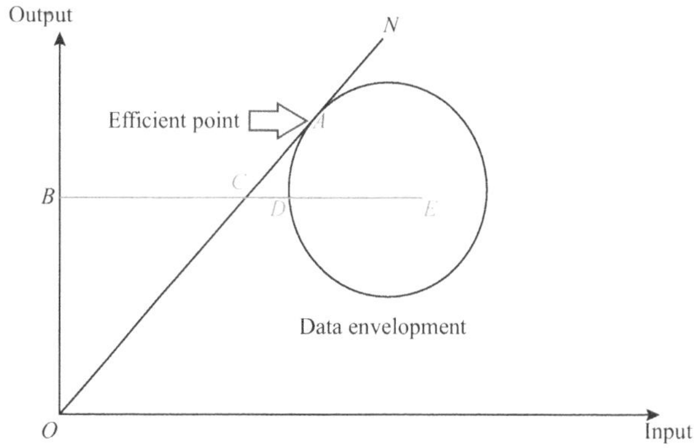

Fig. 7.1 Illustration of the decomposition and measurement of overall efficiency

point A are all equal to 1. Figure 7.1 is just an example of two-dimensional Cartesian coordinate, i.e., input is the only cause for output. If the sample data are collected, they will form a multidimensional sphere in the space. Thus, we can imagine a flat or hyperplane forming the constant return to scale tangent plane as line ON is in two-dimensional condition.

7.1.2.3 Data Selection and Description

Referring to the research of Wu Q and Wu C (2009) and Li and Zhou (2006), this chapter applies DEA to construct a set of energy-efficiency-accessing models: here taking the industrial sector of 30 provinces (autonomous regions, municipalities) (Xizang, Hong Kong, and Macao excluded) in China as DMUs to examine the relationship between energy input and industry output. Chang and Hu (2010) had researched the regional data from 2000 to 2004, and evaluated the energy productivity change in different provinces (autonomous regions, municipalities) of China with the total-factor framework. Similar but different from the former researchers, this model also includes the economic index as output. However, the index is no longer gross domestic product but the gross value of industrial product (GIP) to set the focus of research on the industrial departments of China. According to the real industrial production process, factors of energy, capital, and labor are deemed as input in the DEA model. That is given by the characteristics of the industrial sectors. With the differences in the research purpose, the indicator selection differs (Wu Q and Wu C 2009; Li and Cheng 2008). In this chapter, the indicators and data sources are as follows:

Output:

Gross industry production (GIP) cannot fully reflect the purpose of our research. Since large-scale industrial enterprises are the main consumer of energy and the main provider of provincial or municipal GIP, the gross product value of industrial enterprises above designated size is adopted as the measure of output. The raw data are taken from the *China Statistical Yearbooks (2006–2010).*

Input:

Capital input. Apply the annual average balance of fixed capital and working capital of industrial enterprises above the designated size to measure the capital input.

Labor input. Apply the annual average number of employees of industrial enterprises above the designated size to measure the labor input.

Energy input. Apply the energy consumption of industrial enterprises above the designated size to measure the energy input. Energy consumption calculated in this chapter is the product of Industrial Added Value and energy consumption per unit of Industrial Added Value.

The raw data of the inputs above are also from the *China Statistical Yearbooks (2006–2010).*

7.1.3 Empirical Analysis of Total-Factor Energy Efficiency

7.1.3.1 Simple Statistical Review of Energy Intensity During 2005–2009

Before starting the empirical results of total-factor energy efficiency, we present some simple statistical review of energy intensity in China during the year 2005–2009, here we firstly calculated using the same source of data, to build an overlook of Chinese status in quo. Energy Intensity is calculated using Eq. (7.5), which is a universally accepted method.

$$EI = \frac{\sum_{i=1}^{n} Ei}{GIP} = \sum_{i=1}^{n} \frac{E_i}{GIP_i} \frac{GIP_i}{GIP} = \sum_{i=1}^{n} EI_i \cdot S_i \tag{7.5}$$

In Eq. (7.5), i and n represent provincial units observed and total number of provincial units, EI, E_i, GIP_i, EI_i, and $S_{i,}$ respectively, represent *Energy Intensity*, energy use in the ith provincial unit, the GIP in the ith provincial unit, *Energy Intensity* in the ith provincial unit, and economic output share of the ith provincial unit in the whole nation. The results are listed in Table 7.1.

In Table 7.1, all provincial data are original Energy Intensity without weighing, but the data of eastern, central, western, and total accounts are weighted according to their economic contribution. From the data, we can see the basic paradox of both energy intensity and single-factor energy efficiency indicators: because the gaps among Chinese provincial develop, economic weighing will cause statistical twist. This conclusion can be obviously proved: according to Table 7.1, although the individual energy intensity of each provincial unit in the western area is relatively high, the overall energy intensity level of the western area is the lowest after economic weighing for the purpose of relative energy intensity comparison.

However, in Table 7.1, we can see that in China the energy intensity was decreasing provincially, regionally, and totally during 2005–2009, only with different extents. But we lack the details that how energy use is improved in the Chinese industrial sectors, or whether there is still room for further improvement, etc. Total-factor energy efficiency evaluation based on the DEA method helps to answer these questions.

7.1.3.2 Overall Efficiency Analysis

Table 7.2 lists the result of overall efficiency, pure technical efficiency, and scale efficiency of industrial sectors of 30 provinces (autonomous regions, municipalities) in China.

Based on the overall efficiency results, the industrial department energy efficiency of five provinces of Tianjin, Shanghai, Jiangsu, Shandong, and Guangdong were DEA-effective from 2005 to 2009, which constitutes the frontier of energy efficiency in the interprovincial industrial department. The overall energy efficiency of the

Table 7.1 Energy intensity of industrial sectors in 30 provinces (autonomous regions, municipalities) (2005–2009)

Regions	2005	2006	2007	2008	2009
Beijing	0.360	0.298	0.266	0.203	0.188
Tianjin	0.398	0.381	0.358	0.298	0.261
Hebei	1.264	1.205	1.094	0.880	0.784
Shanxi	2.367	2.144	1.956	1.710	1.662
Inner Mongolia	2.344	2.306	2.127	1.654	1.463
Liaoning	0.900	0.854	0.783	0.647	0.618
Jilin	0.976	0.892	0.760	0.586	0.473
Heilongjiang	1.125	1.051	0.973	0.856	0.550
Shanghai	0.290	0.312	0.248	0.215	0.205
Jiangsu	0.414	0.391	0.341	0.275	0.254
Zhejiang	0.317	0.294	0.273	0.234	0.225
Anhui	1.056	0.912	0.849	0.731	0.629
Fujian	0.392	0.390	0.379	0.321	0.315
Jiangxi	0.858	0.825	0.677	0.531	0.447
Shandong	0.673	0.599	0.560	0.451	0.408
Henan	1.093	1.253	1.244	0.864	0.818
Hubei	1.116	1.069	1.026	0.765	0.716
Hunan	1.004	0.934	0.846	0.613	0.494
Guangdong	0.269	0.274	0.250	0.203	0.191
Guangxi	0.991	0.922	0.865	0.760	0.736
Hainan	1.151	0.936	0.757	0.704	0.685
Chongqing	0.720	0.766	0.765	0.745	0.799
Sichuan	1.227	0.990	0.953	0.829	0.745
Guizhou	1.939	1.884	1.728	1.461	1.475
Yunnan	1.498	1.274	1.150	0.998	1.004
Shaanxi	1.038	1.014	0.950	0.802	0.531
Gansu	1.275	1.308	1.225	1.254	1.064
Qinghai	1.331	1.475	1.445	0.822	1.196
Ningxia	2.933	2.671	2.826	2.531	2.330
Xinjiang	1.349	1.258	1.177	1.211	1.031
Eastern area	0.355	0.340	0.306	0.252	0.233
Central area	0.245	0.228	0.227	0.188	0.170
Western area	0.123	0.104	0.100	0.091	0.084
Total	0.723	0.672	0.633	0.531	0.487

Beijing industrial department increased to 1 and was in the frontier from 2006. So far, the provinces (autonomous regions, municipalities) whose the overall efficiency was DEA effective comprised 20% of all provinces (autonomous regions, municipalities), which are all in the eastern region. According to Table 7.2, only Liaoning, Zhejiang, Fujian, and Guangxi provinces, 4 out of 12, do not meet the overall efficiency above 0.9. In other words, the overall efficiency of 66.7% provinces

Table 7.2 The total-factor energy efficiency of industrial sectors in 30 provinces (2005–2009)

Regions	2005			2006			2007			2008			2009		
	OE	PTE	SE	OE	PTE	SE	OE	PTE	SE	OE	PTE	SE	OE	PTE	SE
Beijing	0.978	0.999	0.979	1.000	1.000	1.000	1.000	1.000	1.000	1.000	1.000	1.000	1.000	1.000	1.000
Tianjin	1.000	1.000	1.000	1.000	1.000	1.000	1.000	1.000	1.000	1.000	1.000	1.000	1.000	1.000	1.000
Hebei	0.871	0.884	0.984	0.853	0.868	0.983	0.878	0.883	0.995	0.843	0.844	0.999	0.902	0.906	0.995
Shanxi	0.531	0.565	0.939	0.502	0.534	0.940	0.536	0.547	0.981	0.422	0.423	0.996	0.503	0.503	0.999
Inner Mongolia	0.666	0.708	0.941	0.643	0.685	0.938	0.747	0.749	0.998	0.848	0.850	0.997	1.000	1.000	1.000
Liaoning	0.775	0.776	0.999	0.765	0.775	0.987	0.792	0.800	0.990	0.855	0.865	0.985	0.869	0.893	0.973
Jilin	0.747	0.787	0.949	0.731	0.779	0.939	0.837	0.856	0.978	0.761	0.773	0.986	0.888	0.897	0.990
Heilongjiang	0.718	0.749	0.958	0.668	0.708	0.944	0.625	0.642	0.974	0.536	0.548	0.979	0.624	0.639	0.976
Shanghai	1.000	1.000	1.000	1.000	1.000	1.000	1.000	1.000	1.000	1.000	1.000	1.000	1.000	1.000	1.000
Jiangsu	1.000	1.000	1.000	1.000	1.000	1.000	1.000	1.000	1.000	1.000	1.000	1.000	1.000	1.000	1.000
Zhejiang	0.942	0.945	0.997	0.960	0.960	1.000	0.928	0.929	0.999	1.000	1.000	1.000	0.883	0.906	0.974
Anhui	0.695	0.755	0.921	0.708	0.772	0.917	0.719	0.750	0.959	0.697	0.706	0.987	0.735	0.740	0.993
Fujian	0.890	0.938	0.949	0.908	0.956	0.949	0.886	0.920	0.963	0.944	0.977	0.966	0.881	0.913	0.965
Jiangxi	0.708	0.837	0.846	0.791	0.921	0.859	0.863	0.948	0.910	1.000	1.000	1.000	0.904	0.975	0.927
Shandong	1.000	1.000	1.000	1.000	1.000	1.000	1.000	1.000	1.000	1.000	1.000	1.000	1.000	1.000	1.000
Henan	0.809	0.840	0.963	0.849	0.879	0.965	0.971	0.991	0.980	0.886	0.893	0.992	0.911	0.928	0.982
Hubei	0.631	0.642	0.982	0.618	0.632	0.977	0.648	0.653	0.992	0.640	0.651	0.984	0.664	0.674	0.985
Hunan	0.758	0.840	0.902	0.781	0.865	0.903	0.847	0.905	0.936	0.899	0.915	0.982	0.859	0.902	0.952
Guangdong	1.000	1.000	1.000	1.000	1.000	1.000	1.000	1.000	1.000	1.000	1.000	1.000	1.000	1.000	1.000
Guangxi	0.650	0.761	0.854	0.698	0.812	0.860	0.705	0.742	0.949	0.755	0.788	0.958	0.704	0.715	0.985
Hainan	0.644	1.000	0.644	0.716	1.000	0.716	0.975	1.000	0.975	0.880	1.000	0.880	0.918	1.000	0.918
Chongqing	0.657	0.779	0.844	0.673	0.800	0.841	0.680	0.765	0.889	0.638	0.670	0.953	0.680	0.742	0.916

(continued)

Table 7.2 (continued)

Regions	2005			2006			2007			2008			2009		
	OE	PTE	SE	OE	PTE	SE	OE	PTE	SE	OE	PTE	SE	OE	PTE	SE
Sichuan	0.630	0.653	0.965	0.639	0.665	0.960	0.679	0.689	0.986	0.663	0.665	0.996	0.720	0.722	0.998
Guizhou	0.534	0.636	0.839	0.541	0.676	0.802	0.549	0.616	0.891	0.456	0.471	0.970	0.539	0.552	0.978
Yunnan	0.679	0.728	0.932	0.647	0.701	0.923	0.658	0.681	0.965	0.615	0.619	0.993	0.643	0.646	0.996
Shaanxi	0.579	0.617	0.939	0.575	0.615	0.934	0.651	0.670	0.972	0.511	0.512	0.999	0.656	0.657	0.999
Gansu	0.631	0.731	0.863	0.609	0.712	0.855	0.668	0.709	0.943	0.525	0.542	0.968	0.615	0.628	0.979
Qinghai	0.571	0.878	0.651	0.597	0.834	0.716	0.634	0.792	0.800	0.519	0.584	0.889	0.631	0.666	0.948
Ningxia	0.546	0.887	0.616	0.542	0.886	0.612	0.562	0.772	0.728	0.464	0.511	0.907	0.557	0.578	0.964
Xinjiang	0.743	0.828	0.898	0.764	0.817	0.934	0.735	0.739	0.994	0.604	0.601	0.990	0.715	0.722	0.990

(autonomous regions, municipalities) in the eastern region was above 0.9, while the overall efficiency of three provinces (autonomous regions, municipalities) out of 9 in the central region and none of the provinces (autonomous regions, municipalities) in the western region reached 0.9. None of the western provinces (autonomous regions, municipalities) can even reach the overall efficiency of 0.8. In all 30 provinces (autonomous regions, municipalities), the overall efficiency of Shanxi was the lowest, with a score of only 0.503. From the above analysis, we can also see that the industrial departments with high energy efficiency were mainly located in the eastern region, while the relatively low ones were located in the central and western regions.

By further analysis of the non-DEA-effective provinces (autonomous regions, municipalities) from the aspects of technology and scale of comprehensive efficiency, we found that the industrial sector in Hainan Province was technically effective but not scale effective (Table 7.2), which indicates that the role of pure technical progress in its output has full utilities, but the scale return has not been formed. This means, in accordance with current output calculation, that it is impossible to decrease its investment. In addition, the pure technical efficiency of Hebei, Zhejiang, Fujian, Jiangxi, Henan, and Hunan industrial sectors reached above 0.9, and those of Liaoning and Jilin were between 0.8 and 0.9. Therefore, the non-DEA-effective provinces (autonomous regions, municipalities), whose pure technical efficiency reached more than 0.8 which accounted for about 30%. Relatively speaking, scale efficiency was considerably higher, with none of the province's industrial departments being less than 0.8. Thus we can see that the industrial departments in China generally rely on the mass resource input of scale production, while technical aspects are relatively underdeveloped.

7.1.3.3 Efficiency Developing Trend Analysis

Generally speaking, the overall efficiency of eastern and central regions was rising while the western region encountered the downfall even since at its beginning. Shanxi is the lowest overall efficiency province in the central region, and even lowest in whole of China at some time points. The reason is that Shanxi is abundant with energy resources, and thus the province did not concern too much about efficiency, and even the processing of other fuels like coke was inefficient. However, no definite trends could be easily identified during western region development. The reason for the phenomenon is that the overall investment environment of the western region is not as ideal as central and eastern regions. Although the west-construction plan of China has been called up by the government for years, the scale of applicable investment and the conversion lag of investment to production are preventing the obvious promotion of overall efficiency.

Pure technical efficiency in the eastern region is relatively stable, and that of the central region is gradually going up while with a similar trend of stableness. However, no optimistic signs can be found in the western region. The pure technical efficiency plummeted from 2007 to 2008 and bottomed in 2008. The rebound was found in 2009, but the height did not exceed efficacy in 2007, and the overall trend is falling. It can be referred that at least until 2009, technological innovation had been ignored by the provinces (autonomous regions, municipalities). Considering the development environment of the western region, it is not hard to understand the trend of the polyline. The western region is still relying on the pattern of investment-driven economic development; technical promotion is not a priority due to the time lag of converting technology into production.

The scale efficiency of all three regions in China was going up during 2005–2009. The western region has rather a high accelerating rate, especially the provinces (autonomous regions, municipalities) of Ningxia and Qinghai which can be referred from Table 7.2. It is easy to conclude that the stableness of overall efficiency in the western region was mainly contributed by massive investment, but not technological promotion, which gave support to our previous discussion. Due to the feature of the extensiveness of the western development map, scale efficiency will be much easier to take effect on overall efficiency than pure technical efficiency caused by technical promotion. It is dangerous if this kind of large-scale-investment-driven economic development continues because it can cause giant energy waste.

7.1.3.4 Scale Benefit Analysis

The industrial sector of Beijing, Tianjin, Shanghai, Jiangsu, Shandong, and Guangdong are at the stage of unchanged scale returns, and the comprehensive efficiency of these six provinces (autonomous regions, municipalities) achieves DEA-effectiveness. This means that the input and output increased in the same proportion. In addition, the industrial sectors of Inner Mongolia achieved unchanged scale returns in 2009. Only the industrial departments of Hebei, Liaoning, Zhejiang, and Hubei, showed decreasing returns to scale. Specifically, the scale returns of Liaoning diminished during 2006–2009, and that of Hubei diminished during 2008–2009. The conditions of decreasing returns to scale indicate that if we increase the input, the output growth rate could be less than the increase of the input ratio. In other words, the output efficiency is quite low. The rest of the industrial sectors in other provinces (autonomous regions, municipalities) are at the stage of decreasing returns to the scale, which means that if the industrial department in these regions increased the quantity of input in the same proportion, a larger proportion of return would be reached. Hence, the analysis also shows that the industrial sectors of most regions in China still fail to achieve economies of scale. Increased investment would bring about larger scale output.

7.1.3.5 Energy Redundancy Analysis

According to the theory of operational research, the reason for DEA-inefficiency is input redundancy and output insufficiency. According to the framework of TFEE and the feature of the DEA method, it is easy to calculate the redundancy of the input factors and the redundancy rate as well. We arranged the data of provincial units whose industrial departments are DEA technologically inefficient. The data shown in Table 7.3 include both the perspectives of input redundancy and output

Table 7.3 The analysis of investment redundancy and output insufficiency for industrial sectors of 21 provinces (autonomous regions, municipalities) in 2009

Regions	Regional belonging (West, Central, East)	Energy investment redundancy/ 10,000 yuan	The rate of energy investment redundancy	Capital investment redundancy/ billion yuan	The rate of capital investment redundancy	Labor force investment redundancy/ 10,000 yuan	The rate of labor force investment redundancy	Output insufficiency of the gross value of industrial output/ billion yuan
Hebei	East	8569.592	0.454	0.000	0.000	0.000	0.000	0.000
Shanxi	Central	4559.705	0.297	0.000	0.000	0.000	0.000	0.000
Liaoning	East	5922.364	0.340	0.000	0.000	0.000	0.000	0.000
Jilin	Central	302.806	0.064	0.000	0.000	0.000	0.000	0.000
Zhejiang	East	0.000	0.000	986.797	0.025	0.000	0.000	0.000
Anhui	Central	590.441	0.071	0.000	0.000	1.929	0.008	0.000
Fujian	East	0.000	0.000	0.000	0.000	48.668	0.128	0.000
Jiangxi	Central	19.358	0.004	0.000	0.000	45.194	0.258	0.000
Henan	Central	9591.969	0.423	0.000	0.000	58.344	0.130	0.000
Hubei	Central	2172.397	0.195	0.000	0.000	0.000	0.000	0.000
Hunan	Central	299.282	0.045	0.000	0.000	44.433	0.184	0.000
Guangxi	East	627.194	0.124	0.000	0.000	2.023	0.016	0.000
Chongqing	West	1113.973	0.206	0.000	0.000	18.787	0.137	0.000
Sichuan	West	2694.153	0.200	0.000	0.000	0.000	0.000	0.000
Guizhou	West	1210.473	0.239	0.000	0.000	0.000	0.000	0.000
Yunnan	West	450.798	0.086	0.000	0.000	0.000	0.000	0.000
Shaanxi	West	442.594	0.098	0.000	0.000	0.000	0.000	0.000
Gansu	West	963.602	0.240	0.000	0.000	0.000	0.000	0.000
Qinghai	West	128.359	0.099	419.790	0.166	0.000	0.000	0.000
Ningxia	West	1229.019	0.361	0.000	0.000	0.000	0.000	0.000
Xinjiang	West	0.000	0.000	454.265	0.071	0.000	0.000	0.000

insufficiency. From Table 7.3, actually, not only energy efficiency but also capital and labor efficiency can be calculated by 1 minus the redundancy rate.

In this chapter, in order to set the alarm bells ringing the discussion of energy efficiency will be substituted with energy redundancy. Besides, Beijing, Tianjin, Inner Mongolia, Heilongjiang, Shanghai, Jiangsu, Shandong, Guangdong, and Hainan, nine in total, are not included in Table 7.3 because there are no signs of redundancy. In other words, among the nine provincial units, not only energy redundancy, but all input redundancy reached a minimum level in 2009.

Moreover, it is not hard to see the advantages of the total-factor energy efficiency indicator compared with those of single-factor energy efficiency and energy intensity. For example, in Table 7.4, we can see that in the year 2009, Hebei has a higher rate of energy redundancy than Shanxi, which means Hebei is less energy efficient than Shanxi. However, Hebei is much less energy intense than Shanxi (Table 7.1), which means Hebei is much more energy efficient than Shanxi. But when considering the realities of two provincial units, as well as their technical efficiency, the results under the framework of total-factor energy efficiency are more convincing, which will be discussed later.

There are different degrees of input redundancy in the DEA technologically efficient provincial units of industrial sectors. It means that there is a waste of resources, particularly energy input redundancy. The highest is Hebei, where the energy input redundancy rate reaches as high as 0.454, which means that the energy input of the industrial sector can also be reduced by 45.4% on the current basis. It is not hard to understand the result with the help of knowing the background. Hebei is famous for its steel industry, and as is well known that, the steel industry is highly energy demanding. Due to the inefficient technology investment, a large amount of energy redundancy can easily occur. Henan, Ningxia, and Liaoning are in second place, and their energy input redundancy rates are, respectively, 0.423, 0.361, and 0.340. In addition, Shanxi reaches more than 0.25. Guizhou and Gansu are only a little bit less than 0.25. Central provinces (autonomous regions, municipalities) as Sichuan and Chongqing are also as high as approximately 0.20.

Comparatively speaking, the status of labor input redundancy and capital investment redundancy are better. In respect of labor inputs, the redundancy rate of Jiangxi Province reaches the highest of 0.258, which means its industrial department labor inputs can reduce up to 25.8%. The redundancy rates of Hunan, Fujian, Chongqing, Henan are, respectively, 0.184, 0.128, 0.137, and 0.130. Anhui and Guangxi are better, with a score of 0.008 and 0.016. The condition of capital investment redundancy is better, as redundancy is only found in Qinghai, Xinjiang, and Zhejiang, appearing redundancy, with rates of 0.166, 0.071, and 0.025, respectively. This also explains from another perspective that there is a serious waste of resources in industrial sectors in China. Labor input also needs to be optimized, but the capital investment is broadly appropriate. In addition, the rightmost column of Table 7.3 also shows that the output of the industrial departments of all provinces (autonomous regions, municipalities) can achieve the basic goals with current inputs. It means, if current input remains, the output cannot be better.

Table 7.4 Regions and provincial units in China

Regions	Provincial units
Eastern region (12 provincial units included)	Beijing, Tianjin, Hebei, Liaoning, Shanghai, Jiangsu, Zhejiang, Fujian, Shandong, Guangdong, Guangxi, Hainan
Central region (9 provincial units included)	Inner Mongolia, Jilin, Heilongjiang, Anhui, Jiangxi, Henan, Hubei, Hunan, Shanxi
Western region (9 provincial units included)	Chongqing, Sichuan, Guizhou, Yunnan, Shaanxi, Gansu, Qinghai, Ningxia, Xinjiang

7.2 Analysis of China's Energy Efficiency from Both Static and Dynamic Perspectives

In the past decades, China's booming economy has driven a rapid increase in energy consumption. According to the statistics, China's GDP accounted for 8.6% of global economic output in 2010, while its energy consumption accounted for 19.3% of the globe. China's energy consumption per unit of GDP was more than twice the world in 2010. Obviously, China's energy efficiency is relatively low, and its current economic growth model with extensive and inefficient energy consumption is unsustainable.

Herein, through calculation and analysis of China's historical performance regarding energy efficiency and revealing the significant difference in energy efficiency between regions and provinces (autonomous regions, municipalities), we try to deepen the understanding of international readers' about the evolution of China's energy utilization at national, regional, and provincial levels. In addition, through analyzing the key factors responsible for the change in energy efficiency and keeping in view the technical progress, productive scale, and management level, we opt to put forward corresponding suggestions for the Chinese government to further improve the energy efficiency from the production technology, management, and scale aspects. We believe that the results and conclusions of this chapter will certainly provide significant references for the Chinese government to make energy policy in the future.

7.2.1 Construction of Global Production Frontier with Undesirable Outputs

DEA is proposed by Charnes et al. (1978), and it is a mathematical procedure using linear programming techniques to assess the efficiency of DMUs. In this chapter, it is used for measuring the energy utilization efficiency in China. According to the main purpose of this book, the calculation was performed based on the global DEA method.

In order to present the changing trend of energy efficiency in time series and make the dynamic model and static model consistent with each other, the global DEA method is utilized. Hence, two definitions of benchmark technology are essential: (i) contemporaneous benchmark and (ii) global benchmark technologies (Oh and Heshmati 2010). Here, x represents the amount of input utilized by DMU, y is the economic output produced by DMU, and b is the undesirable output of DMU. $P(x)$ represents the PPS (production possibility set).

A contemporaneous benchmark technology is defined as

$$Pt(xt) = \{(yt, bt)\text{Ixt can produce}(yt, bt)\} \tag{7.6}$$

where $t = 1, \ldots, T$. The contemporaneous benchmark technology constructs a reference production set, at time t. This set is made from the observations made at that time only (Tulkens and Eeckaut 1995). A global benchmark technology is defined as

$$p^G = p^1 \cup p^2 \cup \cdots \cup p^T \tag{7.7}$$

This global benchmark technology envelopes all contemporaneous benchmark technologies by establishing a single reference PPS from the panel data on inputs and outputs of relevant DMUs (Pastor and Lovell 2005).

The above two definitions of the PPSs are depicted in Fig. 7.2. Taking DMU A, for example, its global efficiency in any period can be stated as

$$\text{Global efficiency} = \|\mathrm{A}'\mathrm{A}''\| / \|\mathrm{AA}''\| \tag{7.8}$$

In our analysis, the efficiency of all DMUs in any period are evaluated using the global reference PG. Hence, all DMUs during the sample period have the same

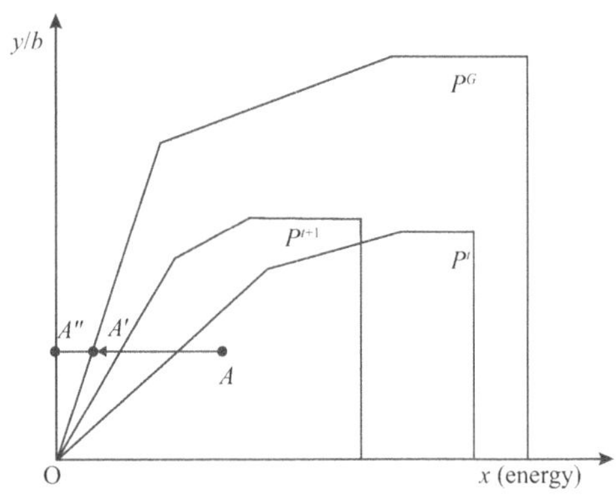

Fig. 7.2 Concept of the global DEA method

reference set. In this way, estimation results obtained by utilizing the global DEA method are more steady compared with those obtained by utilizing a single-phase method.

7.2.2 Non-radial Directional Distance Function and Global Malmquist Productivity Index

In order to present the changing trend of energy efficiency in time series and make the dynamic model and static model consistent with each other, the global DEA method is utilized in this chapter. Combining the global DEA method and energy efficiency evaluation model proposed by Wang et al. (2012), energy efficiency with the assumption of constant returns to scale could be measured as follows:

$$\begin{aligned}
& D^G(x_{i0}^T, y_{i0}^T, b_{i0}^T|CRS) = \min\theta \\
& st. \sum_{t=1}^{T}\sum_{i=1}^{N}\lambda_i^t e_i^t \le \theta e_{i0}^t \\
& \sum_{t=1}^{T}\sum_{i=1}^{N}\lambda_i^t k_i^t \le k_{i0}^t \\
& \sum_{t=1}^{T}\sum_{i=1}^{N}\lambda_i^t k y_i^t \le y_{i0}^t \\
& \sum_{t=1}^{T}\sum_{i=1}^{N}\lambda_i^t k b_i^t \le b_{i0}^t \\
& \lambda_i^t \ge 0, i = 1, 2, \ldots, N, t = 1, \ldots,
\end{aligned} \tag{7.9}$$

where θ represents energy efficiency of $\mathrm{DMU}_{i0}.e_i^t$, k_i^t and l_i^t, respectively, present the energy input, capital stock, and labor input of DMU_i at time t. And y_i^t and b_i^t are the economic output and the undesirable output produced by DMU_i at time t, respectively. λ_i^t is the intensity variable associated with DMU_i at time t for connecting the inputs and outputs by convex combination.

In this case, as shown in Fig. 7.2, DMUs' reference during all periods is PG.

As shown in Fig. 7.3, according to Hu and Wang the radial adjustment and slack of DMU, A in an input-oriented CRS DEA model are ‖BC‖ and ‖CE‖, respectively. Primary energy in China is fossil energy, which is nonrenewable. Hence, the Chinese government has a strong incentive to pursue an extreme reduction in energy consumption even with the sacrifice of other non-energy resources such as labor and capital. Therefore, point $\mathrm{A^e}$ is the projected point on the frontier for DMU, A taking as the target to reach by reducing the radial adjustment ‖BF‖. Compared with the input-oriented CRS DEA model, there is an additional adjustment (‖EF‖) of energy input and sacrifice of radial adjustment (‖GH‖) of other

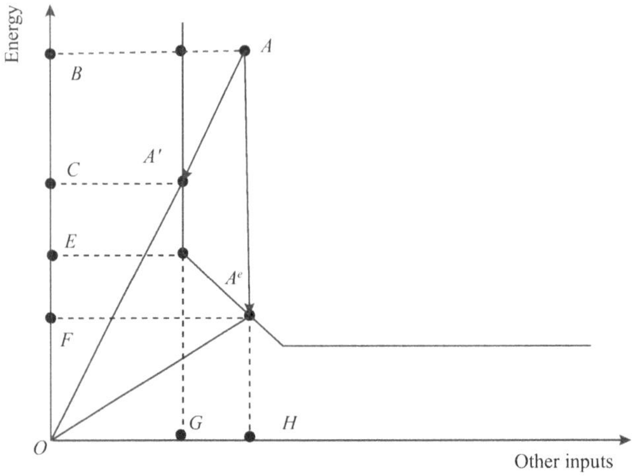

Fig. 7.3 Radial adjustment and slack identified in the energy efficiency evaluation model

non-energy inputs in the static model. Obviously, there would no slack of energy input existing in the energy efficiency evaluation model.

For that reason, the energy efficiency evaluation in this study is performed based on China's provincial panel data. The provincial energy efficiency can be defined as energy efficiency = *a*. The energy efficiency of China's three major regions (i.e., eastern region, central region, and western region) could be estimated according to the following equation:

$$\text{Energy efficiency } a = \sum_{i \in a} e_i \cdot \theta i / \sum_{i \in a} a_i \tag{7.10}$$

where *a* represents region, and *i* represents a provincial unit in region *a*.

Färe et al. (1994) first proposed the DEA-Malmquist index to evaluate the productivity growth and technical progress in industrialized countries. Its core idea is to convert the conventional Malmquist index based on distance function to a new DEA-Malmquist index based on efficiency. Following this proposal, a lot of productivity studies have been done based on the DEA-Malmquist or developed DEA-Malmquist indices. However, as we have discussed in Sect. 7.2.1, most of the existing research utilizes the single-phase benchmark technologies to construct single-phase reference production sets, which are then used for calculating the productivity growth and technical progress. As a result, estimation results from these studies lack stability and may deviate from the practical production activity. Hence, in order to overcome this defect, the global DEA method is utilized here.

Combining with model (4), provincial global Malmquist (GM) index proposed in this chapter could be defined as follows:

$$\mathrm{GM}^{t,t+1}(\mathrm{x}^t, y^t, b^t, \mathrm{x}^{t+1}, y^{t+1}, b^{t+1}|\mathrm{CRS}) = \left[\frac{D^G(\mathrm{x}^{t+1}, y^{t+1}, b^{t+1}|\mathrm{CRS})}{D^G(\mathrm{x}^t, y^t, b^t|\mathrm{CRS})}\right] \quad (7.11)$$

The solution of directional distance function DG (x^t, y^t, bt|CRS), is given by solving model (4). If the GM index is greater than, equal to, or less than 1, it means energy efficiency is increasing, stable, or declining, respectively between periods t and t + 1. Under the assumption of constant returns to scale, the GM index can be further decomposed into three components as follows:

$$\begin{aligned}
&\mathrm{GM}^{t,t+1}(x^t, y^t, b^t, x^{t+1}, y^{t+1}, b^{t+1}|\mathrm{CRS}) \\
&= \left[\frac{\frac{D^G(x^{t+1},y^{t+1},b^{t+1}|\mathrm{CRS})}{D^{t+1}(x^{t+1},y^{t+1},b^{t+1}|\mathrm{CRS})}}{\frac{D^G(x^t,y^t,b^t|\mathrm{CRS})}{D^t(x^t,y^t,b^t|\mathrm{CRS})}}\right] \times \left[\frac{\frac{D^{t+1}(x^{t+1},y^{t+1},b^{t+1}|\mathrm{CRS})}{D^{t+1}(x^{t+1},y^{t+1},b^{t+1}|\mathrm{VRS})}}{\frac{D^t(x^t,y^t,b^t|\mathrm{CRS})}{D^t(x^t,y^t,b^t|\mathrm{VRS})}}\right] \\
&\times \left[\frac{D^{t+1}(x^{t+1}, y^{t+1}, b^{t+1}|\mathrm{VRS})}{D^t(x^t, y^t, b^t|\mathrm{VRS})}\right] = \mathrm{GTCH}^{t,t+1} \times \mathrm{GSCH}^{t,t+1} \times \mathrm{GPCH}^{t,t+1}
\end{aligned} \quad (7.12)$$

where the directional distance function $D^G(x^{t+1}, y^{t+1}, b^{t+1}|\mathrm{CRS})$ is defined on global benchmark technology set P^G under the assumption of CRS (constant returns to scale). $D^t(x^t, y^t, b^t|\mathrm{CRS})$ is defined based on the contemporaneous benchmark technology set P^t under the assumption of CRS (constant returns to scale).

$D^t(x^t, y^t, b^t|\mathrm{VRS})$ is defined based on the contemporaneous benchmark technology set Pt under the assumption of VRS (variable returns to scale). Provincial indices of GM, technical change (GTCH), pure efficiency change (GPCH), and scale efficiency change (GSCH) could be obtained by solving model (7.12). The GTCH reflects the shift in the best practice frontier from t to t + 1, and the product of GPCH and GSCH measures the change in relative efficiency over time. If the GTCH is greater than, equal to, or less than 1, it means the relative distance from technical efficiency to the best practice frontier is increasing, unchanging, closing, respectively. If the GSCH is greater than, equal to, or less than 1, it indicates scale efficiency is increasing, stable, or declining, respectively, between periods t and t + 1. If the GPCH is greater than, equal to, or less than 1, it means pure efficiency is changing, stable or declining respectively, between periods t and t + 1.

Here, a method for calculating the GM indices of China's three major regions (i.e., eastern region, central region, and western region) is given as follows:

Firstly, directional distance functions of three major regions are obtained by solving the following equation:

$$D_a^G(e^t, k^t, l^t, GDP^t, p^t; g|CRS) = \frac{\sum_{i \in a} D_i^G(e^t, k^t, l^t, GDP^t, p^t; g|CRS) \cdot e_i}{\sum_{i \in a} e_i} \tag{7.13}$$

$$D_a^t(e^t, k^t, l^t, GDP^t, p^t; g|CRS) = \frac{\sum_{i \in a} D_i^t(e^t, k^t, l^t, GDP^t, p^t; g|CRS) \cdot e_i}{\sum_{i \in a} e_i} \tag{7.14}$$

$$D_a^t(e^t, k^t, l^t, GDP^t, p^t; g|VRS) = \frac{\sum_{i \in a} D_i^t(e^t, k^t, l^t, GDP^t, p^t; g|VRS) \cdot e_i}{\sum_{i \in a} e_i} \tag{7.15}$$

where a represents region and i represents a provincial unit in region a.

Then, all directional distance functions obtained from the first step are taken into model (7.12). Then GM indices of three main regions could be evaluated by solving model (7.12).

As we have discussed above, utilizing the single-phase reference production sets to evaluate the energy efficiency and index will lead to the result that the estimations from the dynamic model and the static model are not consistent with each other. While with the application of global DEA, the estimations from the dynamic model and the static model are consistent with each other. A brief proof is given as follows: It is shown clearly in model (7.9) and model (7.11) that the changes in energy efficiency between periods t and $t + 1$ exactly equal to GM, the value of in Eq. (7.11). That is to say, based on the global production sets, the states of changing features in both of the dynamic and static models are completely consistent with each other. However, if the evaluation is based on the single-phase reference production sets, then it is obvious that estimations from the dynamic model and the static model will not be consistent with each other. It is not hard to find that the change in energy efficiency between periods t and $t + 1$ is not equal to Mt, $t + 1$. That is to say, based on the single-phase reference production sets, the states of changing features of energy efficiency in both the dynamic and static models are not consistent with each other.

7.2.3 Empirical Results of Static and Dynamic Total-Factor Energy Efficiency in China

For geographical and political reasons, the study area is divided into three regions: the eastern region, the central region and the western region (Table 7.4). Xizang is excluded due to lack of data. Thus, the sample of this study consists of 30 provincial units. The sample period is from 2001 to 2010. In this section, the regional differences and changing trends of energy efficiency are shown in our analysis. Then, an empirical analysis based on the "static-dynamic" perspective is employed to identify the key factors influencing China's energy efficiency. All models are run by MATLAB 7.1.

7.2.3.1 Regional Data and Descriptive Statistics

According to the practical production process, capital stock, energy consumption, and labor are chosen as inputs. The GDP (gross domestic product) is used as a desirable output, and three main environmental pollutants are chosen as undesirable outputs. Data on the labor employment, GDP, and three main environmental pollutants are all collected from *China Statistical Yearbooks (2002–2011)*. GDP data is deflated by GDP deflator and converted into a 2000 constant price. The data of primary energy consumption are collected from *China Energy Statistical Yearbooks (2002–2011)*, and units of all sorts of primary energy are converted into ten thousand tons of standard coal.

The data of capital stock are calculated according to the estimation approach suggested by Zhang et al. (2004). The equation used for capital stock in this work is: $K_j^t = I_j^t + (1-\delta)K_j^{t-1}$ Where K_j^t and K_j^{t-1} represent the capital stock of the *j*th provincial unit at time *t* and *t*–1, respectively. I_j^t is the total volume of investment in fixed assets at time *t*, and δ is the depreciation rate. The capital data is deflated by the price index of investment in fixed assets and converted into a 2000 constant price. The data of investment in fixed assets and the price index of investment is collected from *China Statistical Yearbooks (2002–2011)*.

Here, we have three environmental indicators which are wastewater, waste gas, and solid waste, respectively. Obviously, it is not reasonable for us to take these three environmental indicators all together into our DEA-based model. Hence, in order to obtain a comprehensive environmental evaluation indicator, the entropy method is utilized here. According to the improved entropy method adopted by Wang et al. (2013), all the environmental pollutant indicators are integrated into a comprehensive index *h*. By utilizing the multiple transformation method, the reciprocal of *h* can be used as a positive environmental indicator *b*.

Supposing that there are *m* units to be evaluated and each unit has *n* evaluation indicators, the specific algorithm steps of the entropy method can be organized as follows:

Firstly, the data are standardized as follows:

$$x_{ij}' = \frac{x_{ij} - \min(x_j)}{\max(x_j) - \min(x_j)} \tag{7.16}$$

where x_{ij} represents the *j*th indicator of unit *i*. $\min(x_j)$ is the minimum value of the *j*th indicator for all m units. $\max(x_j)$ represents the maximum value of the *j*th indicator for all *m* units. The standardized value of the *j*th indicator of unit *i* (y_{ij}) can be defined as $y_{ij} = \frac{x_{ij}'}{\sum_{i=1}^{m} x_{ij}'}$.

Secondly, the entropy value e and utility value d of the jth indicator are calculated as:

$$e_j = -k\sum_{i=1}^{m} y_{ij}\ln y_{ij}$$
$$d_j = 1 - e_j \tag{7.17}$$

The utility value of the jth indicator could be calculated as $d_j = 1 - e_j$.
Thirdly, the jth indicator's weight is calculated:

$$w_j = \frac{d_j}{\sum_{j=1}^{n} d_j} \tag{7.18}$$

where w_j presents the weight of the jth indicator in n indicators, $\sum_{j=1}^{n} w_j = 1$. Hence, the comprehensive evaluation indicator of unit i could be calculated as $h_j = \sum_{j=1}^{n} w_{ij}y_{ij}$.

In our study, the sample consists of 30 provinces (autonomous regions, municipalities) for the period of 2001–2010. Hence, there are 300 units in this calculation. Three environmental pollutants are considered in our analysis, so there are 3 evaluation indicators in this evaluation. As the results of our evaluation, the comprehensive indicator h presents the level of environmental pollution. Greater h presents more serious environmental pollution. By utilizing the multiply transformation method, a positive environmental indicator is developed as b = 1/h (Table 7.5).

7.2.3.2 Empirical Results and Discussions

In order to reveal the changing trend of energy efficiency in time series, as well as to comprehensively show the regional differences of energy efficiency, we operate the energy efficiency evaluation model with the application of the global DEA method.

Table 7.5 Summary statistics of inputs and outputs

Index	Unit	Observations	Minimum	Maximum	Mean	STDev
Energy consumption	Ten thousand tons of standard coal	300	520.00	34808.00	9027.41	6387.44
Labor employment	Ten thousand employees	300	240.00	6042.00	2301.35	1522.80
Capital stock	Billion yuan	300	87.50	7217.70	1476.12	1297.72
GDP	Billion yuan	300	29.20	3231.70	644.96	582.58
b	–	300	71.00	16733.00	962.81	2104.63

Energy efficiency in three major regions and China as a whole experienced similar changing trends that energy efficiency firstly decreased during 2001–2005 and then increased during 2005–2010. However, there are some differences in changing trends of energy efficiency in 30 provinces (autonomous regions, municipalities). Concretely, energy efficiency in 66.7% of provinces (autonomous regions, municipalities) experienced a similar changing trend with that in China during the sample period, while energy efficiency in Beijing, Tianjin, Shanxi, Heilongjiang, Shanghai, and Anhui kept increasing throughout the period of 2001–2010. And there are not apparent changing trends of energy efficiency in Liaoning, Guangdong, Chongqing, and Guizhou.

We can see that energy efficiency in China was lower than 0.70 during 2001–2010, indicating that China's current energy efficiency is at a relatively lower level. Among these three major regions, the eastern region was leading in energy efficiency with parameters above 0.70 during the sample period, while energy efficiency in the western region had been encountering the downside ever since the beginning. Energy efficiency in the central region was in the middle. The reason for this phenomenon is that the level of economic development and technology in these three major regions are quite different. Since several decades of reform and opening up, the eastern region, at the forefront of reform and opening up, has actively introduced foreign advanced energy technology, equipment, and management experience through encouraging and increasing international trade and investment. Moreover, the eastern region could make full use of its rapid economic development to make a large number of investments in R&D of high-tech industries and environmental governance. As a result, energy efficiency in the eastern region is at a relatively high level. Compared with the eastern region, the western region lags behind in terms of technology and management. Meanwhile, the investments in environmental governance and protection are not enough in the western region due to its underdeveloped economy. Thus, compared with the eastern region, the western region's energy efficiency is relatively low. Since the situation of the central region is between that of eastern and western regions, energy efficiency in the central region also situates between eastern and western regions.

Guangdong and Fujian, two provinces in the eastern region, had the highest energy efficiency during 2001–2010, whose mean values were 0.9959 and 0.9124, respectively. The mean values of energy efficiency in the eastern provincial units of Zhejiang, Shanghai, Jiangsu, and Hainan were also higher than 0.8. Two provincial units with the lowest energy efficiency were the western provincial unit of Ningxia and the central provincial unit of Shanxi. The corresponding mean values of energy efficiency in these two provincial units were 0.1533 and 0.1920, respectively. The reason is that Ningxia, as one of the poorest provincial units in China, lags behind in terms of technology and management. Shanxi is abundant with energy resources and thus does not concern much about efficiency. Hence, energy efficiency in these two provincial units was at a relatively low level.

According to the changing trend of energy efficiency in China, the sample period of this study could be divided into two stages: 2001–2005 and 2005–2010. In order to make a comprehensive analysis of the key factors influencing China's energy efficiency, we explore the empirical analysis at national, regional, and provincial levels.

At the national level, it can be seen from Table 7.6 that energy efficiency in China decreased by 10.30% from 0.5919 to 0.5310 during 2001–2005. Obviously, technical regression (accumulated GTCH = 0.9106 < 1) contributed the most to this decline. It is not hard to understand the result with the help of knowing the background. During 2001–2005, China's national economy had been steadily growing at a remarkable speed. However, the rapid, extensive economic development also had a high cost in terms of resources and environment. Especially, as China's industrialization and urbanization speeded up during 2001–2005, its energy intensity and several main environmental pollutant emissions increased significantly. During this period, the high dependence on fossil energy resources, backward technology, and outdated equipment finally resulted in technical regress in China. In addition, the decline of scale efficiency (accumulated GSCH = 0.9831 < 1) was another contributor to this decline. It is known that China's roaring economy had been powered for years by high inflated fixed-asset investments which would inevitably have a negative effect on the returns to scale of investment and economy. As a result, the scale efficiency in China decreased during that period.

At the regional level, from Table 7.6, we can see that there were significant differences in the factors influencing energy efficiency among these three regions. Energy efficiency in the eastern region decreased by 9.8% from 0.7194 to 0.6489 during 2001–2005, while scale efficiency, pure efficiency, and technical efficiency in this region had declined by 6.26%, 3.40%, and 0.29%, respectively. Compared with the other two regions, it can be found that the eastern region had the highest energy efficiency, while its performance on GSCH was the worst. The reason for this phenomenon is that the eastern region has the best basis of economic and technology, which makes its energy efficiency stay at a relatively high level. And due to the fact that the eastern region is abundant with funds, the overinvestment in that region would inevitably cause a decline in scale efficiency. During the same period, energy efficiency in the central region dropped by 15.87% from 0.5313 to 0.4470. Technical regression (accumulated GTCH = 0.7739 < 1) was the biggest contributor to the significant decline of energy efficiency in the central region, while

Table 7.6 Global Malmquist indices and the decompositions of three major regions and China during 2001–2005

Region	Energy efficiency		Accumulated GM index and its decomposition during 2001–2005			
	2001	2005	GM	GTCH	GPCH	GSCH
East	0.7194	0.6489	0.9020	0.9961	0.9660	0.9374
Central	0.5313	0.4470	0.8413	0.7739	1.0940	0.9937
West	0.3636	0.3427	0.9425	0.8182	0.9480	1.2150
China	0.5919	0.5310	0.8970	0.9106	1.0020	0.9831

pure efficiency in the central region had been improved (accumulated GPCH = 1.0940 > 1). For that reason, the positive effect of pure efficiency promotion could not offset the negative effect of technical regression (accumulated GTCH = 0.7739 < 1) and the decrease in scale efficiency (accumulated GSCH = 0.9937 < 1) on energy efficiency, energy efficiency in the central region declined during 2001–2005. The western region, whose energy efficiency was the lowest among these three major regions, had the best performance on GSCH. The accumulated GSCH in the western region was as high as 1.2150; this means scale efficiency had increased significantly during 2001–2005. This may be due to the fact that since the implementation of the west-construction plan which has been called up since the year 1999, the poor situation that the western region is lack of investment has been gradually improved. Eventually, the rapid increase of capital investment improved the scale efficiency in this area.

At the provincial level, energy efficiency in 66.67% of provincial units declined during 2001–2005. During the same period, the GTCH indices in 83.33% of provincial units were lower than 1, and likewise, statistics for GPCH and GSCH indices were 16.67% and 36.67%, respectively. This suggests that technical regression was a common problem among 30 provincial units and was the main reason for the decline of energy efficiency in most provincial units during that period.

At the national level, energy efficiency in China increased by 29.64% from 0.5310 to 0.6884 during 2005–2010. Obviously, technical progress (accumulated GTCH = 1.1911 > 1) was the most powerful engine for this significant improvement of energy efficiency. During the same period, scale efficiency and pure efficiency in China increased by 5.41% and 3.26%, respectively.

During 2005–2010, the Chinese central government had set a series of energy saving and emission reduction goals, and local governments at the provincial level were forced to concern these targets during their decision-making. By formulating strict efficiency and emission standards and eliminating backward production capacity, the Chinese government reversed the rising trend of energy intensity and the emissions of several main pollutants. During 2005–2010, the strict policies and standards promoted technical progress. For instance, compared with 2005, the penetration rate of dry quenching technology in the steel industry increased from less than 30% to more than 80%, and the penetration rate of ionic membrane caustic soda technology in caustic soda industry increased from 29 to 84% in China. During the same period, by controlling the excess capacity and adjusting the industrial structure, the scale efficiency and management level in China had been improved.

At the regional level, we can see that there were some differences in the factors influencing energy efficiency among different regions. Energy efficiency in terms of eastern region, central region, and western region increased from 0.6489, 0.4470, and 0.3427 to 0.8525, 0.5657, and 0.4452 during 2005–2010, respectively. Technical progress (accumulated GTCH > 1) in these three regions was the most powerful engine for this sharp promotion of energy efficiency. The only difference among these three regions is that pure efficiency in central and western regions had

Table 7.7 Global Malmquist indices and the decompositions of 30 provinces (autonomous regions, municipalities) during 2005–2010

Region	Energy efficiency		Accumulated GM index and its decomposition during 2005–2010			
	2005	2010	GM	GTCH	GPCH	GSCH
Beijing	0.6172	0.8483	1.3744	1.2541	1.0948	1.0010
Tianjin	0.6204	0.8041	1.2961	1.2671	1.0001	1.0227
Hebei	0.3507	0.4582	1.3065	1.2316	1.0001	1.0608
Shanxi	0.1797	0.2412	1.3422	1.2431	1.0000	1.0797
Inner Mongolia	0.2394	0.3248	1.3567	1.2544	1.0000	1.0816
Liaoning	0.5033	0.6208	1.2335	1.1446	1.0001	1.0775
Jilin	0.4230	0.5677	1.3421	1.2513	0.7521	1.4262
Heilongjiang	0.6119	0.7232	1.1819	1.1819	1.0000	1.0000
Shanghai	0.7899	0.9999	1.2659	1.2659	1.0000	1.0000
Jiangsu	0.7528	1.0000	1.3284	1.3282	1.0001	1.0000
Zhejiang	0.7360	0.9481	1.2882	1.2348	1.0264	1.0164
Anhui	0.5782	0.7721	1.3354	1.2628	1.0993	0.9619
Fujian	0.8237	0.9999	1.2139	1.2138	1.0001	1.0000
Jiangxi	0.6017	0.7842	1.3033	1.2463	1.0000	1.0458
Shandong	0.5365	1.0000	1.8639	1.2438	1.4451	1.0370
Henan	0.4861	0.6151	1.2654	1.2080	1.1189	0.9362
Hubei	0.5942	0.7221	1.2152	0.9334	1.0000	1.3020
Hunan	0.5130	0.6259	1.2201	1.1714	0.9281	1.1222
Guangdong	0.9999	1.0000	1.0001	1.0000	1.0001	1.0000
Guangxi	0.5290	0.6459	1.2210	1.2775	1.0001	0.9556
Hainan	0.8150	0.9125	1.1196	1.1191	1.0005	1.0000
Chongqing	0.4032	0.5847	1.4501	1.2161	0.9322	1.2792
Sichuan	0.4513	0.5987	1.3266	1.2036	1.5478	0.7121
Guizhou	0.2129	0.2655	1.2471	1.1926	0.5846	1.7886
Yunnan	0.3703	0.4614	1.2460	1.2311	0.9845	1.0281
Shaanxi	0.3838	0.4967	1.2942	1.2350	0.9148	1.1455
Gansu	0.2793	0.3492	1.2503	1.1893	0.7449	1.4112
Qinghai	0.2194	0.2547	1.1609	1.1503	1.0002	1.0090
Ningxia	0.1345	0.1645	1.2230	1.1736	1.0001	1.0420
Xinjiang	0.3000	0.3394	1.1313	1.2321	0.9352	0.9819

declined during 2005–2010, while pure efficiency in the eastern region had increased during that period. The main reason is that since the financial crisis in 2008, the export of industrial enterprises in the eastern region has significantly shirked. In order to seek cheaper labor force and raw materials, a large amount of high-tech enterprises have been migrating from the eastern region to central and western regions. This migration would inevitably drive the rapid growth of technology in these two regions. However, for the reason that the poor infrastructure and imperfect market mechanism in central and western regions could not meet the requirements of these high-tech enterprises, the pure efficiency (management or resource allocation efficiency) in these two regions declined during 2005–2010.

At the provincial level, as shown in Table 7.7, energy efficiency in all provinces (autonomous regions, municipalities) had been improved during 2005–2010. Concretely, patents in Jiangsu, whose accumulated GTCH was as high as 1.3282, ranked the first among 30 provinces (autonomous regions, municipalities) in China since the year 2008. It is, therefore, not surprising that Jiangsu performed the best on technical progress in China. Energy efficiency in the Shandong increased by 86.39% from 0.5365 to 1 during 2001–2005, which was the highest score among 30 provinces (autonomous regions, municipalities). Technical progress (accumulated GTCH = 1.2438 > 1) and the improvement of pure efficiency (accumulated GPCH = 1.4451 > 1) were two main contributors to this rapid growth of energy efficiency in Shandong.

7.3 Relationship Between Energy Technology Patents and CO_2 Emissions

In the face of international demand for China's emission reduction, the Chinese Government is committed to fall on the basis of 40–45% target from 2005 to 2020 in the percentage of GDP CO_2 emissions. To achieve the established emission reduction targets, technological progress and technological innovation are indispensable. The Chinese government is actively developing energy technology policies to promote energy conservation and emission reduction through technological innovation. In the past 10 years, good progress has been made, and a large number of patents for energy technology have been obtained. This means there is an improvement on the energy technology innovation level. However, China's energy technology patents have played a role in reducing CO_2 emissions. In this chapter, considering the factors of economic growth, it is important to discuss the relationship between China's energy technology patents and CO_2 emission from the perspective of energy technology innovation output in China's future development of energy technology policy.

7.3.1 Current Situation of Energy Technology Patents and CO_2 Emissions

Under the stress of climate change and resource crises, cutting down GHG emissions and slowing down the process of global warming have received increasing concern worldwide. Along with growing energy consumption and CO_2 emissions, it is crucial for China to enhance energy security and reduce climate change. Many factors determine the CO_2 emissions, including economic scale, population, industrial structure, energy consumption structure, energy efficiency, energy intensity, and level of technology and management (Kaya 1990; Wang and Huang 2008; Xu et al. 2006). In this chapter, we will address this issue from the perspective of energy technology patents.

Patent counts are commonly used to measure the output of innovation activities (Popp 2006; Popp et al. 2009a). Energy technology patents directly reflect the performance of energy technology innovation activities and the development of energy technologies. The increase of patent counts in energy and environmental sectors implies the improvement of energy technology innovation ability (Liu and Sun 2008). However, careful scrutiny of the literature on the main factors determining CO_2 emissions in China shows that almost no study has been executed to explore the impact of energy technology patents. Therefore, it is necessary to explore the relationship between energy technology patents and CO_2 emissions in China. On the one hand, this study could help to get a better understanding of the relationship between energy technology innovation and CO_2 emissions; on the other hand, it could provide references for the Chinese government to make energy technology policy.

Many studies found that energy technology innovation and energy technology R&D could reduce CO_2 emissions (Garrone and Grilli 2010). Technology innovation could help to prevent CO_2 emissions from increasing in the long run (Popp et al. 2009b). And it is promising for developing countries to reduce CO_2 emissions by applying new technologies (Bernstein et al. 2006). Previous studies also reported that increased R&D had a negative influence on energy and emission intensities in developing countries (Fisher-Vanden and Wing 2008). In China, indigenous R&D was negatively related to energy intensity and CO_2 emissions (Ang 2009; Teng 2009).

7.3.2 Models for Relationship Between Energy Technology Patents and CO_2 Emissions

The economic development levels are different in China's provincial units, and it has a direct impact on the development of energy technology and the amount of CO_2 emissions (Wei and Yang 2010; Yu and Qi 2007). So when analyzing the relationship between energy technology patents and CO_2 emissions in China (including its eastern, central, and western regions), the GDP was included as a control variable. We selected the following variables in this research: energy technology patents, CO_2 emissions, and GDP.

7.3.2.1 Date of Energy Technology Patents in China's Provincial units

Energy technology patents in this study include the innovation in energy sectors (e.g., power generation) and energy user sectors (e.g., car motors, heating plants) (van Vuuren et al. 2003). The advances in fossil-fueled technologies (e.g., gas-fired and coal-fired plant technologies) could have a limited effect on CO_2 emissions (Chen et al. 2011; Gnansounou et al. 2004). In contrast, nuclear and renewable energy technologies are typical low-carbon, even zero-carbon generation technologies. If hydropower and new renewable energy technologies are developed and penetrate the power generation sector, investments in fossil-fueled plants will decrease and CO_2 emissions are expected to decrease as well (Lüthi and Prässler 2011). Therefore, we divided the energy technology patents into the patents for fossil-fueled technologies and patents for carbon-free energy technologies, and extracted the energy technology patents relevant to the energy saving and emissions reduction from 1997 to 2008 in China (Popp 2006; Popp et al. 2011). Fossil-fueled technologies include technologies in energy sectors: coal, crude oil, gasoline, diesel, and natural gas relevant to energy saving and emissions reduction (Margolis and Kammen 1999); they also include technologies in energy user sectors: stoves relevant to energy saving and emissions reduction (including the sectors of metallurgy, building, cement, chemical industry, heating plants, generating station, and household), electricity-saving equipment or technology, electric vehicles, car motors, combustion engine, turbine, fuel injection, energy-efficient lighting, and carbon capture and storage (CCS). The carbon-free energy technologies include: solar, wind, ocean, geothermal, hydropower, nuclear, biomass and waste, synthesis gas, hydrogen fuel, biomethane, biodiesel, and ethanol (Johnstone et al. 2010; Popp 2006; Popp et al. 2011; Wang and Chen 2010). The technology fields covered by this study are reported in Table 7.8.

Table 7.8 Fossil-fueled and carbon-free energy technology fields covered

Category	Technology field
Fossil-fueled	Coal, crude oil, gasoline, diesel, and natural gas relevant to energy saving and emissions reduction Stoves relevant to energy saving and emissions reduction (including the fields of metallurgy, building, cement, chemical industry, heating plants, generating station and household) electricity-saving equipment or technology; electric vehicles; car motors, combustion engine, turbine, fuel injection; energy-efficient lighting and carbon capture and storage (CCS)
Carbon-free	Solar, wind, ocean, geothermal, hydropower, nuclear, biomass and waste, synthesis gas, hydrogen fuel, biomethane, biodiesel, and ethanol

Using the above approach, we got the data of patents for fossil-fueled technologies and data of patents for carbon-free energy technology. To organize the data, patents were sorted by their application year. Then, we got the data of energy technology patents of 30 provincial units from 1996 to 2008 in China. We also divided the data of energy technology patents into categories of eastern, central and western regions. Table 7.1 and Fig. 7.1 shows the rising trend of national energy technology patents from 1997 to 2008, as well as a similar trend of eastern, central, and western regions.

7.3.2.2 Date of CO_2 Emissions in China's Provincial units

As there is no direct data of CO_2 emissions in China, most of the previous studies got the data based on the estimates of energy consumption (Xu 2010; Yi et al. 2011). Energy consumption data are obtained from energy balance sheets of all provincial units in China Energy Statistical Yearbook 2009, including 17 types of energy sources. Emission factors refer to the amount of GHG emissions per net calorific value that each energy generates by burning or using (IPCC 2006). Emission factors of each type of energy are obtained from the Intergovernmental Panel on Climate Change (IPCC) guidelines for National Greenhouse Gas Inventories (IPCC, 2006) (Table 7.9). As the main energy consumption for electricity and heating in China is coal, the CO_2 emissions from electricity and heating are estimated based on the energy input of thermal power and heating supply (Zhao et al. 2009). CO_2 emissions per year of each provincial unit are calculated as follows (IPCC 2006; Yi et al. 2011). We firstly calculated the heat of each energy based on the energy consumption and average low calorific value of each energy

Table 7.9 Emission factor of each type of energy

Energy type	Emission factor/(kg CO_2/TJ)	Energy type	Emission factor/(kg CO_2/TJ)
Raw coal	95,700.00	Kerosene	71,500.00
Cleaned coal	95,700.00	Diesel oil	74,100.00
Other washed coal	95,700.00	Fuel oil	77,400.00
Briquettes	95,700.00	Liquefied petroleum gas	63,100.00
Coke	107,000.00	Refinery gas	57,600.00
Coke oven gas	44,400.00	Natural gas	56,100.00
Other gas	44,000.00	Other petroleum products	73,300.00
Crude oil	73,300.00	Other coking products	80,700.00
Gasoline	69,766.67		

and then calculated the CO_2 emissions of each energy based on the heat and emission factor of each energy; finally, we summed up CO_2 emissions of each energy and obtained total CO_2 emissions of each provincial unit.

Before 1996, the energy consumption in Chongqing was contained in Sichuan. In order to accurately reflect the CO_2 emissions of each provincial unit, we chose the data during 1997–2008. We divided the CO_2 emissions data of 30 provinces (autonomous regions, municipalities) during 1997–2008 into the categories of eastern, central, and western regions. As shown in Fig. 7.4, CO_2 emissions of eastern, central, and western regions were increasing gradually during 1997–2008. Especially the CO_2 emissions increased significantly at a high speed during the period of 2002–2008 (Fig. 7.4). This was mainly due to the China's rapid economic growth and a continuous increase in energy consumption. GDP reflects the value of all final products and services of economic activities in a nation or region in a certain time period (a quarter or a year), and it is the symbol of the economic development level of a nation or a region. All provincial GDP data come from China Statistical Yearbooks. We summed up provincial GDP data according to the division of eastern, central, and western China, and then obtained GDP data of three regions. The GDP growth of eastern, central, and western regions in China during 1997–2008. We find that GDP growth rates of the three regions were obviously higher during 2002–2008. In this research, nominal GDP was used for analysis.

In the following parts of this chapter, we will use ETP to represent the energy technology patents, EMS to represent the CO_2 emissions, and GDP to represent the gross domestic product. In order to reduce the volatility of data, we converted the data into the natural logarithm form and named as LETP, LEMS, and LGDP. LETP stands for the natural logarithm of energy technology patent, LEMS stands for that of CO_2 emissions, and GDP stands for the gross domestic product (Figure 7.5).

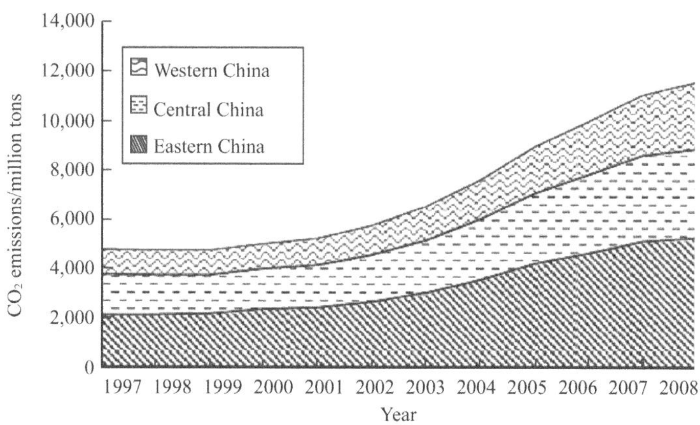

Fig. 7.4 CO_2 emissions in eastern, central, and western China

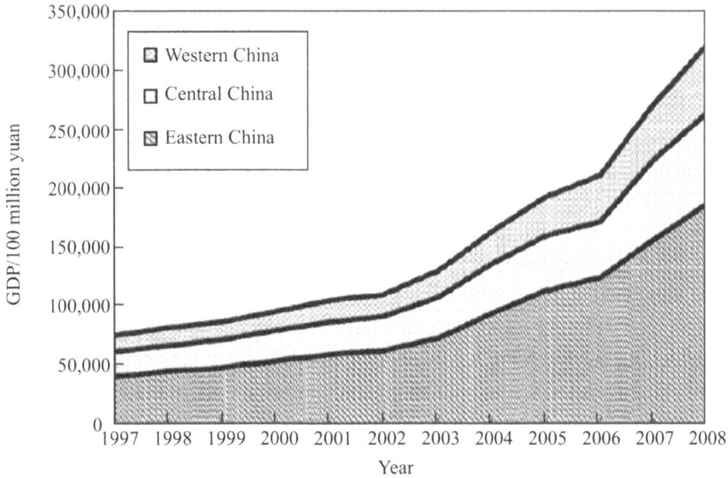

Fig. 7.5 GDP in eastern, central, and western China

7.3.2.3 Model Specification and Methodology

Panel data are also called as pool data, which involve two dimensions: a cross-sectional dimension and a time series dimension. Panel data usually bring researchers a large number of data points, increasing the degree of freedom and reducing the collinearity among explanatory variables, hence improving the efficiency of econometric estimates (Hsiao 2003). With repeated observations of enough cross sections, panel data analysis permits the researchers to study the dynamics of change with short time series. Panel data analysis endows the regression analysis with both a spatial and temporal dimension. A general panel data regression model is written as

$$y_{it} = \alpha + \beta' x_{it} + u_{it}, i = 1, 2, \ldots, N, t = 1, 2, \ldots, T \quad (7.19)$$

where i is the individual dimension and t is the time dimension.

Dynamic panel data were used in this chapter to help to verify the existence of causality between energy technology patents and CO_2 emissions. Firstly, a panel data unit root test was used to inspect the stationarity of the series; secondly, a panel cointegration test was conducted to verify that whether there were long-run relationships among the series; finally, the method of dynamic panel estimation was used to determine the direction of causalities.

Because of the nonstationary nature of time-series data, it is essential to test their stationarity before panel data models are established. The regression of nonstationary panel data will lead to the problem of pseudo-regression. Therefore, firstly the stationarity of panel data has to be examined using a panel unit root test. Panel unit root test is based on the following autoregressive specification:

$$y_{it} = \rho_i y_{it-1} + \delta x_{it} + u_{it}, i = 1, 2, \ldots, N, t = 1, 2, \ldots, T \tag{7.20}$$

where i = 1, 2,..., N represents the provincial units observed over periods t = 1, 2, ...T, X_{it} is the exogenous variable in the model including any fixed effects or individual trends, and ρ_i is the autoregressive coefficient. If $|\rho_i < 1|$, y_i is said to be stationary and has no unit root. Conversely, If $|\rho_i = 1|$, then y_i contains a unit root and is not stationary. U_{it} is the disturbance term.

The panel cointegration tests can be done if panel data are stationary and the corresponding series is integrated with the same order. Cointegration test is used to determine the long-run equilibrium relationship among the series. According to the different characteristics of energy technology patents, CO_2 emission, and GDP data, panel data model with variable coefficients was selected to conduct cointegrating regression:

$$LETP_{it} = \alpha_i + \beta_1 LEMS_{it} + \beta_2 LGDP_{it} + u_{it} \tag{7.21}$$

where α_i is the provincial-unit-specific intercept, the slope coefficients β_1 and β_2 vary from one individual to another allowing the cointegration vectors to be heterogeneous across provincial units.

If there is cointegration among LEPT, LEMS, and LGDP, the Vector Error Correction Model (VECM) can be further established to test causalities. In order to verify the causality between energy technology patents and CO_2 emission, the panel VAR model as Eqs. (7.22)–(7.24) is established:

$$LETP_{it} = \alpha_1 + \sum_{j=1}^{m+1} \beta_{1j} LETP_{it-j} + \sum_{j=1}^{m+1} \Upsilon_{1j} LEMS_{it-j} + \sum_{j=1}^{m+1} \sigma_{1j} LGDP_{it-j} + \eta_{1i} + u_{1it} \tag{7.22}$$

$$LEMS_{it} = \alpha_2 + \sum_{j=1}^{m+1} \beta_{2j} LETP_{it-j} + \sum_{j=1}^{m+1} \Upsilon_{2j} LEMS_{it-j} + \sum_{j=1}^{m+1} \sigma_{2j} LGDP_{it-j} + \eta_{2i} + u_{2it} \tag{7.23}$$

$$LGDP_{it} = \alpha_3 + \sum_{j=1}^{m+1} \beta_{3j} LETP_{it-j} + \sum_{j=1}^{m+1} \Upsilon_{3j} LEMS_{it-j} + \sum_{j=1}^{m+1} \sigma_{3j} LGDP_{it-j} + \eta_{3i} + u_{3it} \tag{7.24}$$

where η_{1i}, η_{2i} and η_{3i} are provincial-unit-specific effects for the ith individual in the panel, and u_{1it}, u_{2i} and u_{3i} are the disturbance terms.

In Eqs. (7.22)–(7.24), the ordinary least squares method will lead to the biased estimation because of the correlation between the lagged dependent variables and provincial-unit-specific effect. To avoid this bias, first differences are taken in the equations. However, any information about the long-run adjustments of data may be omitted in the first difference; therefore, the VAR model should be adopted to express the short-run relationships among the variables. According to Engle–Granger two-step method, as long as there is cointegration among variables, VECM can be deduced from the VAR model. The promise of VECM is that the equilibrium which happens at a time point can be corrected in the next time point, and it is possible to identify different relationships between variables in the long run and short run. Therefore, panel-based VECM as Eqs. (7.25)–(7.27) is established:

$$\lambda_1 ECT_{i,t-1} + \Delta u_{1it} \tag{7.25}$$

$$\lambda_2 ECT_{i,t-1} + \Delta u_{2it} \tag{7.26}$$

$$\lambda_3 ECT_{i,t-1} + \Delta u_{3it} \tag{7.27}$$

ECT is error correction term attained from the residuals of estimated cointegration in Eq. (7.19), reflecting the long-run equilibrium among the variables. The coefficient of error correction term (ECT) denotes the adjusting velocity when equilibrium deviates. Coefficients of the different terms reflect the influence of independent variables on the dependent variables in the short run.

On the right side of Eqs. (7.25)–(7.27), lagged dependent variables have been included. As a result, regressions are inherently correlated with disturbance. In this case, OLS estimation will be biased and inconsistent. To solve the problem, Arellano and Bond (1991) proposed the estimator of first differenced GMM (DIF-GMM) in the panel for the systems (5a)–(5c), using lagged dependent variables in levels as instrumental variables in first differences. Instrumental variables are affected only when disturbances m1*it*, m2*it,* and m_{3it} are not correlated. Blundell and Bond (1998) pointed out that DIFGMM estimation may be easily affected by weak instrumental variables. As a result, the biased estimation may be achieved. To resolve this problem, they proposed the method of System-GMM. We tried to use System-GMM as an estimating method, but it did not work. Therefore, Arellano and Bond (1991) first differenced GMM robust, one-step estimator was adopted to solve Eqs. (7.25)–(7.27). The selection of lag order *mj* for instrumental variables should meet the requirement that the problem of the autocorrelation of residuals can be avoided. AR test is used to help to inspect the autocorrelation of residuals. AR (1) and AR (2) are usually used to conduct the test, and the rule is that the hypothesis of the existence of AR (1) should be rejected, and the hypothesis of the existence of AR (2) should not be rejected. Sargan test of overidentifying restrictions is processed to check the validity of the instruments.

Causalities between variables are identified by verifying the significance of the coefficient of the dependent variables in panel VECM. First, short-run causalities are identified by testing hypothesis as follows:

$$H_0 : \gamma_{1j} = 0, \forall_j = 1, \ldots, m \quad and \quad H_0 : \gamma_{1j} = 0, \forall_j = 1, \ldots, m \quad in \quad Eq. \quad (7.25)$$

$$H_0 : \gamma_{2j} = 0, \forall_j = 1, \ldots, m \quad and \quad H_0 : \gamma_{2j} = 0, \forall_j = 1, \ldots, m \quad in \quad Eq. \quad (7.26)$$

$$H_0 : \gamma_{3j} = 0, \forall_j = 1, \ldots, m \quad and \quad H_0 : \gamma_{3j} = 0, \forall_j = 1, \ldots, m \quad in \quad Eq. \quad (7.27)$$

And then, the significance of ECT coefficients determines the existence of a long-run causality. At last, a simple Wald test can be applied to examine the direction of the causal relationship between the variables.

7.3.3 Empirical Results for the Relationship Between Energy Technology Patents and CO_2 Emissions

All statistics in LLC, IPS, Fisher-ADF, and Fisher-PP and some statistics in Breitung test indicate that null hypotheses of LETP (LETP is analyzed based on patents for fossil-fueled technologies and patents for carbon-free energy technologies), LEMS and LGDP having a unit root are not rejected at the 5% or 10% level, so the series are not stationary. However, the null hypothesis that first difference of each variable has a unit root is rejected at a 10% level, indicating that most of the tests provide supporting evidence that series are integrated in the first order.

The series of LETP, LEMS and LGDP are integrated in the first order for the national, eastern, central and western China. Thus we can proceed to conduct the cointegration tests for fossil-fueled technologies, all statistics in panel PP, panel ADF, group PP, and Fisher tests and some statistics in other tests reject null hypothesis of no cointegration is at the 10% level; on the other hand, for carbon-free energy technologies, all the resulting statistics, except for panel v-statistic, panel rho-statistic and group rho-statistic in Pedroni test, reject null hypothesis of no cointegration is at the 1% level. The above results indicate that LETP, LEMS, and LGDP are cointegrated series. Therefore, there is a long-run relationship between the series of provinces (autonomous regions, municipalities) in the panel for fossil-fueled technologies and carbon-free energy technologies, which means that the series move together in the long-run (Table 7.10).

In order to determine the direction of the causal relationship between the series, VECM (Eqs. (5a)–(5c)) is estimated using the DIF-GMM estimator (Arellano and Bond 1991). Table 7.11 shows the Sargan test results and $m1$ and $m2$ statistics on fossil-fueled technologies. The estimates show that 2 lags or 1 lag is selected to make the disturbance to have no serial correlation in the national and other regions. The results of $m1$ and $m2$ show that significant first-order serial correlation is found

Table 7.10 Statistic values for panel causality tests on fossil-fueled technologies

Dependent variables		Short-run ECT			Long-run ECT
		ΔLETP	ΔLEMS	ΔLGDP	
National	ΔLETP		0.4666[a]	0.0636	0.8913[a]
	ΔLEMS	0.0181[b]		0.1868[a]	0.5998[a]
	ΔLGDP	0.0143	0.5726[a]		0.8442[a]
Eastern	ΔLETP		1.3421[a]	0.1104	1.0150[a]
	ΔLEMS	0.0239[b]		0.1785[a]	0.5329[a]
	ΔLGDP	0.0237	0.5626[a]	0.9077[a]	0.0237
Central	ΔLETP		0.0277	0.0727	0.6418[a]
	ΔLEMS	0.0172		0.0987[b]	0.5308[a]
	ΔLGDP	0.0226	0.8178[a]		1.0247[a]
Western	ΔLETP		0.1293	0.8144[a]	0.9080[a]
	ΔLEMS	0.0061		0.4187[a]	0.7209[a]
	ΔLGDP	0.0471[b]	0.4177[a]		0.5127[a]

Notes [a]Null hypothesis of no causation is rejected at the 1% level
[b]Null hypothesis of no causation is rejected at the 10% level

Table 7.11 Results of DIF-GMM estimation on fossil-fueled technologies

		National	Eastern	Central	Western
The number of lags	Eq. (5b)	2	2	2	2
	Eq. (5c)	1	1	2	2
Validity (Yes or no)	Sargan test	Yes	Yes	Yes	Yes
Significance (Yes or no)	*m*1	Yes	Yes	Yes	Yes
	*m*2	No	No	No	No

in the first differenced residuals, while there is no evidence of second-order serial correlation. The Sargan statistics do not reject the validity of the instruments.

To distinguish the possible different sources of causation, Table 7.11 shows the statistic values of the Wald test of no causality on fossil-fueled technologies. Firstly, we analyzed the short-run causality between LEMS and LETP. On the one hand, there is a positive causal relationship running from LEMS to LETP in the eastern and national levels of China at a 1% significant level, indicating that the increase in CO_2 emissions pushes the increase in patents for fossil-fueled technologies. The reasons are as follows: Economy and technology in eastern China have relatively rapidly developed, and there is a substantial increase in CO_2 emissions because of fossil-fueled dominated energy structure. Because in order to achieve higher energy efficiency and reduce the CO_2 emissions, the development of fossil-fueled technologies for energy saving and emissions reduction is promoted.

On the other hand, there is a positive causal relationship running from LETP to LEMS in the eastern and national levels of China at the 10% significant level, indicating that an increase in patents for fossil-fueled technologies causes an increase in CO_2 emissions. Our result is in line with the findings of Hu and Huang (2008), who found that the current technology did not help to reduce the CO_2 emissions in China. This may be due to several reasons. Most advances in gas-fired and coal-fired plant technologies could have a limited effect on CO_2 emissions (Chen et al. 2011; Gnansounou et al. 2004). The cost and risk are higher while firms are going to employ advanced fossil-fueled technologies, which tend to have a positive externality. Private firms are reluctant to invest in these technologies if no payoff is guaranteed (Popp et al. 2009a). Therefore, some of the present patents for fossil-fueled technologies could not be adopted widely to reduce emissions. In addition, the use of energy-efficient fossil-fueled technologies may have produced irrelevant rebound effects (Sorrell et al. 2009), which may lead to some extent CO_2 emissions increase.

The short-run and positively bidirectional causality between LEMS and LETP is not significant in central and western China. On the one hand, an increase in CO_2 emissions does not significantly push up the increase in patents for fossil-fueled technologies. This may be due to poor technology infrastructure and insufficient investment in energy technology R&D in central and western China, which also impedes the absorption of energy-efficient fossil-fueled technologies (Wei and Yang 2010). On the other hand, patents for fossil-fueled technologies could not curb the CO_2 emission. The result is closely related to low energy efficiency and the low transformation rate of patents in central and western China.

Secondly, we analyzed the short-run causality between LEMS and LGDP, and short-run causality between LETP and LGDP. There is a positive causal relationship running from LEMS to LGDP in eastern, central, western, and national levels of China at the 1% significant level. The main cause of CO_2 emissions is energy consumption, while economic growth needs energy consumption. There is a positive causal relationship running from LGDP to LEMS in eastern, central, western, and national levels of China at the 1 or 10% significant level, which shows that the higher GDP is, the greater CO_2 emissions are. This result is affected by the coal-dominated energy consumption structure. There is also evidence of positive bidirectional causality between LETP and LGDP in western China, which implies that an increase in patents for fossil-fueled technologies improves the GDP output and an increase in GDP promotes an increase in patents for fossil-fueled technologies. However, the linkage between LETP and LGDP is not significant in eastern, central, and national levels of China.

Thirdly, we analyzed long-run causality among three variables. If the coefficient of ECT is not significantly equal to zero, a long-run causality among the variables will exist, and a positive coefficient indicates that the variables deviate from long-run equilibrium. The coefficients of ECT in the eastern, central, western, and national levels of China are positive and significant in Eqs. (5a)–(5c) at the 1% level. These results suggest that the patents for the fossil-fueled technology, CO_2 emissions, and GDP are heavily reliant on each other in the long run, and they all respond to a deviation from the long-run equilibrium in the previous period.

The Sargan test and $m1$ and $m2$ statistics on carbon-free energy technologies. The estimates show that 2 lags or 1 lag are selected to make the disturbance have no serial correlation in the national and other regions. The results of $m1$ and $m2$ show that significant first-order serial correlation is found in the first differenced residuals, while there is no evidence of second-order serial correlation. The Sargan statistics do not reject the validity of the instruments.

To distinguish the possible different sources of causation, Table 7.12 shows the statistic values of the Wald test on carbon-free energy technologies. Firstly, we analyzed the short-run causality between LEMS and LETP. There is a negative causal relationship from LEMS to LETP on carbon-free energy technologies in eastern, central, western, and national levels of China at the 1% or 10% significant level, indicating that an increase in CO_2 emissions does not promote the carbon-free energy technologies. This result may be due to several reasons. Coal accounts for 70.4% of China's energy use and coal in electricity generation account for 83% of China's all power generated in 2007 (Wang and Chen 2010). This unreasonable energy structure has a serious threat to China's energy security. Therefore, the main reason is that the Chinese government has tried to reduce the dependence on fossil-fueled energy and promote the development of carbon-free energy technologies (e.g., solar, hydropower, wind, nuclear, etc.) in order to ensure energy security but not reduce emission. In addition, the renewable energy production and fossil-fueled production could be assumed to be perfect substitutes from the perspectives of consumer demand (Fischer and Newell 2008). There is a positive relationship from CO_2 emissions to fossil-fueled technologies, so a negative relationship may exist from CO_2 emissions to carbon-free energy technologies.

There is a negative short-run causality from LETP to LEMS on carbon-free energy technologies, which is significant in eastern China at the 10% level, indicating that patents for carbon-free energy technologies could reduce the emissions in eastern China. The results are consistent with the previous study (Wei and Yang 2010). Eastern China has a strong economic basis and pays more attention to develop carbon-free energy technologies (e.g., hydropower and new renewable technologies) which have a strong effect on reducing CO_2 emissions. If they penetrate the power sector and other sectors, the CO_2 emissions are expected to decrease. However, patents for carbon-free energy technologies have a limited impact on the emissions reduction in eastern China; when patents for carbon-free energy technologies increase by 1%, then CO_2 emissions will decrease by 0.02%. A possible reason is that

Table 7.12 Results of DIF-GMM estimation on carbon-free energy technologies

Items		National	Eastern	Central	Western
The number of lags	Eq. (5a)	2	2	2	2
	Eq. (5b)	1	2	1	1
	Eq. (5c)	2	2	2	2
Validity (Yes or no)	Sargan test	Yes	Yes	Yes	Yes
Significance (Yes or no)	$m1$	Yes	Yes	Yes	Yes
	$m2$	No	No	No	No

private firms face higher costs and risks, so they are not actively adopting energy efficiency technologies. For example, China has a strong R&D ability, and market competitiveness in the field of solar photovoltaic. China has dominated about 70% market share in the world, but only 3–4% capacity is digested in China, and the rest is sold abroad (Gonsense 2011). However, as a matter of fact, China's R&D investment and ability in many fields of carbon-free energy technologies have a larger gap compared with developed countries (Ma et al. 2003). In addition, the actual time for a patent application to be granted lasting 5–6 years also hinders them from reducing the CO_2 emissions (Liu and Zheng 2008). The negative short-run causality from LETP to LEMS in central and western China is not significant, indicating that the role of patents for carbon-free energy technologies reducing the emissions is not obvious. A possible reason is that R&D investment is not sufficient and energy infrastructures are so imperfect that the energy efficiency technologies cannot get a large-scale application in the central and western China.

Secondly, we analyzed the short-run causality between LEMS and LGDP, as well as LETP and LGDP. There is evidence of short-run and positive bidirectional causality between LEMS and LGDP in eastern, central, western and national levels of China at 1 or 10% significant level, indicating that an increase in CO_2 emissions pushes up the GDP, and an increase in GDP leads to increase in CO_2 emissions. There is a positive short-run causality from LETP to LGDP in the eastern, central, western, and national levels of China at a 10% significant level, which implies that an increase in patents for carbon-free energy technologies contributes to GDP growth. The short-run LGDP has a positive and statistically significant impact on the LETP in the central, western, and national levels of China at a 10% level, indicating that an increase in GDP promotes the increase in patents for carbon-free energy technologies (Table 7.13).

Table 7.13 Statistic values for panel causality tests on carbon-free energy technologies

Dependent variables		Short-run ECT			Long-run ECT
		ΔLETP	ΔLEMS	ΔLGDP	
National	ΔLETP		−0.5344[a]	0.0996[b]	0.9671[a]
	ΔLEMS	−0.0039		0.1423[a]	0.6039[a]
	ΔLGDP	0.0138[b]	0.5907[a]		0.8178[a]
Eastern	ΔLETP		−0.7070[b]	−0.1901[b]	0.8666[a]
	ΔLEMS	−0.0202[b]		0.0887[b]	0.6640[a]
	ΔLGDP	−0.0096[b]	0.5616[a]		0.9036[a]
Central	ΔLETP		−0.5394[b]	0.1572[b]	0.9389[a]
	ΔLEMS	−0.0018		0.0914[b]	0.4432[a]
	ΔLGDP	0.0492[b]	0.8540[a]		1.0850[a]
Western	ΔLETP		−0.5684[b]	0.5237[b]	1.0376[a]
	ΔLEMS	−0.0052		0.3633[a]	0.5298[a]
	ΔLGDP	0.0096[b]	0.4125[a]		0.4406[a]

Notes [a]Null hypothesis of no causation is rejected at the 1% level
[b]Null hypothesis of no causation is rejected at the 10% level

Thirdly, we analyzed the long-run causality among three variables on the carbon-free energy technologies. The coefficients of ECT in eastern, central, western, and national levels of China are positive and statistically significant in Eqs. (5a)–(5c) at the 1% level. These results add extra evidence for the long-run relationship among carbon-free energy technologies, CO_2 emissions, and GDP, and they all respond to a deviation from the long-run equilibrium in the previous period.

7.4 Summary

In Sect. 7.1, we used DEA method and framework of total-factor energy efficiency to establish an energy efficiency evaluation model, in which output was based on gross industrial output value of industrial enterprises above designated size, and input was measured by energy consumption, average annual balance of fixed capital, and working capital, as well as the average annual number of employees. Based on these, we conducted an empirical analysis for the total-factor energy efficiency of industrial sectors of 30 provinces (autonomous regions, municipalities) in China.

Overall, the results of this chapter show that the energy efficiency of China is experiencing developing gaps among regions. During the years 2005–2009, areas where energy efficiency of industrial sectors has attained DEA-efficiency are mainly distributed in the coastal provinces (autonomous regions, municipalities) of the east. The industrial sectors of the east which have attained DEA-efficiency have entered the stage of constant returns to scale, which means that output increases in the same proportion as the input, and its development is saturated. However, the industrial sectors of the central and western areas which have not attained DEA-efficiency are still at the stage of increasing returns to scale, which means that increased input will lead to the larger scale production of output.

Generally speaking, DEA-inefficient industrial sectors in central and western areas have the problem of redundant inputs, particularly energy inputs. Therefore, other than saying that capital, energy, and labor can substitute each other (Smyth et al. 2011), we find that wastage in provinces (autonomous regions, municipalities) is quite serious. However, it also reflects that there is huge energy-saving potentiality in these areas. Moreover, in view of the important position of industry in China's economy, if the central and western regions can effectively reduce the redundant investment of energy in industrial sectors, the overall energy consumption of China can be reduced considerably and remarkably.

The target of energy saving and emission reduction has been emphasized by the Chinese central government in further development. It can be seen that eliminating the backward production will continue in the future to build resource-saving and environmental-friendly society. To realize the plan, it is important to promote efficiency, especially energy efficiency in China. Based on the analysis of this chapter, DEA-inefficient industrial sectors in central and western regions should speed up innovation in changing their modes from extensiveness to intensiveness, in order to comply with the call of energy saving and emission reduction by government, as well as to reduce the energy consumption and contribute to the overall improvement of China's energy efficiency.

In Sect. 7.2, we focused on analyzing the regional energy efficiency in China from both dynamic and static perspectives. In order to reveal the changing trend of energy efficiency in time series and make the dynamic model and the static model consistent with each other, the global DEA method is utilized in this study. In this way, we explore the changing trends of China's energy efficiency from 2001 to 2010, and identify the key factors for the change of energy efficiency during these periods from the aspects of technical progress, productive scale, and management level, so as to find out the potential capacity of China's energy efficiency improvement as well as the corresponding approaches. The estimation results show that: at the national level, China's energy efficiency presented an overall downward trend during 2001–2005, and technical regression and the decline in scale efficiency were due to two main reasons. At the regional level, the eastern region had the highest energy efficiency above 0.70 during the sample period, while energy efficiency in the western region fell far behind since the beginning. At the provincial level, eastern provinces of Fujian and Guangdong performed the best on energy utilization during 2001–2010.

The policy implications of the estimation results are also clear and straightforward. Firstly, at the national level, the Chinese central government should learn from the failure during 2001–2005 as well as the success during 2005–2010 in energy saving and improvement of energy efficiency. It is necessary for the Chinese government to make clear about its current status of energy efficiency and find out the key factors for the changes in energy efficiency during these two periods. Obviously, effective measures and efforts should be taken to accelerate further technical progress. Secondly, at the regional level, western China had the lowest energy efficiency. This indicates that in future the western region is China's promising growth engine of energy efficiency. Hence, preferential policies and more investments in technical innovation and environmental governance should be given to this region. In addition, it is necessary for provinces (autonomous regions, municipalities) to make clear about their performance on technical, scale and management efficiency and then foster strengths, and circumvent the weaknesses.

Section 7.3 empirically studies whether there was a causal relationship between energy technology patents and CO_2 emissions from the perspective of energy technology innovation output. Our findings fill the literature gap of ignoring this important relationship and enrich energy technology innovation theory and CO_2 reduction literature. The 1997–2008 panel data of 30 provinces (autonomous regions, municipalities) were collected, and used to examine the causal relationship between patents for fossil-fueled technologies and CO_2 emissions, between patents for carbon-free energy technologies and CO_2 emissions in eastern, central, western, and national levels of China. Using the dynamic panel data approach, unit root test and co-integration test on patents for fossil-fueled technologies, patents for carbon-free energy technologies, CO_2 emissions, and GDP were conducted to determine the stationarity of series and long-run relationships between the series. Then, the dynamic relationships between the series were examined using DIF-GMM.

We find that there is a long-run causality among patents for fossil-fueled technologies, CO_2 emissions, and GDP, and there is also a long-run causality among patents for carbon-free energy technologies, CO_2 emissions, and GDP. In the short-run: from the perspective of patents for fossil-fueled technologies, there is positive bidirectional causality between LEMS and LETP, and it is significant in the eastern and national levels of China, while not significant in central and western China. From the perspective of patents for carbon-free energy technologies, there is a significantly negative causality from LEMS to LETP in eastern, central, western, and national levels of China; there is a negative causality from LETP to LEMS, and it is significant in eastern China, while not significant in central, western and national levels of China. According to our findings, the patents for fossil-fueled technologies have no effect on emissions reduction. However, patents for carbon-free energy technologies are found to be helpful in reducing CO_2 emissions, which is significant in eastern China, while not significant in central and western China. It is consistent with findings that the effect of domestic innovation on emissions reduction is clearly different in eastern, central, and western China (Wei and Yang 2010).

References

Andrews-Speed, P. 2009. China's ongoing energy efficiency drive: origins progress and prospects. *Energy Policy* 37: 1331–1344.

Ang, J.B. 2009. CO_2 emissions, research and technology transfer in China. *Ecological Economics* 68: 2658–2665.

Arellano, M., and S. Bond. 1991. Some tests of specification for panel data: Monte Carlo evidence and an application to employment equations. *Review of Econometric Studies* 58: 277–297.

Banker, R.D., A. Charnes, and W.W. Cooper. 1984. Some models for estimating technical and scale inefficiencies in data envelopment analysis. *Management Science* 30 (9): 1078–1092.

Bernstein, P.M., W.D. Montgomery, and S.D. Tuladhar. 2006. Potential for reducing carbon emissions from non-Annex B countries through changes in technology. *Energy Economics* 28: 742–762.

Blundell, R., and S. Bond. 1998. Initial conditions and moment restrictions in dynamic panel data models. *Journal of Econometrics* 87: 115–143.

Chang, T.P, and J.L. Hu. 2010. Total-factor energy productivity growth, technical progress, and efficiency change: an empirical study of China. *Applied Energy* 87 (10): 3262–3270.

Charnes, A., W.W. Cooper, and E. Rhodes. 1978. Measuring the efficiency of decision making units. *European Journal of Operational Research* 2: 429–444.

Charnes, A., W.W. Cooper, and B. Golany, et al. 1985. Foundations of data envelopment analysis for Pareto-Koopmans efficient empirical production functions. *Journal of Econometrics* 30 (1–2): 91–107.

Chen, Q., C. Kang, and Q. Xia, et al. 2011. Preliminary exploration on low-carbon technology roadmap of China's power sector. *Energy* 36: 1500–1512.

Färe, R., S. Grosskopf, and M. Norris, et al. 1994. Productivity growth, technical progress, and efficiency change in industrialized countries. *American Economic Review* 84 (1): 66–83.

Fischer, C., and G. Newell. 2008. Environmental and technology policies for climate mitigation. *Journal of Environmental Economics and Management* 55: 142–162.

Fisher-Vanden, K., and I.S. Wing. 2008. Accounting for quality: issues with modeling the impact of R&D on economic growth and carbon emissions in developing economies. *Energy Economics* 30 (6): 2771–2784.

Garrone, P., and L. Grilli. 2010. Is there a relationship between public expenditures in energy R&D and carbon emissions per GDP? An empirical investigation. *Energy Policy* 38: 5600–5613.

Gnansounou, E., J. Dong, and D. Bedniaguine. 2004. The strategic technology options for mitigating CO2 emissions in power sector: assessment of Shanghai electricity-generating system. *Ecological Economics* 50 (1–2): 117–133.

Gonsense. 2011. The energy industry research report. http://www.gosense.cn/baogao_neirong_14.html [2011-10-21].

Hao, H., and J.F. Zong. 2007. The Analysis of System and Methods of Evaluation. Beijing: Economic Science Press.

Herring, H. 1999. Does energy efficiency save energy? The debate and its consequences. *Applied Energy* 63 (3): 209–226.

Hu, J.L., and S.C. Wang. 2006. Total-factor energy efficiency of regions in China. *Energy Policy* 34: 3206–3217.

Hu, C.Z., and X.J. Huang. 2008. Characteristics of carbon emission in China and analysis on its cause. *China Population, Resources and Environment* 18 (3): 38–42.

Hsiao, C. 2003. Analysis of panel data. Cambridge: Cambridge University Press.

IPCC. 2006a. *IPCC guidelines for national greenhouse gas inventories*. Hayama, Japan: The Institute for Global Environmental Strategies.

IPCC. 2006. IPCC guidelines for national greenhouse gas inventories. http://www.ipcc.ch/ipccreports/methodology-reports.htm.

IPCC. 2006. IPCC guidelines for national greenhouse gas inventories. http://www.ipcc-nggip.iges.or.jp/public/2006gl/pdf/2_Volume2/V2_2_Ch2_ Stationary Combustion.pdf.

IPCC. 2006. IPCC guidelines for national greenhouse gas inventories. Hayama: The Institute for Global Environmental Strategies.

Jenne, C.A., and R.K. Cattell. 1983. Structural change and energy efficiency in industry. *Energy Economics* 5 (2): 114–123.

Johnstone, N., I. Haščič, and D. Popp. 2010. Renewable energy policies and technological innovation: evidence based on patent counts. *Environmental and Resource Economics* 45: 133–155.

Kaya, Y. 1990. Impact of carbon dioxide emission control on GNP growth: interpretation of proposed scenarios. Paris: IPCC Energy and Industry Subgroup, Response Strategies Working Group.

Li, L.S., and Y. Zhou. 2006. Can the advance of technology improve energy efficiency? An empirical study based on Chinese industry department. *Management World* 10 (10): 82–91.

Li, S.X., and J.H. Cheng. 2008. Study on the energy efficiency of China and its determinants. *Statistical Research* 25 (8): 18–27.

Liu, F., and Y. Sun. 2008. Effect of innovation, industrial change on energy consumption. *China Population, Resources and Environment* 18: 108–113.

Liu, J., and X.T. Zheng. 2008. Patent system: the incentive role of technological innovation and optimization technique. *Journal of Hubei Administration Institute* 1: 58–60.

Lüthi, S., and T. Prässler. 2011. Analyzing policy support instruments and regulatory risk factors for wind energy deployment—a developers' perspective. *Energy Policy* 39 (9): 4876–4892.

Ma, C., C. Gao, and H. Shi. 2003. Energy R&D investment in China. *Energy Research and Utilization* 3: 3–6.

Margolis, R.M., and D.M. Kammen. 1999. Evidence of under-investment in energy R&D in the United States and the impact of federal policy. *Energy Policy* 27: 575–584.

Miketa, A., and P. Mulder. 2005. Energy productivity across developed and developing countries in 10 manufacturing sectors: patterns of growth and convergence. *Energy Economics* 27 (3): 429–453.

Oh, D.H., and A. Heshmati. 2010. A sequential Malmquist-Luenberger productivity index: environmentally sensitive productivity growth considering the progressive nature of technology. *Energy Economics* 32 (6): 1345–1355.

Pastor, J.T., and C.A.K. Lovell. 2005. A global Malmquist productivity index. *Economics Letters* 80 (3): 266–271.

Patterson, M.G. 1996. What is energy efficiency? Concepts, indicators and methodology issues. *Energy Policy* 24 (5): 377–390.

Patterson, M.G., M.A. West, and V.J. Shackleton, et al. 2005. Validating the organizational climate measure: links to managerial practices, productivity and innovation. *Journal of Organizational Behavior* 26 (4): 379–408.

Popp, D. 2006. International innovation and diffusion of air pollution control technologies: the effects of NO_x and SO_2 regulation in the US, Japan, and Germany. *Journal of Environmental Economics and Management* 51 (1): 46–71.

Popp, D., I. Hascic, and N. Medhi. 2009a. Technology and the diffusion of renewable energy. Draft, the Maxwell School, Syracuse University, Syracuse, NY.

Popp, D., R.G. Newell, and A.B. Jaffe. 2009b. Energy, the environment, and technological change. Cambridge: National Bureau of Economic Research.

Popp, D., I. Hascic, and N. Medhi. 2011. Technology and the diffusion of renewable energy. *Energy Economics* 33: 648–662.

Renshaw, E.F. 1981. Energy efficiency and the slump in labour productivity in the US. *Energy Economics* 3 (1): 36–42.

Sari, R., and U. Soytas. 2009. Are global warming and economic growth compatible? Evidence from five OPEC countries? *Applied Energy* 86 (10): 1887–1893.

Sorrell, S., J. Dimitropoulos, and M. Sommerville. 2009. Empirical estimates of the direct rebound effect: a review. *Energy Policy* 37: 1356–1371.

Teng Y. 2009. Indigenous R&D, technology import and energy intensity: an empirical analysis in panel data of the industrial sectors in China. *Industrial Economics Research* 5 (42): 1–6.

Tulkens, H., and P.V. Eeckaut. 1995. Non-parametric efficiency, progress and regress measures for panel data: methodological aspects. *European Journal of Operational Research* 80 (3): 474–499.

van Vuuren, D., F. Zhou, and B. de Vries, et al. 2003. Energy and emission scenarios for China in the 21st century—exploration of baseline development and mitigation options. *Energy Policy* 31 (4): 369–387.

Wang, Q., and Y. Chen. 2010. Status and outlook of China's free-carbon electricity. *Renewable and Sustainable Energy Reviews* 14 (3): 1014–1025.

Wang, W., and X. Huang. 2008. Decompositions model of factors and substantiation analysis of area carbon dioxide emissions intensity change—take Jiangsu Province as an example. *Ecological Economy* 12: 32–35.

Wang, K., B. Lu, and Y.M. Wei. 2013. China's regional energy and environment efficiency: a range-adjusted mearsure based analysis. *Applied Energy* 112: 1403–1415.

Wang, K., Y.M. Wei, and X. Zhang. 2012. A comparative analysis of china's regional energy and emission performance: which is the better way to deal with undesirable outputs? *Energy Policy* 46: 574–584.

Wei, W., and F. Yang. 2010. Impact of technology advance on carbon dioxide emission in China. *Statistical Research* 27: 36–44.

Wilson, J.K. 1944. Over five hundred reasons for abandoning the cross-inoculation groups of the legumes. *Soil Science* 58 (1): 61–70.

Wu, Q., and C. Wu. 2009. Research on evaluation model of energy efficiency based on DEA. *Journal of Management Sciences* 22 (1): 103–112.

Xu, G. 2010. The research on the relationship for energy consumption, carbon emissions and economic growth in China. Wuhan: Hua Zhong University of Science and Technology.

Xu, G., Z. Liu, and Z. Jiang. 2006. Decomposition model and empirical study of carbon emissions for China, 1995–2004. *China Population, Resources and Environment* 6: 158–161.

Yi, W., L. Zou, and J. Guo, et al. 2011. How can China reach its CO_2 intensity reduction targets by 2020? A regional allocation based on equity and development. *Energy Policy* 39: 2407–2415.

Yu, F., and J. Qi. 2007. Decomposing and analyzing the environmental pollution under the opening economic circumstances: based on the panel data of provinces—autonomous regions and municipals of China from 1990 to 2003. *Statistical Research* 24: 47–53.

Zhang, J., G.Y. Wu, and J.P. Zhang. 2004. The estimation of China's provincial capital stock: 1952–2000. *Economic Research Journal* 10: 35–44. (In Chinese).

Zhao, M., W. Zhang, and L. Yu. 2009. Carbon emission from energy consumption in Shanghai. *Research of Environmental Sciences* 22 (8): 984–989.

Zhou, P., B.W. Ang, and J.Y. Han. 2010. Total factor carbon emission performance: a Malmquist index analysis. *Energy Economics* 32: 194–201.

Chapter 8
Inter-company Cooperation on CO_2 Emission Abatement

Industrial sectors are the largest energy-related CO_2 emission sources in China, which account for more than 70% of total emission (Liu et al. 2010a, b). Increasing concern on climate change has brought great pressure for industrial companies to reduce their CO_2 emissions. A large scale of energy-intensive industrial equipment was required to be abandoned, and many heavy-energy-consuming companies have been forced to shut down due to increased regulations' control. For instance, energy-intensive production of 31.22 Mt iron and 27.94 Mt steel was made obsolete in 2011, which involved more than 150 industrial companies (MIIT 2011).

Thus, industrial companies are trying their best to reduce their CO_2 emissions. Technology innovation in energy efficiency is one of the important ways for energy-intensive industrial companies to reduce their CO_2 emissions. The application of more energy-efficient technologies is useful for emission reduction. Cleaner energy substitution is considered as another choice for industrial companies to solve their heavy CO_2 emission problems. Renewable energy (such as solar power) has been used more and more in the industrial production process.

Although there has been a great improvement in energy efficiency and energy structure in China, the CO_2 emission still increases rapidly every year because of obvious economic growth. Li et al. (2012) estimated that there have to be at least 1651 Mt of CO_2 emissions reduced by 2020 in China in order to meet the national CO_2 reduction goal. Liu et al. (2010a, b) also pointed out that industrial technology innovation could only contribute 12–14% to the national target of reducing 40–45% CO_2 emissions per unit of GDP in 2020 compared with 2005. It is hard to achieve this target merely depending on technology innovation. It is indicated that the single company's practices on CO_2 reduction (e.g., technology innovation by companies themselves) are not enough to satisfy the increasing carbon reduction demand.

This chapter takes the following literature for reference: Zhang B, Wang Z. 2012. Inter-firm collaborations on CO_2 reduction within industrial chains in China: practices, drivers and effects on firms' performances. Energy Economics, 42: 115–131.

Z. Wang and B. Zhang, *Low-Carbon Consumption in China: Residential Behavior, Corporate Practices and Policy Implication*,
https://doi.org/10.1007/978-981-15-2792-0_8

More CO_2 reduction paths need to be introduced as the supplements for China to confront the increasing CO_2 emission pressures.

As a single company's effect on emission reduction is limited, what if we combine multiple companies' powers and collaborate on CO_2 reduction? Industrial companies have already taken measures to collaborate with other companies to deal with environmental problems, such as eco-design cooperation and industrial symbiosis. Some of these measures are also helpful for CO_2 reduction. This study is designed to enrich the current research on the status of intercompany cooperation on CO_2 reduction within industrial chains. Above all, this chapter is focusing on solving the following problems.

How do the industrial companies cooperate on the carbon reduction within the industrial chains?

What kinds of determinants drive or hinder the implementation of this carbon reduction cooperation within the industrial chains?

Does this carbon reduction cooperation have any effect on the industrial companies' environmental and economic performance?

8.1 Overview of Inter-company Cooperation on CO_2 Emission Abatement

Intercompany cooperation is a new option for industrial companies to reduce their CO_2 emissions. Particularly for the companies within same industrial chain, they have similar technical backgrounds. Some of them even have supply–demand business link with each other. The cooperation of CO_2 reduction within these companies is easily conducted and less costly. More and more companies realize the importance of inter-company cooperation in CO_2 reduction. Different modes and types of collaboration have been emerging.

8.1.1 Major Modes of Inter-company Cooperation on CO_2 Emission Abatement

As cooperation is becoming important for solving CO_2 emission problems, the coming question is how to implement cooperation. Intercompany cooperation within the industrial chain is one of the options for industrial companies to solve their environmental problems. With the increasing concern about environmental problems, industrial companies begin to take a fresh look at the effects of their industrial supply chain management on the improvement of environmental performance. Many scholars have begun to recognize the potential capacity of carbon reduction through industrial chains. Zhu and Geng (2013) discussed the drivers and barriers for Chinese manufacturers cooperating with their suppliers and customers to meet the energy saving and emission reduction goals. Closed-loop supply chain became one of the options to deal with environmental and emission problems, which are being considered worldwide (e.g., Korhonen and Snäkin 2005; Lieckens and Vandaele 2007; Pochampally et al. 2009).

There are many different actions related to the intercompany cooperation within industrial chains, such as eco-product design, considering several phases of life cycle (Hugo and Pistikopoulos 2005), product recovery and remanufacturing including the participation of both suppliers and customers (Jayaraman 2006; Luo et al. 2001), reverse logistics which needs the cooperation with customers (Sheu 2008) and closed-loop supply chain which requires all the participants on the industrial chain cooperating with each other (Barker and Zabinsky 2008). Some of these practices also have positive effects on CO_2 reduction. For instance, some energy-saving designs of the products are completed under the cooperation with customers (Zhu and Geng 2013).

From the perspective of the industrial chain, this chapter classified intercompany cooperation into three categories. Considering the differences of cooperation partners, the basic modes of intercompany cooperation on CO_2 reduction through industrial chains can be described as Fig. 8.1. It follows the paths for industrial companies to reduce their CO_2 emissions by cooperation within industrial chains.

(1) From the perspective of vertical extended industrial chain, industrial companies could cooperate with their suppliers and customers on CO_2 reduction. Cooperation with customers or suppliers in using less energy during product transportation is a representative example of this kind of cooperation on CO_2 emission. Recovery of waste energy and product from the customers could not only reduce the cost of purchasing materials, but also save energy by waste energy reuse and decrease the energy consumption needed for the initial production of the saved materials. These recovery practices need cooperation from customers and suppliers. Moreover, the design of energy-saving products would

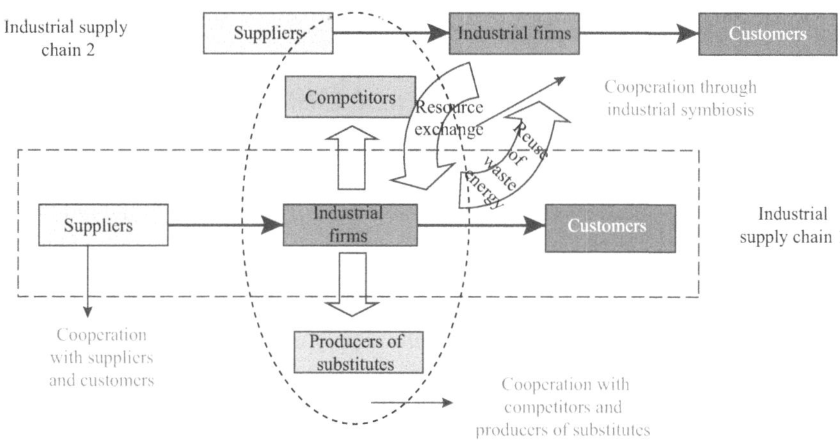

Fig. 8.1 The basic modes of intercompany cooperation on CO_2 reduction

also need to consider the demand of customers and supply capacity of the suppliers. The cooperation with customers and supplies, as a result, becomes an effective way to improve products' energy efficiency. Other cooperation practices with suppliers and customers such as green packaging and cleaner production would also have indirect positive effects on CO_2 reduction.

(2) From the point of view of horizontal expanded industrial chain, another option for industrial companies is to cooperate with their competitors and surrogates. There are some similarities in industrial technology and process between industrial companies and their competitors or surrogates. And CO_2 emission problems, most of the time, are common problems of industrial competitors and producers of substitute since they belong to the same industry and face similar environmental pressures. Therefore, there is a common ground for industrial companies to cooperate with their competitors and surrogates on CO_2 reduction. Besides, the R&D investment in energy efficiency and CO_2 reduction would be a big burden for most industrial companies. The cost could be shared by means of cooperation with competitors who have similar technology demand.

(3) From the perspective of industrial symbiosis, industrial companies in different industrial chains can cooperate with each other by the waste energy and resource exchange. Industrial symbiosis gives a platform for industrial companies to exchange their waste resources with other companies, which in most situations have no direct "supply–demand" relationship with them. This kind of cooperation among several industrial chains based on industrial symbiosis, most of the time, combines the effects of CO_2 reduction. Hashimoto et al. (2010) analyzcd thc CO_2 rcduction through industrial symbiosis for cement companies cooperating with other companies such as iron and steel companies and power plants. This provides a new option for industrial companies to reduce CO_2 emissions through cooperation with the companies in other industrial chains.

8.1.2 Cooperation with Suppliers and Customers on CO_2 Reduction

The energy-saving and emission reduction activities of enterprises in the industrial chain are often constrained by the upstream and downstream enterprises. On one hand, energy conservation and emission reduction of enterprises need clean and low-carbon inputs of raw materials and energy. The strong upstream enterprises in the industrial chain have a strong voice in the supply of raw materials, affecting the energy conservation and emission reduction decisions of enterprises. On the other hand, with the increasing popularity of the concept of energy saving and emission reduction, energy-saving and environmental protection products have gradually become a new in-demand product for the consumers. This consumption demand will lead the downstream enterprises on the industrial chain to put forward low carbon environmental protection requirements for the product manufacturing and design of its upstream enterprises, thus promoting the low carbonization of the whole industrial chain. In addition, in order to realize the aim of logistics and distribution of low carbonization, the cooperation between the upstream and downstream enterprises of the industrial chain can also be carried out in the logistics of products. Through the establishment of a stable contractual relationship, the low-carbon and energy-saving transport mode is chosen, the unnecessary packaging is reduced, the low carbonization of the raw materials procurement and transportation is realized, and the CO_2 emission is reduced. Therefore, there is a good practical basis and development potential in the upstream suppliers and the downstream target enterprises carrying out the cooperation of energy saving and emission reduction, and realizing the consistency of the emission reduction action and performance of the upstream and downstream enterprises in the industrial chain. Next, the chapter makes a detailed analysis of the collaborative emission reduction operation structure based on the vertical extension of the industrial chain.

Based on the vertical extension of the industrial chain, the realization path of collaborative emission reduction can be divided into two levels: logistical integration and technological integration. Logistical integration refers to the management cooperation between enterprises in the industrial chain on the flow of resources generated by their business contacts. It not only requires enterprises to agree on logistics, such as purchasing, transportation, inventory, and maintenance, but also sharing and exchanging information related to logistics (including enterprise inventory information, procurement plan, distribution process, etc.). From the point of view of energy saving and emission reduction, logistics synergy can reduce the use of resources and waste of energy to a certain extent, so as to achieve the overall reduction of the supply chain. For example, upstream and downstream enterprises can reach an agreement to choose more energy-efficient and environmentally friendly distribution mode, or to rationally plan distribution schemes by sharing information of procurement and inventory, reduce the number of distributions, thus reducing transportation costs and resulting energy consumption. It is also possible to reduce unnecessary packaging and achieve resource savings through the conclusion of relevant product packaging agreements, so as to achieve an indirect emission reduction effect.

Technological integration refers to knowledge or technology sharing in the strategic field between the upstream and downstream enterprises in the industrial chain. It is easier to form mutual trust and achieve technical cooperation because there is no conflict between businesses in the upstream and downstream industries. The ecological design of products is one of the effective means for enterprises to carry out energy saving and emission reduction. The realization of this product needs the technical coordination of the upstream and downstream enterprises to some extent. For example, in order to save energy and reduce emissions, it is necessary to modularize the product parts to facilitate the disassembly and recovery of parts. On one hand, it's needed to share the technical cooperation and knowledge with the suppliers of the upstream parts and components. On the other hand, reverse logistics should be designed together with the recycling channels of downstream enterprises.

According to the operation process of the whole industrial chain, the synergistic emission reduction based on the vertical extension of the industrial chain is to combine each link of the whole product life cycle. On the basis of giving full play to the professional advantages of each enterprise, the energy-saving potential in the process of inter-enterprise business exchanges is escalated. The effective integration of resources and information can achieve the effect of energy conservation and emission reduction in enterprises. The recovery and reuse of waste products is an important part of collaborative emission reduction among enterprises. The implementation of this includes logistics collaboration and technological synergy. On one hand, the recycling of abandoned products needs to build a corresponding reverse logistics channel, which requires cooperation between the upstream and downstream enterprises in the recycling of abandoned products. On the other hand, in order to facilitate the recycling of waste products, the recycling of waste products should be taken into account in the design of the product, and the corresponding recycling logistics network and management system designed. The reclaimed waste products can be divided into reuse and recycling according to their usage. The reutilization can be divided into two types: one is to reclaim the recycled products by cleaning, grinding, repairing, and repacking, and the other is to dismantle the waste products and recycle the parts with the value of use. The process of reuse generally involves cooperation between the manufacturer and the seller, and requires cooperation between the enterprises in the recovery of logistics and the discrimination of the use value of the waste products. Recirculation refers to the process of processing the waste products or parts without reusing value, and reducing them to the initial raw materials. This process is mainly related to the coordination between vendors and suppliers.

8.1.3 Cooperation with Competitors and Surrogates on CO_2 Reduction

Energy saving and climate change are not only the problems faced by the individual enterprise but also the common problems faced by all enterprises on the industrial chain. On this point, there is a common interest demand between the enterprises and their competitors. Moreover, the products produced and operated by enterprises and their competitors have the same or similar characteristics in terms of use value. Therefore, enterprises can rely on their own industrial chains to carry out horizontal expansion of energy-saving and emission reduction business, and seek cooperation between competitors and product substitutes in the same industry.

The cooperative emission reduction based on the horizontal expansion of the industrial chain can be divided into two types: one is cooperative emission reduction with the cooperation of energy saving and emission reduction technology, and the other is the cooperative emission reduction with the information service of energy saving and emission reduction. In order to reduce the R&D cost of energy-saving and emission reduction technology, the enterprise and their competitors can carry out R&D cooperation around the non-core benefits of energy-saving and emission reduction technology, realize the optimal allocation of resources and improve the efficiency of research and development. Additionally, cooperative emission reduction can reduce the cost of individual R&D and enhance the enthusiasm of enterprises to participate in energy conservation and emission reduction. However, the coordinated development of energy-saving and emission reduction technology requires a smooth system design to avoid the opportunistic behavior of participating enterprises. Because of the competitive relationship between enterprises and competitors in market possession, it is easy to cause asymmetric information in the process of cooperation, resulting in "hitchhiking" and other speculative behavior, affecting the efficiency of coordinated emission reduction. The relevant contract mechanism needs to be improved. The responsibilities and obligations of the parties to reduce the emission reduction should be made clear, and the system of rewards, punishment and supervision should be improved. The smooth development in the R&D of energy-saving and emission reduction technology must be ensured.

The cooperation of energy saving and emission reduction information service refers to the joint construction of information service platform or strategic alliance by all parties involved, sharing energy saving and emission reduction information related to industry, coping with various energy saving and emission reduction pressures in the external environment of the industry, and providing corresponding technical services. For example, in order to reduce the carbon footprint of their own products, many companies have begun to try to form raw material purchasing alliances with their competitors to improve their bargaining advantages in the procurement of low carbon raw materials and reduce the overall CO_2 emissions of their products. In addition, we can accelerate the exchange and sharing of information between the production links within, or with other enterprises outside the enterprise, through the formation of the platform or the alliance, so as to simplify the process of carbon trading and thus improve the enthusiasm of the enterprises to save energy and reduce emission.

8.1.4 Inter-company Cooperation on CO_2 Emission Abatement Based on Industrial Symbiosis

In the course of production and processing, enterprises will produce some by-products and waste. If we discard or discharge this waste directly, we will waste resources and increase energy consumption and CO_2 emissions indirectly. Although these by-products or waste do not have the use value to the enterprises themselves, it is possible to become the necessary raw materials for the production process of other enterprises through certain treatment. Therefore, if we make reasonable use of these abandoned resources, we can reduce the consumption of resources, so as to achieve the effect of energy conservation and emission reduction. The cooperative emission reduction based on industrial symbiosis refers to the reuse of the abandoned resources of other enterprises through the exchange or transfer of the abandoned resources between the enterprises and other enterprises on the industrial chain. At the same time, the by-products and waste produced in the process of production are transferred to other enterprises, while the cost of their own operation is reduced, the recycling of resources between enterprises is realized, and the purpose of energy saving and emission reduction is achieved. This process is not only limited to enterprises within the industrial chain. As long as there is a demand for the exchange or use of abandoned resources among enterprises in different industrial chains, the collaborative emission reduction of this model can also be realized.

According to the different intergrowth relationships between enterprises, the cooperative emission reduction based on industrial symbiosis can be divided into the type of core enterprise leading synergistic emission reduction and the type of equal cooperative emission reduction. The core enterprise led synergistic emission reduction means that the industrial symbiotic relationship is carried out under the guidance of one or several core enterprises, and other enterprises attach to the core enterprises to carry out synergistic emission reduction. In this process, the core enterprises produce a large number of by-products and waste gas resources (such as waste heat, residual energy, etc.). These waste gas resources are indispensable raw materials for the production and operation of other subsidiary enterprises. The core enterprise led synergistic emission reduction is characterized by the strong dependence of the whole cooperative network on the core enterprises, and the operation of the core enterprise leading the whole cooperative emission reduction is in the absolute dominant position in the negotiation with the subsidiary enterprises. Generally speaking, the core enterprises are large enterprises engaged in petrochemical, smelting, mechanical manufacturing, or energy production. The demand for raw materials and the supply of abandoned resources are sufficient and stable, and have a strong scale advantage. It is often the founder of the core enterprise led cooperative emission reduction relationship, which determines the technical feasibility of the whole cooperative network operation and has an irreplaceable role. Once the operating environment or operation strategy of the core enterprise changes, it will have a great impact on the relationship with the enterprise, and can even lead to the breakdown of the cooperative emission reduction relationship. In order to maintain the stability and security of the cooperative emission reduction relationship, dependent enterprises often need to choose a number of core enterprises to establish a stable cooperative relationship, so as to avoid the risk of dependence on a single core enterprise.

The cooperative emission reduction based on equality and cooperation means that there is no dependency relationship between enterprises in the process of cooperative emission reduction. At the same time, an enterprise will interact with many enterprises to dispose of resources, so as to achieve the synergistic emission reduction effect of many enterprises. The participating enterprises are in a relatively equal position in the process of coordinating emission reduction, and rely on the regulation mechanism of the market to realize the effect of synergistic emission reduction among enterprises on the industrial chain and the value chain. The enterprises participating in the equal type of cooperative emission reduction generally are the majority of the small and medium enterprises, the organizational structure is relatively flexible, and the regulation mechanism of the market should be based on the benefit of reducing the emission reduction. In Canada, for example, there are more than 1300 enterprises in the Burnside Industrial Park, which is mainly composed of small and medium high-tech enterprises engaged in deep processing of products (such as pulp mills, paper mills, construction boards, petroleum refineries, etc.). An enterprise can establish a symbiotic relationship based on the exchange of waste resources with several enterprises to achieve synergistic emission reduction. Their operation mode is based on an equal type of cooperative emission reduction relations. Through the specialized division of labor, the outsourcing of resources and the exchange of resources in all directions, enterprises can form a complex symbiotic network. In this process, the overall effect of resource recycling and collaborative emission reduction is achieved.

In the actual collaborative emission reduction operation process, the distinction between the core enterprise led collaborative emission reduction type and the equal type of collaborative emission reduction is sometimes blurred. On one hand, the cooperative emission reduction system is implemented in some core enterprises, with each core enterprise having a lot of subsidiary enterprises. The subsidiary enterprises rely on the abandoned resources of the core enterprises to carry out production and management. On the other hand, the core enterprises also have the exchange of waste resources, and operate on the basis of the equal type of cooperative emission reduction structure. In addition, there is also an exchange of waste resources between the subsidiary enterprises of core enterprises. The whole cooperative emission reduction system is operated in this nested mode to achieve the improvement of emission reduction efficiency of the system.

8.2 Determinants of Inter-company Cooperation on CO_2 Emission Abatement

After we understand the basic modes of inter-company cooperation on CO_2 emission abatement, it is important to make clear what drives and hinders the implementation of each kind of intercompany cooperation. A conceptual model (Fig. 8.2), as a result, is constructed to solve these problems. These determinants from the model are identified on the basis of institutional theory and resource-based view.

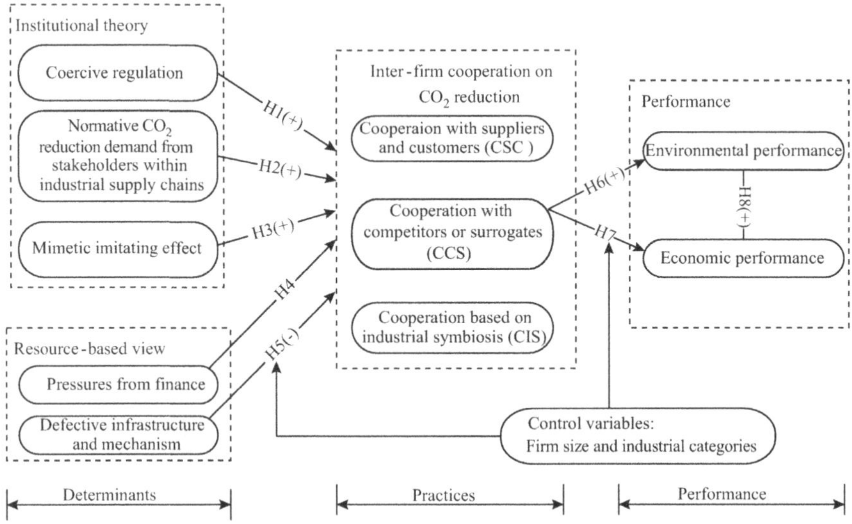

Fig. 8.2 Conceptual model

8.2.1 Hypothesis Modeling for Determinants of Inter-company Cooperation on CO_2 Emission Abatement

Institutional theory is widely used in identifying the external determinants for environmental management practices (e.g., Zhu and Geng 2013; Prajogo et al. 2012; Zhu et al. 2012). DiMaggio and Powell (1983) pointed out that the environmental practices could be driven by three kinds of institutional factors: coercive, normative, and mimetic. Intercompany cooperation on CO_2 reduction is a representative environmental practice, which as a result, can also identify its external determinants from the view of institutional theory.

Coercive driver comes from the compulsive pressure exerted by powerful agencies such as the government (Rivera 2004). Environmental regulation is often considered as the coercive pressure, and its effect on motivating the environmental practices of the companies has been verified by many scholars (e.g., Jones 2010; López-Gamero et al. 2010). Zhang and Cheng (2009) verified that the government of China could pursue CO_2 reduction policy in the long run without impeding economic growth. Correspondingly, the Chinese government has promulgated a series of laws and decrees to regulate the industrial companies' production process on energy use and CO_2 emission, such as Energy Conservation Law (first effective in 1997 and amended in 2007) and Circular Economy Promotion Law (effective from January 1, 2009, and amended in 2018). Many of these regulations promote industrial companies to find more ways to deal with their CO_2 emission problems. Especially when the industrial companies' single CO_2 reduction practice has limited effect, they have to rely on cooperation with each other on CO_2 reduction. From the above analysis, we present the first hypothesis.

Hypothesis 8.1 Environmental regulations motivate intercompany cooperation on CO_2 reduction within the industrial chain.

Normativity refers to the influence from external stakeholders who have business relation with the organization. From the perspective of the industrial chain, suppliers and customers, and even the competitors can be considered as the stakeholders who can influence the industrial companies' environmental behavior (Yin and Ma 2009; Jørgensen et al. 2010). For one thing, the demand for CO_2 reduction for these stakeholders on the industrial chain is the prerequisite of intercompany cooperation. Only when both the industrial companies and their stakeholders (e.g., suppliers, customers or the competitors) need to reduce CO_2 emission, there are chances for them to cooperate with each other on CO_2 reduction. For another, the requirements of carbon reduction from customers and suppliers are also key drivers for industrial companies to implement environmental practices (Zhang et al. 2008), such as intercompany cooperation on carbon reduction. Based on the above discussion, the second hypothesis is as follows.

Hypothesis 8.2 CO_2 reduction demand from stakeholders motivates intercompany cooperation on CO_2 reduction within the industrial chain.

Imitation refers to the imitating effect that individuals tend to learn from the successful practices of other individuals. The imitating effect is also reflected in the implementation of environmental practices (Prajogo et al. 2012). The good performance of some industrial companies' environmental practices often motivates other companies to imitate. Intercompany cooperation on CO_2 reduction also needs several successful cooperation cases to demonstrate the advantage and possibility of cooperating with other stakeholders within the industrial chains on carbon reduction. As a result, other industrial companies can be attracted to imitate these successful experiences and participate in the intercompany cooperation on CO_2 reduction. Then we have the following hypothesis.

Hypothesis 8.3 Imitating effect motivates intercompany cooperation on CO_2 reduction within the industrial chain.

The internal determinants for industrial companies to implement environmental practices are often identified from the resource-based view (e.g., Zhu and Geng 2013; Prajogo et al. 2012). It was verified that the lack of resource and capacity are the main barriers to environmental practices (Ebinger et al. 2006). This chapter also identifies the internal determinants of intercompany cooperation on carbon reduction from the resource-based view.

Financial factors such as short-term cost burden (Sarkis et al. 1997) and unclear benefits (van Hemel and Cramer 2002) are the important determinants for the implementation of environmental practices. However, the effects of financial pressures on intercompany cooperation on CO_2 reduction are a bit more complicated. On one hand, the cooperation increases the operating cost of industrial companies, which impedes the implementation of intercompany cooperation on

CO_2 reduction. On the other hand, the increasing carbon reduction cost forces the industrial companies to cooperate with each other to share the heavy financial burden of carbon reduction. From this perspective, financial pressures play a positive role in motivating intercompany cooperation on CO_2 reduction. From the above analysis, the following hypotheses are given.

Hypothesis 8.4 Financial pressures motivate intercompany cooperation on CO_2 reduction within the industrial chain.

Financial pressures impede intercompany cooperation on CO_2 reduction within the industrial chain.

Defective infrastructure and mechanism are a big problem that influences the intercompany cooperation. Lopéz (2008) pointed out that the mechanism of cost-risk sharing is the key factor for successful cooperation. It is more complicated to design the cost-risk sharing mechanism for intercompany cooperation on carbon reduction, because of its variety of cooperative practice patterns. Trust is another important factor that determines the success of cooperation (Okamuro 2007). Industrial companies prefer to cooperate with trusted participants on carbon reduction, especially those who are familiar with or have cooperative experience with before. The trust mechanism can reduce the risk of opportunistic behavior such as the "free riders", which could be quite common during the intercompany cooperation on carbon reduction. The defective safety mechanism also impedes the implementation of intercompany cooperation on CO_2 emission. Some industrial companies worry about the leakage of their key information during the cooperation. Thus, the fifth hypothesis is put forward.

Hypothesis 8.5 Defective infrastructure and mechanism impede intercompany cooperation on CO_2 reduction within the industrial chain.

8.2.2 Questionnaire Survey Design and Data Collection

We designed a questionnaire survey to investigate the intercompany cooperation on CO_2 emission abatement. The processes of questionnaire development and data collection are described in this section.

(1) Questionnaire development

Four sections are designed to constitute the questionnaire: (1) Intercompany cooperation on CO_2 reduction; (2) Determinants; (3) Performance; and (4) Basic information. The measurement structure is organized as the conceptual model shown in Fig. 8.2. The measurement items in each section are developed on the basis of the previous studies. There are five measurement items in the section of intercompany cooperation on CO_2 reduction. Three measurement items are to evaluate the carbon reduction cooperation with suppliers and customers, which refers to the studies of Zhu et al. (2007), Carter and Carter (1998) and Chan and Lau (2001). One measurement item is designed to evaluate the carbon reduction cooperation with competitors and surrogates, which refers to the study of Poyago-Theotoky (2007). And the last measurement item is for cooperation based on industrial symbiosis, which refers to the study of Zhu and Geng (2013). This is because there are various patterns of cooperation with suppliers and customers, such as the cooperation on eco-design, energy-saving transportation, and waste product recovery. Contrarily, the other two kinds of cooperation are simple. All the items in this section are measured according to the implementation experience of the cooperation, by using a five-point Likert-type scale (1—not considering it; 2—plan to consider it; 3—just implemented for less than one year; 4—implemented for 1–3 years; 5—implemented for more than 3 years).

There are 15 measurement items to evaluate the determinants in the second section of the questionnaire. Each determinant is in response to three measurement items. The three items of *Regulation* are learned from the study of Zhu et al. (2007). The dimension of *CO_2 reduction demand from stakeholders within industrial supply chains* is evaluated referring to the studies of Jørgensen et al. (2010), Yin and Ma (2009) and Zhu et al. (2007). The items of *Imitating effect* are designed referring to Zhu and Geng (2013), Prajogo et al. (2012). The items for evaluating *Financial pressure* are developed on previous studies (Sarkis et al. 1997; van Hemel and Cramer 2002; Zhang et al. 2012). And defective infrastructure and mechanism is measured by three items referring to the studies of Okamuro (2007), Lopéz (2008), Bonte (2008). Questions for these items are answered by the agreement on the items' description with the five-point Likert-type scale (1—strongly disagree; 2—tend to disagree; 3—neither agree nor disagree; 4—tend to agree; and 5—strongly agree).

The 8 measurement items are designed in the third section of performance: 4 items for evaluating environmental performance and 4 items for economic performance. These items are designed mainly by referring the experience of Zhu et al. (2010). Questions for each item are developed by asking about companies' performance improvement in recent 3 years, which are also answered using a five-point Likert-type scale (1—none, 2—some but insignificant, 3—some but slightly significant, 4—significant, and 5—highly significant). A brief explanation at the beginning of each section is given to avoid the misunderstanding of the items.

In order to validate the measurement structure of the questionnaire, we made a pilot test by distributing the questionnaire draft to 10 senior managers who were in charge of environmental management issues from industrial companies in China. We asked them to complete the questionnaire and provide comments on the understandability of the questionnaire items and how to make further improvements. Minor modifications to the wording of items were made during the pilot test.

(2) Data collection and sampling characteristics

Energy-intensive industrial companies are targeted as the source of data collection in this chapter. Energy-intensive industries are characterized by high energy to output ratios. They are the industries that have large amounts of energy consumption but relatively low-value output (Mongia et al. 2001). Accordingly, six industries[1] are chosen out of 37 identified sectors according to the criterion that the total primary energy consumption is over 100 million tons SCE[2] and energy consumption out of the output value is more than 50 thousand tons SCE/billion yuan, as shown in Fig. 8.3.

Convenience sampling was used in this chapter by investigating the MBA/EMBA from the School of Management and Economics at Beijing Institute of Technology. Similar data collection methods were also conducted by some precedents (e.g., Zirger and Maidique 1990), due to the difficulties in obtaining data. Christmann and Taylor (2001) and Luo (2001) also completed their empirical studies in China by surveying MBA/EMBA students or participants in workshops. Besides, the MBA/EMBA students are at the same time the middle or senior managers of companies, or have the working experience in management. They often have board knowledge about the companies' daily operation.

The whole survey was undertaken over a period of five months from April to August 2011. Out of a total of 693 questionnaires distributed to representatives, a total of 258 usable responses from energy-intensive industrial companies were collected.[3]

Table 8.1 shows the description of respondent companies in terms of industrial type and company size using employment levels. The company size of industrial companies is often characterized by the number of full-time employees (Dean and Snell 1991). The company size of the respondents ranged from under 100 to over 3000 employees with the majority of companies falling into middle-sized companies classified between 500 and 3000 employees. These five types of industries represent the energy-intensive consumers and heavy carbon emitters.

(3) Reliability and validity test

To further verify the reliability of the responses, Cronbach's alpha coefficients and item-total correlations are introduced in this chapter. As shown in Table 8.2, the majority of Cronbach's alpha coefficients of measurement items are above 0.90, and the lowest one reaches 0.889. The threshold value of reliability was recommended above 0.70 by Nunnally and Bernstein (1994). This suggests a good internal consistency of the questionnaire. Besides, the item-total correlations of all measurement items are at a high level (above 0.70), which suggests that each item has a close correlation to their corresponding higher level constructs. Most of the values decreased compared with the Cronbach's alpha coefficients before deleting the items. This further indicates the good reliability of the questionnaire. Although the

[1]Six industries are Petroleum Processing and Coking (PPC), Smelting and Pressing of Nonferrous Metals (SPNM), Smelting and Pressing of Ferrous Metals (SPFM), Raw Chemical Materials and Chemical Products (RCMCP), Electric Power, Steam and Hot Water (PESP), Nonmetal Mineral Products (NMP).

[2]SCE is unit of energy, which is short for standard coal equivalent.

[3]Although 667 questionnaires were received, most were from the industries other than the six energy-intensive industrial firms.

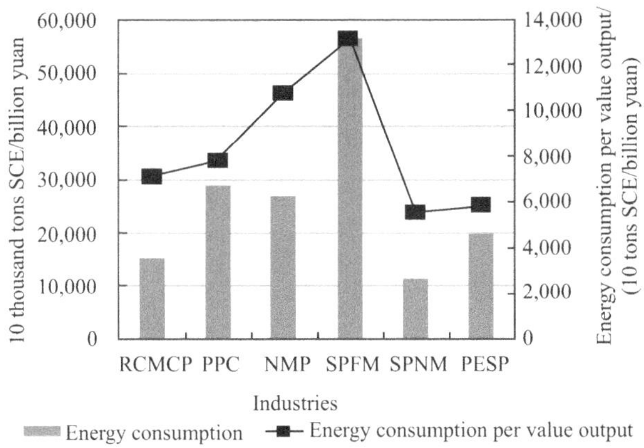

Fig. 8.3 Energy intensive industries: some indicators. *Source* NBSC (2009)

Table 8.1 Profile of the respondents participating in the survey

Characteristics of respondents		Number	Percentage/%	Cumulative percentage/%
Number of employees	<100	41	15.9	15.9
	[100–500)	64	24.8	40.7
	[500–1000)	43	16.7	57.4
	[1000–3000)	65	25.2	82.6
	>3000	45	17.4	100
Total		258	100	–
Industrial type	PPC & RCMCP[a]	79	30.6	30.6
	PESP	23	8.9	39.5
	NMP	47	18.2	57.8
	SPFM	60	23.3	81.0
	SPNM	49	19.0	100
Total		258	100	

[a]PPC and RCMCP have a strong industrial relationship with each other. Many firms of the respondents involved in both industrial businesses. Therefore, we combined the two industries into one type

Cronbach's alpha coefficients increased after the three items (L3, M3, J1) were deleted, the increased values are too minor to be considered. And the previous Cronbach's alpha coefficients are high, we still confirmed the three items reliability in the measurement structure.

Table 8.2 Reliability test and descriptive statistics

Factors	Items	Item-total correlations	Cronbach's alpha after the items were deleted	Cronbach's alpha
Regulation (REG)	L1	0.929	0.931	0.960
	L2	0.944	0.918	
	L3	0.876	0.970	
CO_2 reduction demand from stakeholders within industrial chains (CDI)	I1	0.773	0.854	0.889
	I2	0.780	0.850	
	I3	0.806	0.822	
Imitating effect (IE)	M1	0.872	0.892	0.932
	M2	0.898	0.871	
	M3	0.813	0.938	
Financial pressure (FP)	E1	0.781	0.903	0.910
	E2	0.882	0.817	
	E3	0.802	0.886	
Defective infrastructure and mechanism (DIM)	J1	0.901	0.970	0.967
	J2	0.954	0.931	
	J3	0.929	0.950	
Environmental performance (EP)	EP1	0.795	0.881	0.908
	EP2	0.850	0.860	
	EP3	0.815	0.873	
	EP4	0.716	0.907	
Economic performance	CP1	0.783	0.881	0.907
	CP2	0.868	0.855	
	CP3	0.743	0.898	
	CP4	0.777	0.883	

The confirmatory factor analysis (CFA) was introduced to test the construct validity of the questionnaire. First, the factors were extracted using the orthogonal rotation, following the hypothesis construction in Fig. 8.2. The results were shown in the left part of Fig. 8.4. It was shown that the estimated standard coefficients of each item were all above 0.8, which suggests a good measurement structure. However, some fitting indices were not as good as the estimated standard coefficients, seen in Table 8.3. Root mean square error of approximation (RMSEA) was 0.083, which is a bit lower than the good fitting criterion of below 0.08 suggested by Steiger (1990). Goodness of fit index (GFI) was 0.88, which does not reach the threshold of 0.90 suggested by Huang (2005) as well.

Modification indices (MIs) suggest that there is a certain co-variation relation between the factor *Regulation* and *CO_2 reduction demand from stakeholders within industrial chains*. In order to verify this relationship, we conducted another CFA using oblique rotation. The results were shown in the right part of Fig. 8.4. It was shown that the estimated standard coefficient between the two factors was 0.43, which means there are certain but not very significantly important relations between the two factors. However, the fitting indices after the modified CFA become much better. Both the RMSEA and GFI pass the threshold of the criterion, as seen in Table 8.3. It is an understandable explanation of the co-variation relation between the two factors, since regulation could play a positive role in forcing the other stakeholders within the supply chain to address their CO_2 emission problems.

The Chi-square values in both two CFA models are a bit high, however, the normed Chi-square indices (χ^2/df) are low enough to fit the criterion of below 5 suggested by Hou et al. (2004). Above all, we confirm the acceptable validity of the questionnaire structure.

8.2.3 Modeling for the Influence Towards Inter-company Cooperation on CO_2 Emission Abatement

After the data collection by the survey, we used Multiple Linear Regression (MLR), Binary Choice Model (BCM), and Ordinal Choice Regression (OCR) to analyze three kinds of intercompany cooperation on CO_2 emission, respectively.

(1) MLR for determinants of carbon reduction cooperation with suppliers and consumers

There are three measurement items for evaluating cooperation with suppliers and consumers on carbon reduction. Factor analysis was introduced to validate the measurement construction using the principal component methods. Table 8.4 shows the factor matrix with varimax rotation. One theoretical factor was extracted using the Kaiser criterion (eigenvalues > 1), with the explanation power of the inherent variations reaching 70.912%. Cronbach's alpha reached 0.789, which confirms the reliability of the factor.

We used multiple linear regression to discuss the relationship between cooperation with suppliers and consumers on carbon reduction and its determinants. The dependent variable is from the calculation of factor score in the above factor analysis, seen as follows:

$$ZC_i = \varpi_1 ZC_{1i} + \varpi_2 ZC_{2i} + \varpi_3 ZC_{3i} \tag{8.1}$$

where ZC_i refers to the score of cooperation with suppliers and consumers on carbon reduction for the No. i response. ϖ_j refers to the loading coefficients for No. j item on the factor ZC.

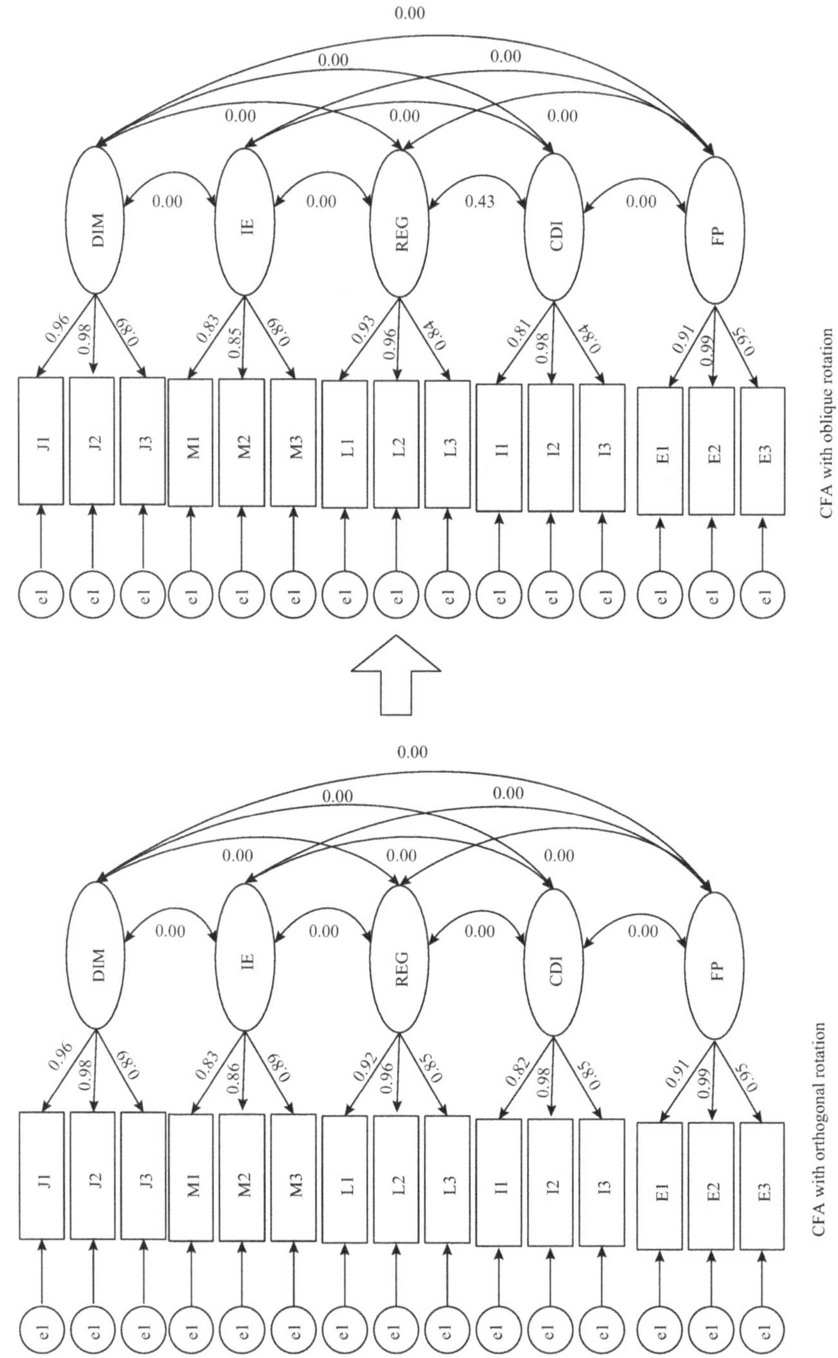

Fig. 8.4 CFA results with orthogonal rotation and oblique rotation

Table 8.3 The fitting indices of CFA

	χ^2	df	Fitting indices					
			χ^2/df	RMSEA	GFI	NFI	CFI	IFI
Criterion			<5	<0.08	>0.9	>0.9	>0.9	>0.9
CFA with orthogonal rotation	250.31	90	2.781	0.083	0.88	0.935	0.957	0.958
CFA with oblique rotation	203.408	89	2.285	0.071	0.90	0.947	0.970	0.970

Table 8.4 The factor matrix on cooperation with suppliers and consumers on carbon reduction

Indicator	Measurement item	Component
		1
Cooperation with suppliers and consumers on carbon reduction (Cronbach's alpha = 0.789)	ZC_1	0.856
	ZC_2	0.903
	ZC_3	0.761

The following functional form is the MLR in this chapter:

$$ZC = \beta_0 + \beta_1 REG + \beta_2 CDI + \beta_3 IE + \beta_4 FP + \beta_5 DIM + \beta_6 control + \varepsilon \quad (8.2)$$

where each determinant (REG, CDI, IE, FP and DIM) identified in Sect. 8.2.1 is also evaluated by factor scores through the EFA to form a single indicator factor. Company size and industrial type are used as the control variables, and industrial type is introduced in the format of the dummy variable. $\beta_0 - \beta_6$ are the estimated coefficients of this model; ε is the regression error term.

(2) BCM for determinants of carbon reduction cooperation with competitors and surrogates

From the distribution of the variable HC, most scores are below 3 or at 3. This means most companies either have not implemented carbon reduction cooperation with competitors or surrogates (below 3), or just conducted the cooperation within a year (at 3). Only a few responses get the score at 4 or 5, which means carbon reduction cooperation with competitors or surrogates has been implemented for more than 2 years. As a result, we transformed the score structure into a "0–1" model. The scores below 3 were transferred to 0, which means the company did not conduct carbon reduction cooperation with competitors or surrogates. And the scores above 2 were transferred to 1, which means the company has conducted carbon reduction cooperation.

Then a binary choice model (BCM) was introduced to discuss the determinants for carbon reduction cooperation with competitors or surrogates. The following specification was used:

$$HC^* = \beta_0^{hc} + \beta_1^{hc} REG + \beta_2^{hc} CDI + \beta_3^{hc} IE + \beta_4^{hc} FP + \beta_5^{hc} DIM + \beta_6^{hc} control + \varepsilon^{hc} \tag{8.3}$$

$$P(HC = 1) = 1 - F(-\mathrm{HC}^*) \tag{8.4}$$

$$s.t. \quad 0 \leq \mathrm{P}(HC_i = 1|\mathrm{x_i}) \leq 1, \quad \mathrm{x_i} = \mathrm{RGE}, CDI, IE, FP, DIM \tag{8.5}$$

where HC^* is latent and continuous measure of carbon reduction cooperation with competitors or surrogates; HC is the observed variable of carbon reduction cooperation with competitors or surrogates; β^{hc} is the vector of parameters to be estimated; ε^{hc} is the random error term. F is the probability distribution function assumed proactively; P is the possibility value between 0 and 1.

(3) OCR for determinants of carbon reduction cooperation based on industrial symbiosis

The observed variable of IS is characterized by how long has carbon reduction cooperation based on industrial symbiosis been implemented. Five tiers from 1 to 5 have been given to evaluate the implementation. As a result, ordinal choice regression (OCR) was introduced to discuss the determinants of carbon reduction cooperation based on industrial symbiosis. The following specifications were used:

$$IS^* = \beta_0^{is} + \beta_1^{is} REG + \beta_2^{is} CDI + \beta_3^{is} IE + \beta_4^{is} FP + \beta_5^{is} DIM + \beta_6^{is} control + \varepsilon^{is} \tag{8.6}$$

$$IS = \begin{cases} 1, & IS^* \leq c_1 \\ 2, & c_1 < IS^* \leq c_2 \\ 3, & c_2 < IS^* \leq c_3 \\ 4, & c_3 < IS^* \leq c_4 \\ 5, & c_4 < IS^* \end{cases} \tag{8.7}$$

$$\begin{gathered} P(IS = 0) = F(c_1 - IS^*) \\ P(IS = 1) = F(c_2 - IS^*) - F(c_1 - IS^*) \\ P(IS = 2) = F(c_3 - IS^*) - F(c_2 - IS^*) \\ P(IS = 4) = F(c_4 - IS^*) - F(c_3 - IS^*) \\ P(IS = 5) = 1 - F(c_4 - IS^*) \end{gathered} \tag{8.8}$$

where IS^* is latent and continuous measure of carbon reduction cooperation with competitors or surrogates; IS is the observed variable of carbon reduction cooperation with competitors or surrogates; β^{is} is the vector of parameters to be estimated; c is the vector of unobserved threshold value; ε^{is} is the random error term; F is the probability distribution function assumed proactively.

8.2.4 Empirical Analysis for the Influence of Inter-company Cooperation on CO_2 Emission Abatement

(1) Determinants of carbon reduction cooperation with suppliers and consumers

Table 8.5 shows the regression results of determinants' effects on carbon reduction cooperation with suppliers and consumers. Variance inflation factors (VIFs) were calculated to check multi-collinearity among independent variables. The largest VIF score in all of the regression models in Table 8.5 is 2.495, which passes the threshold of below 10.0 suggested by Mason and Perreault (1991). In order to avoid the heteroscedasticity during the regression, the ordinary least squares method after white robust standard deviation correction was used by using EVIEWS 6.0. Model 1 has only introduced company size as the control variable. Model 2 considered both company size and industrial type as the control variables. Model 3 was to verify the interaction effect between REG and CDI on carbon reduction cooperation with suppliers and consumers. And the F values of the three models are all high enough to reject the null hypotheses that the independent variables are not associated with the dependent variable.

All three models show that carbon reduction demand from the stakeholders in the supply chain is the main driver for carbon reduction cooperation with suppliers and consumers. If the carbon reduction pressures of the suppliers and consumers are strong, the industrial companies are more easily getting into carbon reduction cooperation with them. Financial pressure is the significant determinant to impede the carbon reduction cooperation with suppliers and consumers. The cooperation might increase the operation cost of the companies, which makes them unwilling to cooperate with each other. The defective infrastructure and mechanism also play a negative role in carbon reduction cooperation with suppliers and consumers. Since the notion of carbon reduction cooperation with suppliers and consumers is relatively new, companies do not dare to try this kind of cooperation until there are some mature and trustworthy mechanisms. However, the imitating effect does not come into effect since the successful cases of carbon reduction cooperation are still few.

Table 8.5 MLR results for determinants of carbon reduction cooperation with suppliers and consumers

Factors	Regression results			Model 1 VIF	Model 2 VIF	Model 3 VIF
	Model 1	Model 2	Model 3			
REG	−0.017 (−0.197)	−0.023 (−0.272)		1.192	1.063	
CDI	0.219*** (3.170)	0.233*** (3.410)	0.239* (1.800)	1.025	1.058	1.058
IE	0.126 (1.501)	−0.112 (1.362)	−0.110 (−1.348)	1.136	1.327	2.560
FP	−0.144** (−2.107)	−0.154** (−2.357)	−0.158** (−2.419)	1.147	1.152	1.152
DIM	−0.116*** (−2.935)	−0.121** (−3.030)	−0.120** (−3.023)	1.033	1.052	1.050
REG × CDI			−0.002 (−0.105)		2.495	2.495
Company size	−0.016 (−0.324)	−0.034 (−0.720)	−0.034 (−0.704)	1.041	1.203	1.064
PPC & RCMCP		0.491*** (3.022)	0.490*** (3.028)		1.864	1.864
PESP		0.454* (1.651)	0.453 (1.648)		1.358	1.358
NMP		0.113 (0.587)	0.110 (0.575)		1.663	1.662
SPFM		0.538*** (3.824)	0.535*** (3.091)		1.778	1.778
β_0	−0.382 (−0.750)	−0.575 (−1.165)	−0.636 (−1.264)			
F statistic	4.270***	4.030***	4.020***			
R^2	0.093	0.140	0.140			

Notes ① *Means $p < 0.1$; **means $p < 0.05$; ***means $p < 0.01$; ② The data at the upside of the cell are the standard coefficients; the data at the bottom of the cell are the t statistic values after robust standard error modifying

From Model 2, we can see the performance of carbon reduction cooperation with suppliers and consumers for the industries of PPC and RCMCP, PESP and SPFM are better than NMP. And the SPFM gets the highest score.

It is shown by Model 3 that the interaction effect of REG and CDI on carbon reduction cooperation with suppliers and consumers is not significant. The environmental regulation does not have a direct positive effect on promoting the carbon reduction cooperation with suppliers and consumers, and has no indirect effect on this kind of cooperation by influencing other companies' carbon reduction practices as well.

(2) Determinants of carbon reduction cooperation with competitors or surrogates

The BCM results for determinants of carbon reduction cooperation with competitors or surrogates are shown in Table 8.6. Both Logit model and Probit model were introduced in this study. The estimated results of the significance for each determinant are not significantly different among the models. However, Hosmer–Lemeshow (H–L) statistic values show that the fitting degree of the Probit model is better than the Logit model. The VIF values of the independent variables in each model are all below 10, which shows no multi-collinearity problem during the estimation. In order to avoid heteroscedasticity during the regression, the estimation method of the generalized linear model (GLM) was used by the application of EVIEWS 6.0. Company size was considered as the only control variable in Model 4. Model 5 considered both company size and industry type as the control variables. Model 6 considered whether the interaction of REG and CDI had some effects on carbon reduction cooperation with competitors or surrogates. And LR statistic values of the three models are all high enough to reject the null hypotheses that the independent variables are not associated with the dependent variable.

Carbon reduction demand from the stakeholders in the supply chain is also the main driver for carbon reduction cooperation with competitors or surrogates. The carbon reduction pressures from suppliers and customers could promote the industrial companies to conduct carbon reduction practices. Cooperation with competitors or surrogates is one of the options for carbon reduction. The defective infrastructure and mechanism are also the barriers that impede carbon reduction cooperation with competitors or surrogates. The environmental regulation and imitating effect also play no significant role in this kind of cooperation.

Different from the cooperation with suppliers and customers, financial pressure has a positive effect on carbon reduction cooperation with competitors or surrogates. Carbon reduction cost becomes an increasing expenditure for industrial companies since climate change gets more and more attention from the public in recent years. In order to save the expenditure of carbon reduction, many industrial companies choose to cooperate with each other, for example, cooperation on carbon reduction R&D.

Company size became a significant factor after the industrial type was introduced in the model. It is indicated that the smaller companies are willing to cooperate with others on carbon reduction R&D for each industry. This might be attributed to the fact that smaller companies have more financial difficulties by conducting carbon reduction practices. The industry of PESP gets the highest score on the performance of carbon reduction cooperation with competitors or surrogates.

Table 8.6 BCM results for determinants of carbon reduction cooperation with competitors or surrogates

	Logit model			Probit model			Model 4 VIF	Model 5 VIF	Model 6 VIF
	Model 4-L	Model 5-L	Model 6-L	Model 4-P	Model 5-P	Model 6-P			
REG	−0.170 (−1.066)	−0.107 (−0.702)		−0.104 (−1.196)	−0.071 (−0.778)		1.192	1.063	
CDI	0.637*** (3.449)	0.687*** (3.376)	0.753*** (2.830)	0.393*** (3.584)	0.416*** (3.515)	0.463** (2.978)	1.025	1.058	1.058
IE	−0.179 (−1.066)	−0.204 (−1.075)	−0.203 (−1.069)	−0.103 (−1.005)	−0.129 (−1.153)	−0.128 (−1.150)	1.136	1.327	2.560
FP	0.285** (2.174)	0.308** (2.173)	0.309** (2.178)	0.168** (2.137)	0182** (2.173)	0.182** (2.177)	1.147	1.152	1.152
DIM	−0.201** (−2.127)	−0.226** (−2.195)	−0.226** (−2.200)	−0.125** (−2.148)	−0.138** (−2.262)	−0.138** (−2.265)	1.033	1.052	1.050
REG * CDI			−0.022 (−0.607)			−0.015 (−0.700)		2.495	2.495
Company size	−0.111 (−1.125)	−0.193* (−1.745)	−0.193* (−1.745)	−0.066 (−1.091)	−0.112* (−1.707)	−0.113* (−1.711)	1.041	1.203	1.064
PPC & RCMCP		1.837*** (4.105)	1.839*** (4.111)		1.118*** (4.236)	1.119*** (4.241)		1.864	1.864
PSEP		3.523*** (4.643)	3.529*** (4.653)		2.067*** (5.108)	2.071*** (5.119)		1.358	1.358
NMP		1.216** (2.530)	1.215** (2.529)		0.758*** (2.618)	0.756*** (2.614)		1.663	1.662
SPFM		1.775*** (3.824)	1.779*** (3.832)		1.090*** (3.920)	1.092*** (3.930)		1.778	1.778
LR value	36.565	73.971	73.844	36.602	74.104	73.988			
McFadden R^2	0.103	0.208	0.208	0.103	0.208	0.208			
H−L statistic	8.412	19.795	21.427	8.267	12.769	17.814			

Notes ① ***Means $p < 0.01$; **Means $p < 0.05$; *Means $p < 0.1$; ② The data at the upside of the cell is the standard coefficient; the data at the bottom of cell is the t statistic value after robust standard error modifying

From Model 6, we can see that the interaction effect of REG and CDI is also not significant in the implementation of carbon reduction cooperation with competitors or surrogates. There is also no indirect effect on this kind of cooperation from environmental regulation.

(3) Determinants of carbon reduction cooperation by industrial symbiosis

The OCR results for the determinants of carbon reduction cooperation on the basis of industrial symbiosis are shown in Table 8.7. Both the Logit model and Normal model were introduced in this study according to the difference in pre-hypothesized probability distributions of F in equation (8.8). The estimated results of the significance for each determinant are not significantly different among the models, which shows the robustness of the estimated results. The estimated threshold values c_1–c_4 are verified significantly (except c_1 in Model 7), and are increased hierarchically. This also indicates good estimations by OCR.

There are no multi-collinearity problems during the estimation as well, since the VIF values of the independent variables in each model are all below 10. GLM estimation methods were used in this OCR model by the application of EVIEWS 6.0, where the coefficient covariance matrix was estimated by quadratic hill climbing. Model 7 was the estimation only considering company size as the control variable. Both company size and industrial type were considered in Model 8 as the control variables. Model 9 considered whether the interaction of REG and CDI had some effects on carbon reduction cooperation by industrial symbiosis. LR statistic values of the three models are all high enough to reject the null hypotheses that the independent variables are not associated with the dependent variable.

Carbon reduction demand from the stakeholders in the supply chain is also the main determinant that promotes carbon reduction cooperation by industrial symbiosis. And the main barrier that impedes this kind of carbon reduction cooperation is also attributed to the defective infrastructure and mechanism. However, different from the above two kinds of carbon reduction cooperation, financial pressure plays no significant role in carbon reduction cooperation by industrial symbiosis. Although there might be some transaction costs during the cooperation by industrial symbiosis, resource purchasing expenditure or waste emission cost would also decrease by the exchange of waste resources. The financial pressure from this kind of cooperation could be offset by these potential benefits.

From Model 9, there are also no significant effects on the implementation of carbon reduction cooperation by industrial symbiosis from the interaction of REG and CDI. It is indicated that environmental regulations also attach no importance to carbon reduction cooperation by industrial symbiosis from both direct and indirect prospective.

Table 8.7 OCR results for determinants of carbon reduction cooperation with competitors or surrogates

	Logistic model			Normal model			Model 7 VIF	Model 8 VIF	Model 9 VIF
	Model 7-L	Model 8-L	Model 9-L	Model 7-N	Model 8-N	Model 9-N			
REG	−0.090 (−0.694)	−0.049 (−0.372)		−0.022 (−0.295)	−0.003 (−0.039)		1.192	1.063	
CDI	0.871*** (5.320)	0.870*** (5.300)	0.913*** (4.149)	0.513*** (5.488)	0.513*** (5.453)	0.517*** (4.130)	1.025	1.058	1.058
IE	0.131 (0.925)	0.117 (0.814)	0.117 (0.813)	0.097 (1.139)	0.082 (0.939)	0.082 (0.939)	1.136	1.327	2.560
FP	0.064 (0.537)	0.047 (0.396)	0.048 (0.404)	0.033 (0.483)	0.028 (0.407)	0.029 (0.410)	1.147	1.152	1.152
DIM	−0.312*** (−3.772)	−0.315*** (−3.741)	−0.315*** (−3.743)	−0.199*** (−4.020)	−0.200*** (−3.979)	−0.199*** (−3.978)	1.033	1.052	1.050
REG * CDI			−0.013 (−0.411)			−0.001 (−0.065)		2.495	2.495
Company size	−0.133 (−1.512)	−0.197** (−2.179)	−0.197** (−2.184)	−0.074 (−1.424)	−0.103* (−1.950)	−0.103* (−1.950)	1.041	1.203	1.064
PPC & RCMCP		1.255*** (3.555)	1.255*** (3.557)		0.726*** (3.489)	0.726*** (3.487)		1.864	1.864
PSEP		1.493*** (3.238)	1.492*** (3.235)		0.898*** (3.258)	0.898*** (3.257)		1.358	1.358
NMP		0.821** (2.126)	0.822** (2.129)		0.530** (0.293)	0.530** (2.293)		1.663	1.662
SPFM		1.011*** (2.730)	1.012*** (2.735)		0.588*** (2.664)	0.588*** (2.665)		1.778	1.778
c_1	1.268 (1.376)	1.903** (1.975)	2.068** (2.153)	0.888 (1.599)	1.279** (2.241)	1.291** (2.280)			
c_2	2.413***	3.119***	3.284***	1.571***	1.997***	2.009***			

(continued)

Table 8.7 (continued)

	Logistic model			Normal model			Model 7 VIF	Model 8 VIF	Model 9 VIF
	Model 7-L	Model 8-L	Model 9-L	Model 7-N	Model 8-N	Model 9-N			
	(2.620)	(3.222)	(3 393)	(2.835)	(3.494)	(3.534)			
c_3	2.940*** (3.179)	3.683*** (3.783)	3.848*** (3 948)	1.894*** (3.407)	2.340*** (4.075)	2.352*** (4.115)			
c_4	3.629*** (3.873)	4.404*** (4.460)	4.569*** (4 624)	2.311*** (4.108)	2.773*** (4.768)	2.786*** (4.814)			
LR statistic	66.087	82.531	82.562	68.362	84.201	84.204			
Pseudo R^2	0.083	0.103	0.103	0.085	0.105	0.105			

Notes ① ***Means $p < 0.01$; **Means $p < 0.05$; *Means $p < 0.1$; ② The data at the upside of the cell is the standard coefficient; the data at the bottom of cell is the Z statistic value after robust standard error modifying

Company size also played a negative role in the implementation of carbon reduction cooperation by industrial symbiosis after the industrial type was introduced in OCR. It is indicated that waste resources exchange for carbon reduction is easier to implement by smaller companies.

8.3 Performance of Inter-company Cooperation on CO_2 Emission Abatement

After we identify the determinants of inter-company cooperation on CO_2 reduction, it is important to know whether the collaboration can help improve corporate performance. Both economic performance and environmental performance are considered in our analysis.

8.3.1 Hypothesis Modeling for the Performance of Inter-company Cooperation on CO_2 Emission Abatement

Whether intercompany cooperation on CO_2 emission can improve the performance of industrial companies is a basic problem discussed in our analysis. Various types of performance measures have been applied in evaluating sustainable industrial chains. But combining economic performance with environmental performance is most frequently accepted by many scholars (e.g., Sheu et al. 2005; Lu et al. 2007; Frota Neto et al. 2008). This division of performance is continued to be used in our analysis.

It is indicated that cooperation with suppliers and customers is instrumental in improving the environmental performance (Darnall et al. 2008). Collaborative industrial companies within the supply chain reducing CO_2 emissions are helpful for reducing unnecessary energy use and waste emissions. Some researches have also found the direct effect of intercompany cooperation within supply chains on enhancing the environmental performance, but the performance improvements are not always significant (Zhu et al. 2005). From the above analysis, we present the sixth hypothesis.

Hypothesis 8.6 Intercompany cooperation on CO_2 emission improves the environmental performance of industrial companies.

It is still a controversial problem of whether environmental practices within the industrial chain can bring economic benefits or not (Gil et al. 2001). Some practices of intercompany cooperation on CO_2 emission within the industrial chain, for instance, recovery of waste energy, can directly reduce the resource purchasing expenditure, but might increase the operating costs during the intercompany cooperation. From the above analysis, the following hypotheses are given.

Hypothesis 8.7 Inter-company cooperation on carbon emission improves the economic performance of industrial companies. Inter-company cooperation on carbon emission has negative effect on the economic performance of industrial companies.

The positive effect of improving environmental performance on companies' economic performance has been discussed by many scholars (e.g., Henri and Journeault 2010; Nishitani et al. 2012). Good environmental performance can improve the social image of the industrial company, which makes the industrial company get more support from society and government. The economic performance, as a result, could be improved due to the benefits of the related social image improvement. Based on the above analysis, we give Hypothesis 8.8.

Hypothesis 8.8 Environmental performance improvement due to the intercompany cooperation on CO_2 emission is positively associated with the companies' economic performance.

8.3.2 Methodology for the Performance of Inter-company Cooperation on CO_2 Emission Abatement

Both environmental performance and economic performance were measured by 4 items. We also used factor analysis to validate the measurement construction with the principal component methods. The factor matrix is shown in Table 8.8. The values of Cronbach's alpha reach 0.908 and 0.907, respectively, with the explanation power of the inherent variations reaching 78.468 and 78.666%. The final factor scores are used as the value of dependent variables as well, as seen as follows:

$$EP_i = \omega_1 EP_{1i} + \omega_2 EP_{2i} + \omega_3 EP_{3i} \tag{8.9}$$

$$CP_i = \omega_1^* CP_{1i} + \omega_2^* CP_{2i} + \omega_3^* CP_{3i} \tag{8.10}$$

Table 8.8 The factor matrix on cooperation with suppliers and consumers on carbon reduction

Environmental performance	Component	Economic performance	Component
	1		1
EP_1	0.887	CP_1	0.887
EP_2	0.920	CP_2	0.935
EP_3	0.900	CP_3	0.849
EP_4	0.834	CP_4	0.875

where EP_i and CP_i refer to the scores of environmental performance and economic performance for the No. i response. ω_j and ω_j^* refer to the loading coefficients for No. j item on the EP and CP.

We also developed two MLR models to discuss the effects of carbon reduction cooperation on environmental performance and economic performance, as seen as follows:

$$EP = \beta_0^{ep} + \beta_1^{ep} HC + \beta_2^{ep} ZC + \beta_3^{ep} IS + \beta_4^{ep} control + \varepsilon^{ep} \tag{8.11}$$

$$CP = \beta_0^{cp} + \beta_1^{cp} HC + \beta_2^{cp} ZC + \beta_3^{cp} IS + \beta_4^{cp} control + \varepsilon^{cp} \tag{8.12}$$

And in order to verify Hypothesis 8.8, another MLR model was introduced as follows:

$$CP = \beta_0^{cpe} + \beta_1^{cpe} HC + \beta_2^{cpe} ZC + \beta_3^{cpe} IS + \beta_4^{cpe} control + \beta_5^{cpe} EP + \varepsilon^{cpe} \tag{8.13}$$

where HC, ZC and IS are the types of carbon reduction cooperation within supply chains described in Sect. 8.2.1; company size and industrial type are also used as the control variables; β^{ep}, β^{cp} and β^{cpe} are the vectors of parameters to be estimated; ε^{ep}, ε^{cp} and ε^{cpe} are the random error terms.

8.3.3 Empirical Analysis for the Performance of Inter-company Cooperation on CO_2 Emission Abatement

Table 8.9 shows the MLR results for the effects of carbon reduction cooperation on industrial companies' performance. Both the effects on environmental performance and economic performance were tested in this study. No multi-collinearity problem arose during the estimation, since the VIF values of the independent variables in each model are all below 10. And the ordinary least squares method after white robust standard deviation correction was used to avoid the heteroscedasticity by the application of EVIEWS 6.0.

Model 10 only considered company size as the control variable. Model 11 considered both company size and industrial type as the control variables. Model 12 introduced environmental performance in the regression as an independent variable to verify whether it has some effects on economic performance. And F statistic values of the three models are all high enough to reject the null hypotheses that the independent variables are not associated with the dependent variable.

Carbon reduction cooperation by industrial symbiosis has a positive effect on industrial companies' environmental performance. However, the effects of the other two kinds of carbon reduction cooperation are not significant. This might be attributed to the fact that most industrial companies just began to realize the

Table 8.9 MLR results for effects of carbon reduction cooperation on industrial companies' performance

	Environmental performance		Economic performance			VIF		
	Model 10-e	Model 11-e	Model 10-b	Model 11-b	Model 12	Model 10	Model 11	Model 12
IS	0.289*** (4.426)	0.262*** (3.865)	0.182** (2.555)	0.155** (2.209)	0.022 (0.375)	3.554	3.758	3.963
HC	−0.265 (−1.296)	−0.081 (−0.371)	−0.083 (−0.419)	0.107 (0.522)	0.148 (0.941)	3.431	3.860	3.862
ZC	0.027 (0.412)	0.041 (0.630)	0.089 (1.458)	0.114* (1.942)	0.093 (1.617)	1.122	1.149	1.151
Company size	−0.045 (−0.987)	−0.027 (−0.563)	−0.091** (−1.997)	−0.070 (−1.490)	−0.056 (−1.275)	1.018	1.056	1.057
PPC and RCMCP		−0.321* (−1.854)		−0.471*** (−2.671)	−0.308** (−2.336)		2.004	2.030
PSEP		−0.620*** (−2.666)		−0.589** (−2.502)	−0.274 (−1.258)		1.550	1.587
NMP		−0.198 (−1.040)		−0.118 (−0.602)	−0.017 (−0.116)		1.687	1.695
SPFM		−0.340* (−1.788)		−0.379** (−2.001)	−0.206 (−1.331)		1.903	1.928
EP					0.509*** (6.668)			1.057
β_0	−0.599*** (−2.842)	−0.402* (−1.810)	−0.233 (−1.085)	−0.010 (−0.045)	−0.402* (−1.810)			
F-statistic	10.330***	6.094***	7.900***	5.351***	6.094***			
Adjusted R^2	0.127	0.137	0.097	0.119	0.137			

Notes ① ***Means $p < 0.01$; **Means $p < 0.05$; *Means $p < 0.1$; ② The data at the upside of the cell is the standard coefficient; the data at the bottom of cell is the Z statistic value after robust standard error modifying possibility of carbon reduction cooperation with suppliers, customers, competitors, or surrogates. As a result, their promotion of environmental performance improvement has not come into effect. Carbon reduction cooperation by industrial symbiosis, gets a higher score than the other two kinds of cooperation. Its effect on performance improvement, as a result, is more significant.

For economic performance, carbon reduction cooperation by industrial symbiosis still plays a positive role. Cooperation with suppliers and customers also has a positive effect after the industrial type energy-related in the regression as a control variable. Carbon reduction R&D cooperation with competitors attaches limited importance to the improvement of industrial companies' economic performance. This, for one thing, is due to the low implementation level of carbon reduction R&D cooperation. For another thing, carbon reduction R&D cooperation also needs some expenditure, which offsets the improvement on economic performance.

From Model 12, we can see that the environmental performance of industrial companies has a positive effect on their economic performance. However, the positive effects of the carbon reduction cooperation by industrial symbiosis decline to the point of no significant effect after the environmental performance was introduced to the regression. This indicates that most effects of carbon reduction cooperation by industrial symbiosis on economic performance improvement are from its positive effects on environmental performance improvement. Its direct effect on environmental performance is too minor to be considered. Comparatively, the effect of carbon reduction cooperation with suppliers and customers does not decline that much like the cooperation by industrial symbiosis.

8.4 Summary

To achieve the sustainable development in response to the pressures from climate change, it is an important option for Chinese industrial firms to implement carbon reduction cooperation based on their industrial chains. This chapter characterized the types of carbon reduction cooperation through industrial chains and then developed a conceptual model to identify its determinants and the effects on firms' performance.

The results show that carbon reduction cooperation through the industrial chain is totally at an elementary stage. Many industrial firms have just realized the effectiveness of reducing CO_2 emissions through this kind of cooperation. Carbon reduction demand from stakeholders in the industrial chain is the main driver for the carbon reduction cooperation. And defective infrastructure and mechanism are the main barrier that impedes the inter-company cooperation on CO_2 reduction. Hypothesis 8.2, 8.4 are verified as a result. Financial pressure plays a negative role in carbon reduction cooperation with suppliers and customers, but has a positive effect on R&D of carbon reduction cooperation with competitors and producers of substitutes. However, its effects on carbon reduction cooperation through industrial symbiosis are not significant. Hypothesis 8.4 is partially verified.

Environmental regulation has no direct effects on carbon reduction cooperation, and also does not have any indirect effects by influencing other stakeholders' carbon reduction awareness on the industrial chain. As a result, Hypothesis 8.1 is not verified in this chapter. This somewhat reflects that current environmental regulatory system in China has not effectively put enough pressure on industrial firms' carbon reduction practices. There are few strict regulations and higher enforcement levels for industrial firms to care about their CO_2 emission problems.

Imitating and demonstrating effects on CO_2 reduction have not come into work at present in China. Hypothesis 8.3 is not verified either. This might be largely due to the starting step for carbon reduction cooperation for Chinese industrial firms as a whole. There are not many successful cases of carbon reduction cooperation at present.

Hypothesis 8.6 is partially verified as well. Carbon reduction cooperation through industrial symbiosis attaches significant importance to the environmental performance of industrial firms. However, carbon reduction cooperation with suppliers and customers, and R&D cooperation with competitors play a limited role in the improvement of environmental performance.

The improvement of environmental performance could promote industrial firms' economic performance. Hypothesis 8.8 is verified. Carbon reduction cooperation through industrial symbiosis also has positive effects on the economic performance of industrial firms, but the effects are mostly independent of its effects on the improvement of environmental performance. Carbon reduction cooperation with suppliers and customers could also improve the economic performance of industrial firms, but the effects are smaller than cooperation through industrial symbiosis. The positive effects of cooperation with competitors and producers of substitutes have not come out yet. Hypothesis 8.7 is also partially verified.

There is still a great potential to enrich this study although some valuable findings have been gained. Firstly, the respondents of the questionnaire survey in this study are MBA or EMBA students who have working experience in the industrial firms. Further investigation would be conducted by surveying the actual industrial firms to better understand their performance of carbon reduction cooperation. Moreover, some case studies about carbon reduction cooperation could be made to further verify the conclusions in this study.

References

Barker, T.J., and Z.B. Zabinsky. 2008. Reverse logistics network design: A conceptual framework for decision making. *International Journal of Sustainable Engineering* 1 (4): 250–260.

Bonte, W. 2008. Inter-firm trust in buyer–supplier relations: Are knowledge spillovers and geographical proximity relevant? *Journal of Economic Behavior & Organization* 67: 855–870.

Carter, C.R., and J.R. Carter. 1998. Interorganizational determinants of environmental purchasing: Initial evidence from the consumer products industry. *Decision Sciences* 29 (3): 28–38.

Chan, R.Y.K., and L.B.Y. Lau. 2001. Explaining green purchasing behavior: A cross-cultural study on American and Chinese consumers. *Journal of International Consumer Marketing* 14 (2/3): 9–40.

Christmann, P., and G. Taylor. 2001. Globalization and the environment: Determinants of firm self-regulation in China. *Journal of International Business Studies* 32 (3): 439–458.

Darnall, N., G.J. Jolley, and R. Handfield. 2008. Environmental management systems and green supply chain management: Complements for sustainability? *Business Strategy and the Environment* 17 (1): 30–45.

Dean, J.W., and S.A. Snell. 1991. Integrated manufacturing and job design: Moderating effects of organizational inertia. *Academy of Management Journal* 34 (4): 776–804.

DiMaggio, P.J., and W.W. Powell. 1983. The iron cage revisited: Institutional isomorphism and collective rationality in organizational fields. *American Sociological Review* 48 (2): 147–160.

Ebinger, F., M. Goldbach, and U. Schneidewind. 2006. Green supply chains: A competence-based perspective. In *Greening the supply chain*, ed. J. Sarkis, 251–270. Berlin: Springer.

Frota Neto, J.Q., J.M. Bloemhof-Ruwaard, J.A. Nunen, et al. 2008. Designing and evaluating sustainable logistics networks. *International Journal of Production Economics* 111 (2): 195–208.

Gil, A.M.J., J.B. Jimenez, and J.C. Lorente. 2001. An analysis of environmental management, organizational context and performance of Spanish hotels. *Omega* 29 (6): 457–471.

Hashimoto, S., T. Fujita, Y. Geng, et al. 2010. Realizing CO_2 emission reduction through industrial symbiosis: a cement production case study for Kawasaki. *Resources, Conservation and Recycling* 54 (10): 704–710.

Henri, J.F., and M. Journeault. 2010. Eco-control: The influence of management control systems on environmental and economic performance. *Accounting, Organizations and Society* 35 (1): 63–80.

Hou, J.T., Z.L. Wen, and Z.J. Cheng. 2004. Structural equation model and its application. Beijing: Science and Education Press. ISBN: 978-7-50412-816-4 (in Chinese).

Huang, F.M. 2005. Structural equation model: Theory and application. China's Tax Press (in Chinese).

Hugo, A., and E.N. Pistikopoulos. 2005. Environmentally conscious long-range planning and design of supply chain networks. *Journal of Cleaner Production: Recent Advances in Industrial Process Optimization* 13 (15): 1471–1491.

Poyago-Theotoky, J.A. 2007. The organization of R&D and environmental policy. *Journal of Economic Behavior & Organization* 62: 63–75.

Jayaraman, V. 2006. Production planning for closed-loop supply chains with product recovery and reuse: An analytical approach. *International Journal of Production Research* 44 (5): 981–998.

Jones, C. 2010. Exploring new ways of assessing the effect of regulation on environmental management. *Journal of Cleaner Production* 18: 1229–1250.

Jørgensen, M.S., U. Jørgensen, K. Hendriksen, et al. 2010. Environmental management in Danish transnational textile product chains. *Management Research Review* 33 (4): 357–379.

Korhonen, J., and J.P. Snäkin. 2005. Analysing the evolution of industrial ecosystems: Concepts and application. *Ecological Economics* 52 (2): 169–186.

Li, H., R. Raeside, T. Chen, et al. 2012. Population ageing, gender and the transportation system. *Research in Transportation Economics* 34 (1): 39–47.

Lieckens, K., and N. Vandaele. 2007. Reverse logistics network design with stochastic lead times. *Computers & Operations Research* 34 (2): 395–416.

Liu, W., et al. 2010a. Development framework and scientific foundation for low-carbon economy in China. Commercial Press (in Chinese).

Liu, W.B., Meng, W., Li, X.X., et al. 2010b. DEA models with undesirable inputs and outputs. *Annals of Operations Research* 173: 177–194.

Lopéz, A. 2008. Determinants of R&D cooperation: Evidence from Spanish manufacturing firms. *International Journal of Industrial Organization* 26: 113–136.

López-Gamero, M.D., J.F. Molina-Azorín, and E. Claver-Cortés. 2010. The potential of environmental regulation to change managerial perception, environmental management, competitiveness and financial performance. *Journal of Cleaner Production* 18 (10–11): 963–974.

Lu, L.Y.Y., C.H. Wu, T.C. Kuo, et al. 2007. Environmental principles applicable to green supplier evaluation by using multi-objective decision analysis. *Sustainable Design and Manufacture, International Journal of Production Research* 45 (18–19): 4317–4331.

Luo, Y., M. Zhou, and R.J. Caudill. 2001. An integrated e-supply chain model for agile and environmentally conscious manufacturing. *IEEE/ASME Transactions on Mechatronics* 6 (4): 377–386.

Luo, Y.D. 2001. Toward a cooperative view of MNG host government relations: building block and performance implications. *Journal of International Business Studies* 32 (3): 401–419.

Mason, C., and W.D. Perreault. 1991. Collinearity, power, and interpretation of multi-ple regression analysis. *Journal of Marketing Research* 28 (3): 268–280.

MIIT. 2011. Issued notice about the elimination of backward production capacity of industrial objectives and tasks in 2011. http://finance.sina.com.cn/china/bwdt/20110711/104810125706.shtml [2011-07-12] (in Chinese).

NBSC. 2009. China Energy Statistical Yearbook 2009. Beijing: China Statistics Press.

Nishitani, K., S. Kaneko, H. Fujii, et al. 2012. Are firms' voluntary environmental management activities beneficial for the environment and business? An empirical study focusing on Japanese manufacturing firms. *Journal of Environmental Management* 105: 121–130.

Nunnally, J.C., and I.H. Bernstein. 1994. *Psycometric theory*, 3rd ed. New York: McGraw-Hill.

Okamuro, H. 2007. Determinants of successful R&D cooperation in Japanese small businesses: The impact of organizational and contractual characteristics. *Research Policy* 36: 1529–1544.

Pochampally, K.K., S. Nukala, and S.M. Gupta. 2009. *Strategic planning models for reverse and closed-loop supply chains*. Boca Raton: CRC Press.

Prajogo, D., A.K.Y. Tang, and K.H. Lai. 2012. Do firms get what they want from ISO 14001 adoption? An Australian perspective. *Journal of Cleaner Production* 33: 117–126.

Mongia, P., K. Schumacher, and J. Sathaye. 2001. Policy reforms and productivity growth in India's energy intensive industries. *Energy Policy* 29: 715–724.

Rivera, J. 2004. Institutional pressures and voluntary environmental behavior in developing countries: Evidence from the Costa Rican hotel industry. *Society & Natural Resources* 17 (9): 779–797.

Sarkis, J., A. Presley, and D. Liles. 1997. The strategic evaluation of candidate business process reengineering projects. *International Journal of Production Economics* 50 (2–3): 261–274.

Sheu, J.B. 2008. Green supply chain management, reverse logistics and nuclear power generation. *Transportation Research Part E: Logistics and Transportation Review* 44 (1): 19–46.

Sheu, J.B., Y.H. Chou, and C.C. Hu. 2005. An integrated logistics operational model for green-supply chain management. *Transportation Research Part E: Logistics and Transportation Review* 41 (4): 287–313.

Steiger, J.H. 1990. Structural model evaluation and modification: An interval estimation approach. *Multivariate Behavioral Research* 25: 173–180.

van Hemel, C., and J. Cramer. 2002. Barriers and stimuli for ecodesign in SMEs. *Journal of Cleaner Production* 10: 439–453.

Yin, H., and C. Ma. 2009. International integration: A hope for a greener China? *International Marketing Review* 26 (3): 348–367.

Zhang, B., J. Bi, Z. Yuan, et al. 2008. Why do firms engage in environmental management? An empirical study in China. *Journal of Cleaner Production* 16: 1036–1045.

Zhang, X.P., and X.M. Cheng. 2009. Energy consumption, carbon emissions, and economic growth in China. *Ecological Economics* 68 (10): 2706–2712.

Zhang, B., Z.H. Wang, J.H. Yin, et al. 2012. CO_2 emission reduction within Chinese iron & steel industry: practices, determinants and performance. *Journal of Cleaner Production* 33: 167–178.

Zhu, Q., Y. Geng, and K.H. Lai. 2010. Circular economy practices among Chinese manufacturers varying in environmental-oriented supply chain cooperation and the performance implications. *Journal of Environmental Management* 91 (6): 1324–1331.

Zhu, Q., and Y. Geng. 2013. Drivers and barriers of extended supply chain practices for energy saving and emission reduction among Chinese manufacturers. *Journal of Cleaner Production* 40: 6–12.

Zhu, Q., J. Cordeiro, and J. Sarkis. 2012. International and domestic pressures and responses of Chinese firms to greening. *Ecological Economics* 83: 144–153.

Zhu, Q., J. Sarkis, and K.H. Lai. 2007. Green supply chain management: Pressures, practices and performance within the Chinese automobile industry. *Journal of Cleaner Production* 15 (11–12): 1041–1052.

Zhu, Q., J. Sarkis, and Y. Geng. 2005. Green supply chain management in China: Pressures, practices and performance. *International Journal of Operations and Production Management* 25 (5): 449–468.

Zirger, B.J., and M.A. Maidique. 1990. A model of new product development: An empirical test. *Management Science* 36: 867–883.

Chapter 9
Low-Carbon Policies in China

9.1 Major Low-Carbon Policies in China

At present, climate change is a common human challenge: to mitigate climate change, it has become the consensus of governments all over the world to promote the transformation to a low-carbon economy. Scholars at home and abroad have reached a certain consensus on the importance of low-carbon policies in promoting the transformation to a low-carbon economy, and believe that low-carbon policies not only play a key role in low-carbon technology innovation and international low-carbon technology transfer, but also change energy systems and promote the achievement of greenhouse gas stabilization goals.

In 2003, the UK White Paper on Energy first proposed the concept of a low-carbon economy as a new economic model that tries to reduce energy and resource consumption through technological and institutional innovation. At the same time, it can reduce greenhouse gas emissions to the maximum extent and realize sustainable socioeconomic development without affecting economic growth. In recent years, especially since the 11th Five-Year Plan, the Chinese Government has implemented a series of policies to promote China's transition to low-carbon economic development. In November 2009, the Chinese Government announced that China's CO_2 emissions per unit GDP would be reduced by 40–45% from 2005 to 2020, which will further accelerate the pace of China's development as a low-carbon economy.

On August 10, 2010, the National Development and Reform Commission issued a Circular referring to a pilot scheme for low-carbon provinces (autonomous regions, municipalities) and cities, which will cover Guangdong, Liaoning, Hubei, Shaanxi,

---

This chapter quotes from the following literature: Wang Z, Zhang B, Zeng H. 2016. The effect of environmental regulation on external trade: empirical evidence from the Chinese economy. Journal of Cleaner Production, 114: 55–61. Wang Z, Zhang B, Zhang Y. (2012). Determinants of public acceptance of tiered electricity price reform in China: evidence from four urban cities. Applied Energy, 91 (1): 235–244.

Z. Wang and B. Zhang, *Low-Carbon Consumption in China: Residential Behavior, Corporate Practices and Policy Implication*,
https://doi.org/10.1007/978-981-15-2792-0_9

and Yunnan (autonomous regions, municipalities), and eight cities (Tianjin, Chongqing, Shenzhen, Xiamen, Hangzhou, Nanchang, Guiyang, and Baoding).

On March 16, 2011, the fourth session of the Eleventh National People's Congress approved *The Outline of the Twelfth Five-Year Plan* for Economic and Social Development of the People's Republic of China, in which Chap. 21 proposes the idea of "exploring the establishment of low-carbon product standards, and a marking and certification system, establishing and improving the accounting system of greenhouse gas emissions, and gradually establishing a CO_2 emission trading market."

On October 29, 2011, the General Office of the State Development and Reform Commission issued a notice on the pilot scheme for CO_2 emissions trading, formally approving such work in Beijing, Shanghai, Tianjin, Chongqing, Hubei, Guangdong, and Shenzhen.

On December 1, 2011, the State Council issued the "Twelfth Five-Year Plan for Controlling Greenhouse Gas Emissions", which defines the overall requirements and main targets for China's control of greenhouse gas emissions by 2015. It is suggested that comprehensive use of various measures should be taken to control greenhouse gas emissions, and a number of typical low-carbon provinces (autonomous regions, municipalities), cities, parks, and communities should be trialed, so as to enhance the capability of greenhouse gas emission control in a holistic manner.

On October 15, 2013, the General Office of the State Development and Reform Commission issued a notice on the issuance of the first batch of greenhouse gas emission accounting methods and reporting guidelines for 10 industrial enterprises (on a trial basis). The industries include power generation, power grid, iron and steel, chemical industry, electrolytic aluminum, magnesium smelting, flat glass, cement, ceramics, civil aviation, etc. The purpose of this notice is to establish and improve the greenhouse gas statistical accounting system and to establish a CO_2 emissions trading market.

On September 19, 2014, the National Development and Reform Commission issued the National Plan on Climate Change (2014–2020), which sets out the main objectives for the response to climate change by 2020. On November 12, 2014, China and the US jointly issued a Joint Statement on Climate Change: for the first time, China and the US announced their respective targets for action after 2020. Both sides indicated that they were responding to the great threat to humanity posed by global climate change, and will work together for the common good. Both sides agreed to reduce greenhouse gas emissions, with the US promising to reduce emissions by 26% by 2025 and China to stop increasing carbon dioxide emissions by 2030.

On January 11, 2016, the General Office of the State Development and Reform Commission issued a notice effectively starting the national CO_2 emissions trading market, with the aim of ensuring the start of national CO_2 emissions trading and implementing a carbon trading system in 2017. The notice proposes a list of enterprises to be included in the national CO_2 emissions trading system and requires the accounting, reporting, and verification of the historical CO_2 emissions of enterprises to be included.

9.2 Environmental Regulation Policy and Its Performance

9.2.1 Overview of Environmental Regulations in China

Differing from previous studies (Lai and Hu 2008; Long et al. 2013), the environmental regulations discussed here were not limited to those compulsory regulations issued by the government. In China, there have been many laws and regulations, as well as administrative commands, enacted to ensure that enterprises comply with environmental requirements. For instance, China has enacted both Cleaner Production Promotion Law (effective from January 1, 2003, and amended in 2012) and Circular Economy Promotion Law (effective from January 1, 2009, and amended in 2018) (Zhu and Geng 2013). Although it is said that China's accession to the WTO did not guarantee better environmental conditions even after the government adopted stronger regulations (Long et al. 2013), Command-and-control (C & C) is still the most popular way to regulate environmental problems for industrial production and its corresponding international trade.

Market-based regulations: the most popular forms of market-based regulations include: pollution tax, pollution control subsidy, tradable pollution permission, and deposit refund. These regulations often do not have compulsive effects on the pro-environmental practices of industrial firms, however, the cost and benefit of firms can be affected directly thereby. It is worthwhile to point out that market-based regulations (i.e, tradable pollution permission or pollution taxes) are often more efficient than C & C regulations (Bruneau 2005).

Reluctant regulation: different from market-based regulation, reluctant regulation comes from the business stakeholders such as customers, competitors, and suppliers, instead of the government. In the former two kinds of environmental regulations, government determines the rules. Even in the market-based regulation such as carbon market, the government is also responsible for in carbon quota allocation and the regulations of carbon trading.

Other differences between market-based regulations and reluctant regulations come from various sources, particularly in China. Firstly, China has been suffering from "green trade barriers" during international trade. International standards such as ISO 14000 dictate the acceptance of strengthened environmental regulation in China (Chang 2002). Moreover, pro-environment requirements from suppliers or customers provide regulation of the producers' environmentally friendly production protocols (Zhang et al. 2012). Zhu et al. (2007) pointed out that environmental protection requirements from supply chains were more popular in international trade. Chinese manufacturers are acquiring increasing green certification to meet the environmental regulations imposed by foreign customers (Zhu et al. 2007).

9.2.2 Environmental Regulation Theoretical Model

A theoretical model based on panel data was established to analyze the relationship between environmental regulation and international trade. In reality, different sections were not expected to develop at the same pace. Regression results of fixed effect, random effect, and random coefficient models also reflected the fact that different sections exhibited different development patterns. Generalized least squares estimation (FGLS) was used as the estimation method to deal with the relationship between environmental regulation level (ERL) and overall indicators of international trade: with regard to the regression method for different trade sectors, Zellner (1962) proposed SUR which is more efficient than ordinary least squares (OLS) for panel data models, especially when there are heteroscedasticities or autocorrelations in residuals.

Environmental regulation level (ERL) is another independent variable in this regression model. Both Xu (2000) and Busse (2004) pointed out that, because of the difficulties in acquiring environmental regulation data and the resulting unsatisfactory data quality, it is not easy to analyze environmental regulation in an empirical manner. Expenditure in reducing pollution was used in previous studies for ERL quantification (Copeland and Taylor 2005; Levinson and Taylor 2008); however, due to a lack of statistical data, there are alternative indicators for indirectly estimating ERL in China. Xiong and Xu (2007) pointed out that rates of pollution abatement could reflect the effect of environmental regulation and lead to similar results. Taking the Chinese political system into consideration, it was also appropriate to adopt abatement rates of discharged waste and pollutants as instrumental variables to assess ERL.

According to Copeland and Taylor (2005), a bilateral exchange rate is the relative price of one currency to another. The fluctuation of exchange rates will directly change the costs and benefits of international buyers and sellers. Tsen (2011) stated that the exchange rate was a significant determinant of international trade volume. As a result, the exchange rate was taken into consideration as a control variable in this model.

Above all, according to the panel data regression model of Wang et al. (2012), the model applied in this chapter can be described as follows:

$$\begin{aligned} \mathrm{IT}_{it} &= \alpha_i + \beta_1 \mathrm{ERL}_{it} + \beta_2 \mathrm{GDP}_{it} + \beta_3 \mathrm{ER}_{it} + u_{it} \\ i &= 0, 1, \ldots, 9; t = 1985, 1986, \ldots, 2010 \end{aligned} \tag{9.1}$$

where t is the time dimension; i represents the nine types of trade categories involved in China's international trade; ERL represents the environmental regulation level; GDP represents the economic development level; ER is the exchange rate; u_{it} is the disturbance term; β_1, β_2, and β_3 are estimated coefficients for each independent variable; α_i is the estimated parameter which is independent of the independent variables.

9.2.3 Data Sources and Preprocessing

It was not difficult to obtain exchange rate data from the global exchange rate market, however, the data acquired directly from the exchange market are usually subject to real-time fluctuations. For the purposes of this research, annual data were used. In the *China Statistical Yearbooks*, average exchange rates for the Chinese yuan against the main foreign currencies are published annually by The People's Bank of China. Here, the trade value of commodities was counted in USD. As a result, the annually averaged exchange rate of USD to yuan was used here.

According to the explanation in Sect. 9.3.1, the ERL data came from the *China Statistical Yearbooks* since 1985: however, the statistical standard for atmospheric data before 1990 was different from that after 1990. Therefore, only after some calculations could comparable statistical data for the purification rate of atmospheric pollution be obtained. Data before 1990 were complemented by deductively calculating a one-order random walk process. After regression, the purification rate of atmospheric pollution was 1.07% higher than in the previous year. The significance level was greater than 0.001, with R^2 as high as 0.9998. The data for the three series are shown in Fig. 9.1.

To calculate the overall ERL, a principal component analysis (PCA) was adopted. This method was chosen because PCA can determine, in an objective manner, the contributions of individual factors. As a result, the ERL was quantified by calculating the PCA summary of the abatement rate of wastewater, the purification rate of atmospheric pollution, and the recycling rate of solid waste with weights of 0.5717, 0.5773, and 0.5830, respectively, which reflects 96.56% of the information in the original series.

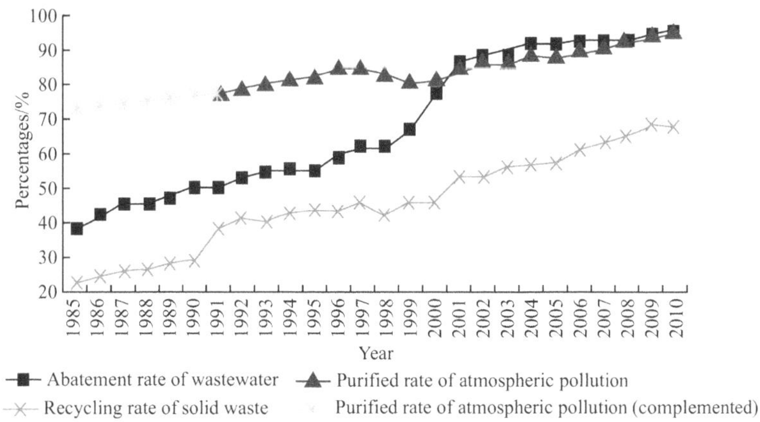

Fig. 9.1 The trends for abatement rates of discharged waste and pollutants

The Standard International Trade Classification (SITC) is used to categorize the different products imported or exported in the course of international trade. Compared with the Harmonised Commodity Description and Coding System (HS), which is the most popular system for the management of statistics relating to import and export through customs, SITC is more often used as an analytical tool. In the most recent version of SITC (Revision 4), the Introduction, Clause 9 states that "… in 1993, the Statistical Commission endorsed the use of HS at the national level in compilation and dissemination of international merchandise trade statistics; and in 1999, the Commission confirmed its recognition of SITC as an analytical tool." Besides, although HS provides more detailed information, its frequent updating has led to inconsistencies in statistical items, however, the STIC system is relatively stable, which makes comparison among annual datasets possible. Therefore, SITC is used in this analysis. The statistics of import and export under the framework are published annually in the *China Statistical Yearbooks.*

Time series data usually contain an inherent trend and are, therefore, nonstationary. As a result, it was necessary to test for stationary points before implementing any panel data model. Any regression using nonstationary panel data of different integration orders would be problematic and even spurious. As a result, before a panel data model was used, a panel unit-root test should be done. The test results are summarized in Table 9.1.

Table 9.1 Panel unit-root tests

Items	Lags	LLC	IPS W-stat	ADF-Fisher	PP-Fisher
Import and export under the framework of SITC	Level data	6.45	9.27	16.56	13.25
	Firsst difference	−16.96***	−17.15***	382.01***	701.96***
Import with the exchange rate, environmental regulation, and GDP	Level data	8.5159	13.0862	4.8564	5.0182
	First difference	−3.3170***	−4.3805***	84.0277***	139.458***
Export with exchange rate, environmental regulation, and GDP	Level data	5.1424	7.5237	10.6514	10.7245
	First difference	−11.1804***	−9.5089***	146.665***	182.789***
Net export with exchange rate, environmental regulation, and GDP	Level data	4.5687	8.9363	4.3767	4.2043
	First difference	−5.8472***	−6.2948***	108.149***	169.057***

Notes Probabilities for Fisher tests are computed using an asymptotic Chi-square distribution. All other tests assume asymptotic normality; *** means the rejection of unit-root null hypothesis at a significant level above 0.999

Both common unit-root process and individual unit root process indicate the fact that all data series are the same I(1) series which was tolerated and deemed significant at the 0.99 confidence level. If the corresponding series were nonstationary but integrated at the same order, a panel cointegration test can be carried out. The purpose of the cointegration test was to determine the long-term static equilibrium among different series. As a result, it is necessary to carry out a cointegration test before launching the regression to ensure that the regression is appropriate. The results of the panel cointegration test are listed in Table 9.2 where it may be seen that there is cointegration between dependent and independent variables.

9.2.4 Discussion of Environmental Regulation Policies

There were two stages to the empirical analysis: one was to identify the influence of determinants on the overall trading volume, including the simple category of primary goods and industrial goods; the other was to explore the effect on the SITC categories. The regression results are listed in Tables 9.3 and 9.4, with their significance levels and corresponding standard deviations. To state these more clearly, and especially to facilitate comparability between the results and their corresponding trading values, both the Prais–Winsten and Cochrane–Orcutt formulations were used to reflect the robustness of the regression results.

Table 9.3 shows that the economy was still the most important, significant factor influencing Chinese international trade. This was identical to the findings of many previous studies. International trade is the great engine powering economic growth, and rapid economic growth provides better market opportunities and attracts more foreign direct investment (FDI). This, otherwise, mobilizes the development of international trade.

Table 9.2 The panel co-integration test

Statistics	Import, exchange rate, environmental regulation level, and GDP increment	Export, exchange rate, environmental regulation level, and GDP increment	Net export, exchange rate, environmental regulation level, and GDP increment
Panel v-statistic	2.153719** (0.0156)	2.007685** (0.0223)	1.342843* (0.0897)
Panel rho-statistic	−0.497625(0.3094)	−0.345444(0.3649)	−0.248197(0.4020)
Panel PP-statistic	−2.334267*** (0.0098)	−1.881285** (0.0300)	−1.342807* (0.0897)
Panel ADF-statistic	−3.880980*** (0.0001)	−2.064729** (0.0195)	−3.343747*** (0.0004)

Notes p values are in parentheses. *Significant at 10%; **Significant at 5%; ***Significant at 1%

Table 9.3 Regression results of the overall trading volume

Indicator	Prais–Winsten (FGLS)			Cochrane–Orcutt (FGLS)		
	Exchange rate	ERL	ΔGDP	Exchange rate	ERL	ΔGDP
Import	−204.20** (79.99)	3890.81*** (1109.04)	0.3818*** (0.06)	−195.27** (86.72)	3985.19*** (1165.81)	0.3780*** (0.07)
Export	−196.81 (114.79)	4709.40*** (1485.25)	0.4218*** (0.08)	−185.48 (127.44)	4828.52*** (1620.88)	0.4189*** (0.08)
Net export	−1.17 (49.96)	647.77 (634.91)	0.0579* (0.03)	−5.00 (56.32)	583.92 (731.85)	0.0583* (0.03)
Primary goods import	−88.15** (39.50)	1128.53* (547.26)	0.1238*** (0.03)	−89.27** (42.50)	1115.78* (574.40)	0.1242** (0.03)
Primary goods export	−9.55* (4.94)	155.40* (75.39)	0.0184*** (0.005)	−10.01* (5.14)	149.48* (77.57)	0.0187*** (0.005)
Primary goods net export	79.04** (35.27)	−983.13* (478.96)	−0.1049*** (0.03)	79.44* (38.25)	−978.62* (505.98)	−0.1051*** (0.03)
Industrial goods import	−112.27** (47.18)	2786.24*** (634.48)	0.2525*** (0.04)	−99.23* (51.69)	2906.49*** (671.61)	0.2481*** (0.04)
Industrial goods export	−185.41 (110.35)	4538.80*** (1420.46)	0.4018*** (0.08)	−173.94 (122.64)	4669.28*** (1558.67)	0.3992*** (0.08)
Industrial goods net export	−82.98 (72.53)	1766.03* (929.92)	0.1569*** (0.05)	−85.74 (80.78)	1733.46 (1027.22)	0.1575*** (0.05)

Notes ***, ** and * stand for significant level of 0.01, 0.05, and 0.1, respectively

Table 9.4 Results of seemingly unrelated regression

Indicator	Import			Export			Net export		
	Exchange rate	ERL	GDP	Exchange rate	ERL	GDP	Exchange rate	ERL	GDP
SITC0	−2.28(1.72)	22.28(28.20)	0.0050 (0.002)***	−2.04(1.50)	53.79(24.55)**	0.0106(0.002) ***	0.24(1.09)	31.51(17.83)*	0.0056(0.001)***
SITC1	−0.58(0.21)***	4.15(3.41)	0.0007 (0.000)***	0.95(0.24)***	−4.29(3.91)	0.0005(0.0002) **	1.53(0.19)***	−8.44(3.12)***	−0.0002(0.0002)
SITC2	−37.69(14.32)***	337.60 (234.14)	0.0732 (0.015)***	−1.78(0.84)**	−3.88(13.73)	0.0031(0.0009) ***	35.91(13.79)***	−341.49(225.48)	−0.0701(0.014)***
SITC3	−0.58(1.18)	11.30(19.22)	0.0033 (0.001)***	0.50(0.14)***	−6.08(2.37)***	0.0003(0.0001) **	1.07(1.09)	−17.38(17.77)	−0.0030(0.001)***
SITC4	−35.54(13.85)***	361.38 (226.47)	0.0673 (0.014)***	−6.66(2.18)***	91.44(35.61) ***	0.0056(0.002) **	28.88(12.30)**	−269.94(201.15)	−0.0616(0.013)***
SITC5	−18.06(6.02)***	397.78(98.47) ***	0.0424 (0.006)***	−10.34(5.80)*	129.03(94.76)	0.0318(0.006) ***	7.72(3.42)**	−268.76(55.86)***	−0.0106(0.004)***
SITC6	−1.70(5.16)	335.01(84.35) ***	0.0305 (0.005)***	−24.26(17.22)	370.91(281.56)	0.1044(0.018) ***	−22.55(16.03)	35.90(262.04)	0.0740(0.017)***
SITC7	−66.31(20.26)***	1429.04 (331.30)***	0.1741 (0.021)***	−131.98(40.27)***	1869.31 (658.44)***	0.2884(0.041) ***	−65.68(25.59)***	440.27(418.48)	0.1143(0.026)***
SITC8	−16.62(5.55)***	256.78(90.82) ***	0.0423 (0.006)***	−8.76(18.19)	755.8(297.4)**	0.1241(0.019) ***	7.86(14.19)	499.02(231.98)**	0.0818(0.015)***
SITC9	−9.54(3.51)***	−33.39(57.42)	0.0043 (0.004)	−13.85(3.49)***	3.69(57.07)	−0.0017(0.004)	−4.31(2.45)*	37.08(40.14)	−0.0060(0.003)**

Notes: ① * means $p<0.1$; ** means $p<0.05$; *** means $p<0.01$. ② The data at the upside of the cell is the standard coefficient; the data at the bottom of the cell is the t statistic value after robust standard error modifying

Moreover, environmental regulation also plays a significant role in international trade. As shown in Table 9.3, the ERL affects Chinese international trade departments with only one exception in the net export data series. This result refuted previous worries that stiffer environmental regulation might damage the development of international trade. Comparatively speaking, environmental regulation has positive effects on both imports and exports. This may be attributed to the fact that hard environmental regulation motivates industrial firms to provide green, pro-environmental products. These value-added products meet increasing demand from international markets (Zhu et al. 2007). Although the positive effects of environmental regulation on net exports are not statistically significant, it indicated that environmental regulation's positive effects on imports do not lag behind its positive effects on exports. There is a relative balance of positive effects on both imports and exports.

According to Table 9.3, there are apparent differences between the effects of environmental regulation on primary goods and industrial goods. For primary goods, imports were more positively motivated by strict environmental regulations, compared to exports. As a result, an interesting phenomenon arose whereby environmental regulation was shown as having negative effects on the net export of primary goods. It indicated that primary products were inclined to rely on imports from the global market to meet the requirements of environmental regulations. This was because the production of primary goods in China is often a heavy polluter and causes environmental problems. Comparatively speaking, the net exports of industrial goods were positively influenced by environmental regulation: the export of industrial goods benefitted more from the stimulus from strict environmental regulation than did China's imports. With increasing concern about environmental protection, there is an increasing global demand for green products and cleaner production. This brings new opportunities for many industrial firms in China. As the "world's factory", many Chinese industrial firms have upgraded their production process to adopt more strict environmental criteria such as ISO 140001 (Zhu and Sarkis 2006). Under environmental regulation, there is an increasing trend for the migration of international trade from primary goods to manufactured goods.

In Table 9.4, the effects of factors influencing Chinese international trade are listed by SITC. The basic assumptions for the scenario were the same as those for Table 9.3. According to Table 9.4, economic development remained the most important motivational factor no matter whether considering imports, exports, or net exports for almost all international trade sectors, however, the effects of environmental regulation differed across various SITCs. Within the scope of imports, there were four sectors from STIC-5 to SITC-8, each of which was positively motivated by environmental regulation. These STICs were mostly from heavy industrial sectors, such as chemicals and manufactured goods. The import of other primary products such as food and animals was not significantly influenced by environmental regulation.

From the perspective of export, environmental regulation had positive effects on both STIC-0 and STIC-4. Since these two sectors were food related, which were among the most environmental-friendly products, stricter environmental regulation promoted their export. The export of SITC-7 and SITC-8 goods was also motivated by environmental regulation. An interesting phenomenon was that their imports were also positively affected by environmental regulation. Since SITC-7 and SITC-8 represent manufactured products, a possible explanation was that the production of manufactured goods in China was undergoing an upgrade and a transformation. The manufacture of heavily polluting products is migrating to other countries, and environmental-friendly products are produced to satisfy global demand; however, the export of SITC-3 goods (i.e., mineral fuels, lubricants, and related materials) had a negative relationship with environmental regulation. Mining production has a deleterious effect on the environment: strict environmental regulations imposed some restrictions on the development of SITC-3.

Considering both imports and exports, it was seen that only the chemical industry had to bear significant losses on their net exports under the imposition of strict environmental regulation in China. The net exports of most other industries either benefited from environmental regulation regardless of or suffered no adverse effects under the imposition of environmental regulation. Therefore, it was indicated that environmental regulation would not decrease the overall international trade.

9.3 Public Acceptance of Low-Carbon Policies

9.3.1 Tiered Electricity Price Reform in the Household Sector

Tiered electricity price (TEP) reform is a planning policy for household electricity conservation in China. Based on TEP, several price hierarchies are provided, and additional charges will be levied if electricity consumption exceeds the upper bound of each hierarchy, yet, the optimal level of each price tier and the institutions for implementing TEP need further research, which is set on the basis of considering residents' affordability and willingness to pay. Therefore, this chapter is aimed at exploring determinants of public willingness to accept TEP and the determination of the acceptable range of premium.

After several years of brewing, TEP reform in China's household sector was put on the agenda with the promulgation of a consultation draft on October 9, 2010. Corresponding pilot schemes have been conducted in several provinces (autonomous regions, municipalities) such as Sichuan, Fujian, and Zhejiang since 2004 (Chen 2009). There are several reasons for the Chinese Government to promote TEP reform.

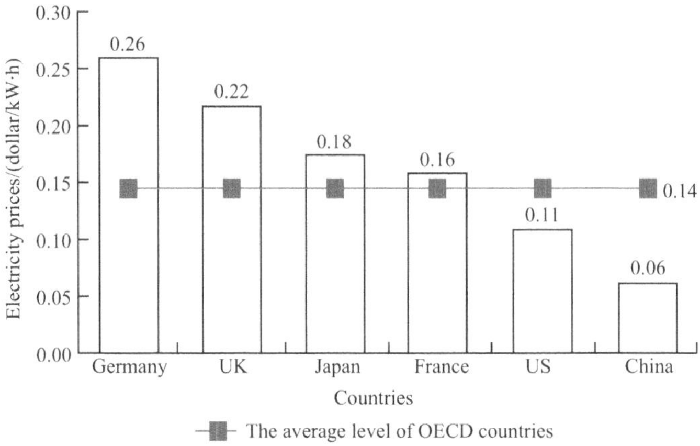

Fig. 9.2 The electricity price of some countries in 2007. Note: The data is referred to IEA (2009)

Firstly, household electricity prices are relatively low (Fig. 9.2). The household electricity price is only $0.06 (USD) per kW·h in China, which is much lower than the price in the US, France, Germany, or other OECD countries. This low price requires a significant subsidy for household electricity use every year, however, the large amounts of high electricity use in the household sector are attributed to its consumption by high-income residents (Liu and Yang 2010). This indicates that rich residents share more subsidized household electricity consumption, which is unfair. To preserve social fairness and stability, it is necessary to conduct TEP reform. Secondly, the low price also puts inadequate pressure on household electricity conservation and results in plenty of high electricity consumption in the household sector. The implementation of TEP could raise the cost of high consumption and reduce high electricity use. Thirdly, a low electricity price imposes a financial burden on government, along with rising power generation costs. Directly raising electricity prices to households is difficult, as it is unfair to low-income residents and would be opposed by most of them. TEP reform allows the government to increase electricity revenues from the household sector, while maintaining subsistence consumption among low-income residents.

TEP reform is a reform policy that requires discriminated electricity price tiers for different amounts of electricity consumption in the household sector. Additional tariffs would be added to the price for electricity consumption exceeding the upper bound of each tier. It is designed to induce electricity conservation with higher prices paid for greater electricity consumption. According to the consultation draft of the TEP reform in China, there are two proposed TEP reform schemes, both of which have three price tiers for household electricity consumption (Fig. 9.3).

According to the first scheme, the price of the first tier stays the same as the present household electricity price, and the upper bound of electricity consumption of this tier is 110 kW·h. The range of electricity consumption in the second tier is between 110 and 210 kW·h. The price of this tier is at least 0.05 yuan higher than

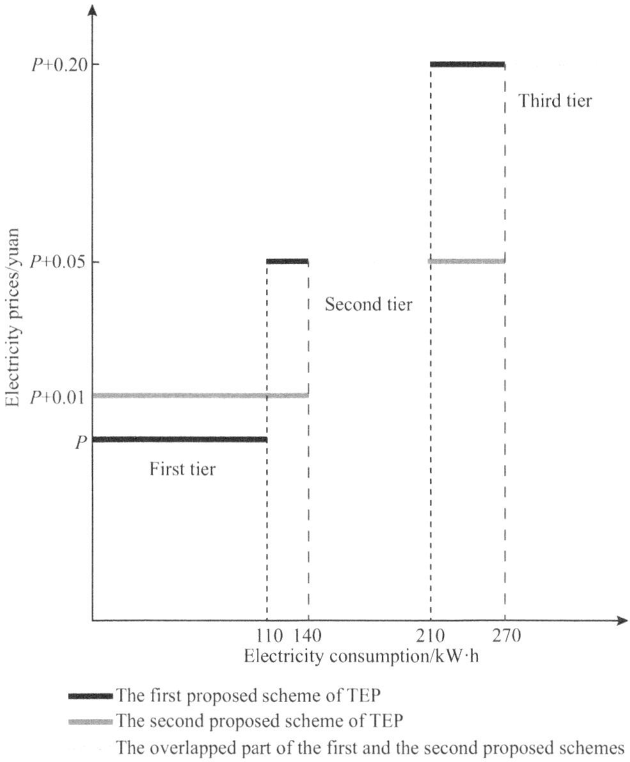

Fig. 9.3 Two proposed schemes of TEP reform by NDRC of China

the present price. The lower bound of the third tier is 210 kW·h, and the price is at least 0.2 yuan higher than the present price. The second TEP reform scheme differs slightly from the first: the lower bound of the second scheme rises to 140 kW·h, and the price of the second tier is at least 0.01 yuan higher than the present price. The required prices in the second tier and third tier are the same as those in the first scheme, but the frontiers of the two tiers are changed. The second tier ranges from 140 to 270 kW·h, and the lower bound of third tier rises to 270 kW·h.

Generally speaking, the first tier is set to satisfy low-income residents' electricity demand for subsistence (Wang 2010). The price of this tier, as a result, should be designed as the acceptable price of the low-income residents, which could be calculated as follows (Zhang et al. 2010):

$$p = \frac{I \times R}{q} \tag{9.2}$$

p: an acceptable electricity price for low-income residents; I: subsistence security standards; R: the proportion of acceptable electricity expenditure in the total expenditure of low-income residents; q: the upper bound electricity consumption of the first tier. The price of the first tier is the lower of the present minimum price and the acceptable price of low-income residents.

The quota electricity consumption in the second tier is designed for normal household electricity demand (Zhang et al. 2010): however, the design of tariff rates in this tier is more complicated than in the first tier. It should be designed to cover the power cost (including the cost of power generation, transmission, distribution, retail, etc.) or to gain reasonable benefits while considering its perceived acceptability to the vast majority (Zhang et al. 2010). As a result, it is necessary to identify the publicly acceptable tariff rate of the second tier before the implementation of TEP reform.

The electricity consumption in the third tier is much higher than that in the other two tiers. It is reasonable to set much higher prices in this tier aiming at inducing a reduction in otherwise high electricity consumption.

There are some typical differences between China's TEP and TEP in some other countries: firstly, the disparity in tariff between the highest and lowest tiers is much smaller than that in other counties. According to consulting draft of TEP reform, the electricity price in the third tier is about 1.2 times that of the first tier: however, the corresponding proportion between the highest tier and the lowest tier in South Korea and California is 11.7 and 1.8, respectively (Borenstein 2013; Liu and Yang 2010). This shows that the marginal prices between high electricity consumption and sustainable electricity consumption are small at present in China. This design can mitigate residents' desire to instigate a boycott (especially among high-income residents), therefore, it is important to take income into account when researching public acceptance of TEP in China. Secondly, increasing block rates are often designed to be revenue neutral in most of the world. This means that the first price tier falls below the average household electricity price, however, the proposed tariff rate of the first tier in China is maintained equal to, or even greater (the second scheme) than the present electricity price. It seems to be in discord with the revenue-neutral paradigm. As a result, residents (especially those with low incomes) may bear an economic burden arising from the increased price caused by TEP reform. It is necessary to identify their attitudes toward TEP reform and explore measures with which to raise their acceptance thereof.

9.3.2 Research Hypotheses

Residential acceptance plays a significant role in the successful implementation of TEP reform, therefore, it is important to identify the drivers and barriers thereto. The theory of planned behavior (TPB) provides a framework for systematic investigation of factors affecting behavioral choices (Ajzen 1991): many studies on pro-environmental behavior or household energy conservation behavior are based on the TPB framework. TPB assumes that behavioral intention is determined by attitude, perceived behavioral control, and subjective norms (Ajzen 1991). It plays a reference role in our research since the residential acceptance of TEP reform is also the result of a reasoned process of weighing various factors related to their external environment and internal attitudes. Therefore, we construct a hypothesis framework mainly based on TPB. Meanwhile, some additional variables are incorporated to enrich the content of our hypotheses.

Attitudes, according to TPB, refer to the degree to which a person has a favorable or unfavorable evaluation of a behavior (Abrahamse and Steg 2009). Residents' attitudes toward environmental protection (pro-environment) and energy-saving would play a significant role in their preferences toward TEP reform. This is attributed to the fact that residents who concern themselves more with the energy crisis and its environmental consequences would be more in favor of electricity conservation (Ek and Soderholm 2010; Samuelson 1990). TEP reform aims to promote household electricity conservation. It obliges residents to sacrifice personal economic benefit for the collective good (e.g, energy saving and environmental protection). Residents concerning themselves more with environmental protection or energy saving, therefore, would be more likely to accept TEP reform. These analyses lead to our first hypothesis:

Hypothesis 1 Residential awareness of pro-environment and energy saving needs would be positively related to their level of acceptance of TEP reform. Perceived behavioral control provides another theoretical lens to account for factors that can influence residential acceptance of TEP reform. It reflects the perceived ease or difficulty of engaging in TEP reform. Cost would be one of the obvious behavioral controls that residents could perceive when they participate in TEP reform. Various costs are imposed by TEP reform: direct financial costs, effort, or time. These costs can be classified into two types: economic cost and life quality loss. Residents often concern themselves greatly with the financial cost when engaging in energy consumption (Banfi et al. 2008; Scarpa and Willis 2010; Zografakis et al. 2010). It is clearly shown that the premiums in the second and third tiers would increase expenditure on electricity consumption. Moreover, the implementation of TEP reform may disturb lifestyle-related aspects of electricity use; less use of air-conditioners or washing machines, for instance, has to be considered if residents would like to reduce their average cost of electricity consumption. The resulting discomfort would negatively influence residents' acceptance of TEP reform because comfort benefits are highly preferred by them (Banfi et al. 2008). Based on the above discussion, the following hypothesis is posited:

Hypothesis 2 The economic cost and life quality loss would be negatively related to the level of residential acceptance of TEP reform. Based on TPB, subjective norms refer to the perceived social pressures to perform, or refrain from, a behavior. Accordingly, individual acceptance of TEP reform might be influenced by socio-environmental factors. It is indicated that other people's attitudes and behaviors in electricity saving may influence an individual's electricity saving activities. Moreover, the prevailing social atmosphere for environmental protection or energy conservation is helpful to an individual's electricity-saving behavior. In recent years, for instance, there have been many educational activities available for disseminating the importance of, and skills in, electricity conservation in China. Such activities would also activate residential enthusiasm toward accepting TEP reform. Besides, corresponding policies might have shaped a potential pressure on an individual's acceptance of TEP reform, therefore, we have the following hypothesis:

Hypothesis 3 The pressure from the social environment would be positively related to the level of residents' acceptance of TEP reform. Beyond the TPB framework, several studies indicate that previous experience has a direct effect on intention or behavior (Bentler and Speckart 1979; Fredricks and Dossett 1983). Macey and Brown (1983) reported that experience was the best predictor of conservation behavior. Typically, it can affect electricity conservation behavior in two ways: residents who are used to electricity saving are more likely to accept new electricity conservation policies. For another, it was also indicated that residents who had suffered brownouts (power cuts, or outages) would express greater concern about electricity savings (Feng et al. 2010). Information is often introduced as an important determinant in the analysis of residential energy-saving behaviors (Banfi et al. 2008; Cameron 1985; Fredricks and Dossett 1983). Holding more information about electricity conservation skills and energy efficiency technology would be helpful for residents to engage in electricity saving (Bentler and Speckart 1979). This facilitates residential electricity-saving behavior, which helps to mitigate the negative impacts of increased electricity charges caused by TEP reform. Then we posit the following two hypotheses:

Hypothesis 4 The experience of electricity saving or shortfall would be positively related to the level of residential acceptance of TEP reform.

Hypothesis 5 Information about electricity conservation skills or energy efficiency technology would be positively related to the level of residential acceptance of TEP reform. There are debates on whether income levels would be related to household electricity saving. Many surveys indicate that income level is a determinant of energy saving, with poor people participating more in energy-saving activities than rich people (Bentler and Speckart 1979; Brandon and Lewis 1999; Samuelson 1990); while others indicate no significant relationship between income level and energy saving (Brandon and Lewis 1999). However, there should be some relationship between the income level and public acceptance of TEP reform, as the rich might pay little heed to additional electricity expense, therefore, we introduce the income level as a control variable in our research model.

9.3.3 Questionnaire Development and Data Collection

Survey data were collected from four cities (Beijing, Shanghai, Tianjin, and Chongqing) in China to test the research hypotheses. The four cities are all municipalities directly under the control of the Chinese Central Government, and would be at the forefront of TEP reform in China. The residents in these four cities

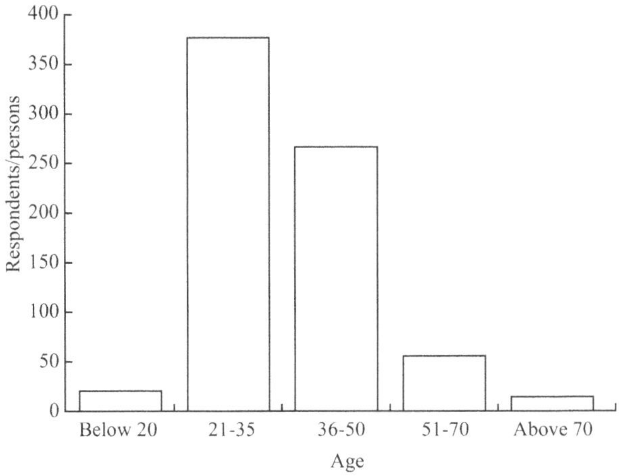

Fig. 9.4 Age profile of respondents

could be representative of residents in other urban areas of China. Data collection was undertaken in two phases including a pilot test and a random survey, whose results were eventually aggregated.

Pilot test: to customize the questionnaire to a Chinese context, we initially conducted a pilot test to validate and refine the measurement instrument. The pilot test was conducted in two residential quarters (Tian Tongyuan and Fang Zhuang) that are the largest and second-largest residential quarters in Beijing by the resident population. Based on the suggestions from 50 respondents in the two residential quarters, we made minor modifications to the wording of the questionnaire.

Random survey: we conducted a random survey in some residential quarters of Beijing, Shanghai, Tianjin, and Chongqing from October 2009 to May 2010. An on-site survey was carried out in Beijing, with 238 usable questionnaires received. The questionnaire survey in the other three cities was carried out by post. The questionnaires were posted to the neighborhood committees in some large residential quarters of these cities. The neighborhood committees were put in charge of questionnaire delivery and collection. Some 1200 questionnaires were despatched, and 649 responses were received. Then we eliminated those responses with missing values on any measurement items of independent variables or attitudes toward TEP reform. After incomplete responses were removed, 531 usable responses remained for subsequent analysis. We performed an independent-samples T-test to compare the data characteristics of the on-site survey and postal survey. The results show that there is no difference (at the 5% level of significance) between the two groups in questionnaire responses on demographic items and the items measuring attitudes toward TEP. The sample description is given in Table 9.5.

Table 9.5 Description of samples

City	Quantities of issued questionnaires	Quantities of responded questionnaires	Incomplete questionnaires	Effective rate/%
Beijing	400	238	12	56.50
Tianjin	400	204	26	44.50
Shanghai	400	217	48	42.25
Chongqing	400	228	52	44.00
Total	1600	887	138	41.8125

Sample aggregation: overall, 749 usable responses were collected. To identify whether or not, there were differences in sample characteristics among these four cities, we performed a one-way analysis of variance (ANOVA). The results are listed in Table 9.6. It is indicated that respondents' acceptance of TEP as a whole is not significantly different in these four cities ($F = 0.756$, Sig. = 0.519). The other sample characteristics show no statistical discrepancy at a 5% significance level, except for the income level. It is, therefore, reasonable to combine the samples collected from the four cities for data analysis.

The demographic information of respondents was well distributed and was representative of gender, age, and income level, although there were some sample biases. 51.23% of Respondents were male, and 48.77% were female. The gender ratio of respondents is similar to prevailing conditions in China (1.06:1). Figures 9.4 and 9.5 show that no single age group, or income group, dominated among respondents.

There are, however, certain biases compared with the actual population in China: survey participants aged 20–50 accounted for 86.52% of the sample. The respondents aged below 20 and above 70 only accounted for 3.47% and 2.27% thereof, respectively. This would not influence the representative nature of the sample because people below 20 years of age do not often own a house and often live with their parents. People aged over 70 are often cared for by the younger members of the family. The income level is divided into five tiers. Since all the respondents are from four urban cities in China, the income levels are above the average level across

Table 9.6 One-way analysis of variance for sample

Variable	Mean squares		*F* value	Significance
	Between groups	Within groups		
ALTP[a]	0.442	0.585	0.756	0.519
Gender	0.314	0.365	0.862	0.461
Education	0.411	0.683	0.602	0.614
Age	1.877	1.075	1.746	0.156
Income	17.755	1.616	10.990	0.000

Note ALTP refers to respondents' attitudes toward TEP reform

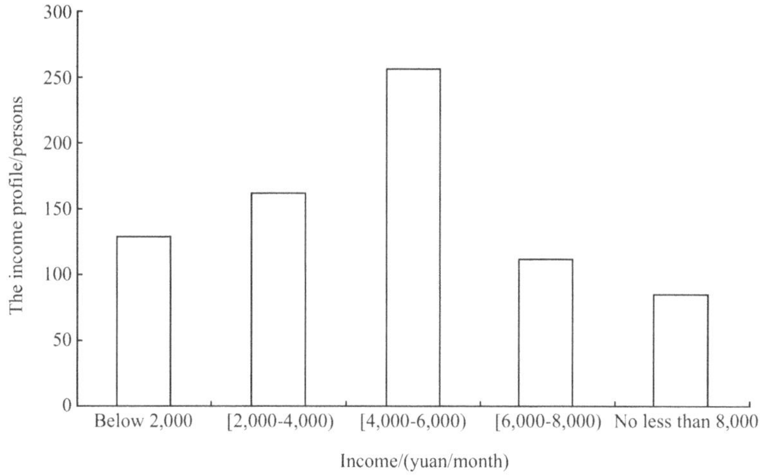

Fig. 9.5 Income profile of respondents

China. The number of respondents with an income of between 4000 and 6000 yuan accounts for 34.45% of the total, and that of respondents with an income of below 2000 or above 8000 yuan account for 17.09% and 11.48%, respectively.

The content of the questionnaire covers three main sections: individual demographic information, dependent item, and determinants of residential acceptance of TEP reform. The five dimensions of determinants of residential acceptance of TEP reform in this study are evaluated by 15 measurement items. As seen in Table 9.7, each construct is measured by three items. The measurement items of awareness of energy saving and pro-environment were developed based on a previous study (Nnorom et al. 2009). Three measurement items of experience were adopted from one study (Zografakis et al. 2010). The items for evaluating information and cost were developed based on previous studies (Banfi et al. 2008; Ek and Söderholm 2010; Tsagarakis et al. 2011). Three measurement items of social–environmental impacts referred to the studies of Abrahamse and Steg (2009), Thøgersen and Grønhoj (2010), and Zhu and Geng (2013). The survey results of the pilot test also provided a reference for the formation of measurement items.

The target respondents answered the questions using a five-point Likert-type scale (e.g., 1 = not at all important, 2 = not important, 3 = not thinking about it, 4 = important, to 5 = extremely important): the question designed to assess household willingness to accept TEP reform is "How much additional payment do you think is reasonable to impose on the present electricity price in the second tier of TEP?" Here respondents selected their willingness in four increasing ranks of choices.

Table 9.7 Descriptive statistics and CFA results for factors

Independent structure*	Items	Mean	Std. dev.	Completely standardized loading	*t* values
Awareness of energy-saving and pro-environment (Cronbach's Alpha = 0.706)	Awareness of energy crisis	2.85	0.966	0.71	18.29
	Realization of global climate change	2.84	0.906	0.75	19.34
	Awareness of environmental protection	2.72	0.824	0.60	15.57
Cost (Cronbach's Alpha = 0.743)	Additional payment caused by TEP reform	2.13	0.872	0.79	19.90
	Discomfort from electricity conservation caused by TEP reform	2.10	0.745	0.75	19.05
	Cost or time wasted for changing energy efficiency appliances because of TEP reform	1.88	0.825	0.58	15.23
Information (Cronbach's Alpha = 0.724)	Knowledge of methods for electricity saving	2.55	0.747	0.67	17.50
	Realization of policy or regulation in electricity using	2.48	0.857	0.66	17.24
	Improvement of technology for electricity conservation	2.32	0.774	0.72	18.98
Social environmental impacts (Cronbach's Alpha = 0.726)	Supportive policy	2.29	0.863	0.62	15.33
	Regulatory pressures	2.77	0.831	0.70	17.17
	Influence from friends' electricity saving behavior	2.66	0.894	0.69	17.01
Experience (Cronbach's Alpha = 0.712)	Experience of electricity shortfall	2.17	1.006	0.64	16.89
	Habits of electricity saving in daily lives	1.99	0.966	0.83	22.04
	Experience of participating in pro-environmental activities	2.62	1.092	0.58	15.06

* X^2 = 210.21; NFI=0.95; NNFI=0.96; GFI=0.97; RESEM=0.045

Before testing the hypotheses, we testified whether our framework of determinants would adequately fit the data collected. A confirmatory factor analysis (CFA) was performed to validate the measurement properties. All the measurement items were forced to load on their corresponding factor, with no correlation with other factors. Table 9.7 lists the results of CFA. It is shown that all measurement items have a reasonably high, and significant, loading with t values greater than 2.0 toward their respective factors. The standard loadings range from 0.58 to 0.83. These results support the convergent validity of the latent constructs, and substantiate the measurement properties. Moreover, data reliability was checked: Table 9.7 lists Cronbach's *alpha* values for the five determinants. The high values of Cronbach's alpha ($\alpha > 0.70$) suggest that all five latent factors fit the data reasonably well.

9.3.4 Determinants of Public Acceptance of TEP Reform

Respondents selected their willingness to accept TEP reform from four alternatives: (1) would not like the implementation of TEP; (2) would accept TEP reform if the premium added to electricity prices was below 0.05 yuan; (3) would accept TEP reform if the premium added to electricity prices was below 0.1 yuan; (4) would accept TEP reform even if the premium added to electricity prices were above 0.1 yuan. Since the response variable is ordinal and has more than two levels, the ordered regression model with maximum likelihood estimation is suitable to testify our hypotheses. Yet there is a choice between ordered logistic regression and ordered probit models. It is indicated that the Logit model is a better choice if the response decision is made based on the maximization of utility (Borsch-Supan 1990). Considering that residents' willingness to accept TEP reform mainly depends on the expected utility from the saving behavior, the Logit model was selected. The following specification was used:

$$y_i^* = \beta x_i + \varepsilon_i \tag{9.3}$$

where y_i: the latent and continuous measure of willingness to accept TEP reform, x_i: the vector of observations for the five dimensions of determinants presented in Table 9.7: β: the vector of parameters to be estimated, e_i: a random error term. The observed and coded discrete willingness variable y_i is determined from the model as follows:

$$y_i = \begin{cases} 1 & \text{if } -\infty < y_i^* \leq \mu_1 \text{(Would not like to accept TEP reform)} \\ 2 & \text{if } \mu_1 < y_i^* \leq \mu_2 \text{ (Would accept TEP reform with the premium} \leq 0.05) \\ 3 & \text{if } \mu_2 < y_i^* \leq \mu_3 \text{ (Would accept TEP reform with the premium} \leq 0.1) \\ 4 & \text{if } \mu_3 < y_i^* \leq \infty \end{cases} \tag{9.4}$$

where μ_i represents the thresholds to be estimated along with the parameter vector b. The probabilities of y_i in different coded value are defined as follows in our ordered logit model:

$$\begin{aligned} P(y_i = 1) &= F(\mu_1 - \beta x_i) \\ P(y_i = 2) &= F(\mu_2 - \beta x_i) - F(\mu_1 - \beta x_i) \\ P(y_i = 3) &= F(\mu_3 - \beta x_i) - F(\mu_2 - \beta x_i) \\ P(y_i = 4) &= 1 - F(\mu_3 - \beta x_i) \end{aligned} \tag{9.5}$$

where $P(y_i = k)$ refers to the probability that individual i adjusts his attitude toward TEP reform at the level of k; $F(\bullet)$ refers to the probability-distribution function of ε_i.

9.3.5 Discussion of Public Acceptance of Low-Carbon Policies

9.3.5.1 Drivers and Barriers to Respondents' Acceptance of TEP Reform

The ordered regression results are listed in Table 9.8. To identify the model-fitting information in this regression model, a χ^2-test was conducted on the -2log-Likelihood plot between the intercept only model and pre-hypothesized model, as it is usually considered as a critical statistic to detect incorrect model specification such as non-linearity in the predictors or missing predictors. The results (χ^2 = 100.490, sig. = 0.000) indicate that it is reasonable to reject the null hypotheses that the independent variables are not associated with the dependent variable. The Pearson χ^2 and deviance χ^2 are 2524.528 and 1535.507, respectively, which evinces goodness of fit in our model. A pseudo-R^2 is also estimated in our analysis. The results (Cox and Snell = 0.126; Nagelkerke = 0.141; McFadden = 0.060) support the explanatory power of the integral estimate. Besides, multi-collinearity was further checked among independent variables. The variance inflation factor (VIF) for all independent variables ranged from 1.024 to 1.401, well below the upper limit of 10.0 suggested by Mason and Perreault (1991). This means that multi-collinearity should not be a serious concern in our regression. The Wald statistics show that the coefficients are significantly different from zero, and then we can assume that the predictors are making a significant contribution to the

Table 9.8 Model estimation results

Indicator	Estimate	Std. dev.	Wald	*df*	Sig.	VIF
Threshold						
$[y_i = 1]$	1.496	0.496	9.086	1	0.003	
$[y_i = 2]$	4.372	0.522	70.147	1	0.000	
$[y_i = 3]$	6.173	0.550	126.033	1	0.000	
Dependent variables						
Awareness of energy saving and pro-environment	0.529	0.109	23.359	1	0.000	1.088
Cost	−0.064	0.109	0.100	1	0.554	1.103
Information	0.270	0.125	4.609	1	0.031	1.230
Social environmental impacts	0.554	0.111	24.804	1	0.000	1.287
Experience	0.064	0.100	0.409	1	0.523	1.170
Income level						**1.01**
[INC = 1]	−0.315	0.259	1.477	1	0.244	0
[INC = 2]	−0.943	0.310	9.221	1	0.002	
[INC = 3]	−0.793	0.251	9.940	1	0.002	
[INC = 4]	−0.293	0.281	1.090	1	0.297	
[INC = 5]	0			0		

Notes INC refers to respondents' income level; INC = 5 is set to be the reference item with the parameter being set to zero

prediction of the outcome. H1 posits that residential awareness of pro-environment and energy-saving influences their acceptance of TEP reform positively. From Table 9.8, we can see that the coefficient is 0.529 ($p < 0.01$), thus supporting H1. H3, which states that the pressure from the social environment could promote residential acceptance of TEP reform, is also confirmed ($b = 0.554$, $p < 0.01$). The support for H5, which posits the positive effect of information on residential acceptance of TEP reform, is weak but still significant at the 5% level.

The driving effect of experience, mentioned in H4, is not, however, significant in our estimation: we can observe the negative effect of cost, but the effect is not significant at the level of 0.05. This provides limited support for H2. Some statistical significances in the income level indicate that the respondents with different incomes present discrepant acceptance levels of TEP reform. Socio-environmental impacts have the largest influence on public acceptance of TEP reform ($b = 0.554$, sig. = 0.000). For one thing, policy regulation is an important aspect of external pressures. China is a country with more centralized power systems, and the household electricity price is under the stringent government regulatory control. The Chinese Government has attached great importance to household electricity price reform and has done much preparatory work before the implementation of TEP. These processes could exert an influence on the residential understanding of TEP, and gradually eliminated their innate resistance thereto. For another thing, several national strategies for responding to climate change and constructing conservation-minded society in China begin to form a social atmosphere conducive to environmental protection and energy conservation. Such an atmosphere is helpful for the implementation of TEP. Typically, this result is also in accordance with Ek and Söderholm, who believe that individuals' energy-saving behavioral intentions are influenced by other residents' behaviors around them (Ek and Söderholm 2010). Therefore, some respondents' positive attitudes toward TEP reform would promote other respondents living around them to accept TEP reform.

Awareness of energy-saving and pro-environment needs also plays a significant role in public acceptance of TEP reform. This is in accordance with previous studies (Lindenberg and Steg 2007; Sardianou 2007; Viklund 2004). Along with the increasingly deteriorated environment in China, more residents are aware of the environmental status of their surrounds. Many residents have begun to express concern about climate change due to more frequent extreme weather events (e.g, disastrous weather events with freezing rain and snow in Southeast china in February 2008, and continuous high temperatures during Summer in 2010) in recent years. All this environmental awareness exerts potential pressure on energy conservation. As a result, respondents who are more concerned with the energy crisis or climate change problems would prefer the implementation of TEP; however, the overall level of pro-environmental awareness in China still lags behind that in more developed countries. The mean value of awareness of energy saving and pro-environment reported in Table 9.7 is only 2.85, falling short of the average value of 2.50. It is indicated that there is great potential to improve environmental awareness in China. Some educational campaigns about serious global concerns such as climate change or the energy crisis may raise public environmental awareness and motivate respondents' acceptance of TEP reform.

Furthermore, respondents holding more information about electricity conservation skills are more likely to accept the implementation of TEP. This is because this specific knowledge could efficiently reduce their electricity consumption and offset the negative impacts caused by increased cost due to the implementation of TEP. Technology improvement for electricity efficiency in household appliances is an important channel for household electricity saving. Respondents who master the corresponding information may control their household electricity consumption within the upper bound of the second tier in TEP. These indicate that high levels of information pertaining to electricity saving is a powerful factor affecting residential acceptance of TEP reform.

Contrary to our hypothesis, the negative effect of the cost is not significant. This may due to the fact that there has been a low level of electricity prices in China for a long time. The financial cost resulting from the TEP reform brings limited pressure on residents' normal lives. In addition, the increased price that the respondents were asked to consider was too small (e.g. 0.05–0.1 yuan/kW·h) to make respondents realize the real bill shock caused by TEP reform. It is thus difficult to form cost pressure. This is in accordance with the argument of Ito (2010). For the cost of life quality, few negative impacts would be brought about. Electrically efficient appliances have been pervasive in China. It is easy to offset the discomfort or inconvenience by adopting these advanced, electricity-efficient technologies.

The contradiction between high electricity demand and insufficient supply ability in China means that many residents have experiences of electricity shortages and power brownouts; however, the results show that these experiences did not efficiently drive respondents' acceptance of TEP reform. A possible reason is that electricity policy in China is inclined to satisfy the demand for household electricity use first. This principle guarantees that the electricity shortfall in household sectors would not last long enough to disturb respondents' daily lives, which as a result could not form enough pressure toward acceptance of TEP reform. The habits of electricity saving also seem to play no special role in the acceptance of TEP reform. This might be attributed to the fact that daily energy-saving habits are often driven by cost saving (Banfi et al. 2008), while the proposed TEP in China raises little cost pressure on electricity consumption. Cost effects on the acceptance of TEP reform are not significant in our results, therefore, there are no significant differences in the acceptance of TEP between respondents with, and without, electricity-saving habits.

Compared with respondents whose incomes are above 8000 yuan/month, there are statistically significant negative impacts on the endorsement of TEP for two groups of respondents (INC2 and INC3). It is indicated that respondents opposed to the implementation of TEP are primary the middle-income group with incomes of between 2000 and 8000 yuan/month. In particular, respondents with incomes of between 2000 and 4000 yuan are more against TEP reform because $|\beta_{INC2}| > |\beta_{INC3}|$; however, respondents with incomes of above 8000 yuan per month show much interest in the implementation of TEP. Besides, it is an unimagined finding that the respondents with the lowest income also present little resistance to TEP reform.

9.3.5.2 Estimation of the Acceptable Premium in the Second Tier of TEP

The electricity price in the second tier of TEP covers the electricity demand in residents' normal lives. As a result, respondents care more about the price mechanism in this tier and show different attitudes toward the premium. It is necessary to identify a generally accepted ratio for the premium before the implementation of TEP. The statistical analysis of responses to the question "How much additional payment do you think is reasonable to impose on the present electricity price in the second tier of TEP?", reveals that about 56.55% of respondents think it would be reasonable to have an increase within 0.05 yuan of the electricity price in the second tier. There are still 19.25% of respondents who would not like to adopt TEP. Only 5.35% of respondents would accept an additional payment exceeding 0.1 yuan/kW·h.

Considering the pressure of financial cost brought by TEP reform on residents, the acceptable additional payment in the second tier would also be discrepant for

Table 9.9 Estimation of the acceptable premium for respondents with different incomes

Indicator	ALTP = 1(%)	ALTP = 2(%)	ALTP = 3(%)	ALTP = 4(%)
[INC = 1]	11.24	57.96	23.95	6.85
[INC = 2]	43.98	49.32	5.53	1.17
[INC = 3]	58.62	37.55	3.17	0.65
[INC = 4]	10.78	57.41	24.66	7.15
[INC = 5]	11.28	58.00	23.90	6.82

Notes ALTP refers to respondents' attitudes toward TEP reform; INC refers to respondents' income level

respondents with different incomes. We further estimate this using our ordinal regression model: the sample was divided into five groups according to the income level. The average values of the independent variables in each group were brought into the estimated ordinal regression model (Table 9.9).

According to our estimation, respondents with incomes of between 2000 yuan/month and 8000 yuan/month are less willing to accept TEP, compared with other respondents. 43.98% of respondents with incomes of 2000–4000 yuan/month and 58.62% of respondents with incomes of 4000–8000 yuan/month would accept the present electricity price with no additional payment. Few respondents in these two groups (1.17% and 0.65%, respectively) would accept a level of additional payment above 0.1 yuan/kW·h. The respondents in the remaining three groups showed more support for TEP. Respondents (57.96%) with incomes of below 2000 yuan/month would accept additional cost within 0.05 yuan/kW·h. Most respondents with an income of above 8000 yuan/month also prefer a level of extra payment within 0.05 yuan/kW·h. Any level of additional payment above 0.1 yuan/kW·h failed to garner widespread acceptance among the respondents in these three groups, but there were many more supporters for this compared with the two groups with incomes of between 2000 yuan/month and 8000 yuan/month.

9.3.6 Summary

This chapter focuses on the residential acceptance of TEP reform and corresponding determinants of that acceptance. Results show that the majority of respondents prefer to accept a premium in the second tier of TEP of below 0.05 yuan/kW·h: residents with different income levels show significantly different attitudes toward TEP reform. Most respondents who are against TEP reform belong to the middle-income group (between 2000 and 8000 yuan/month). Low-income respondents and the high-income group seem more willing to accept a higher premium. For one thing, this might be attributed to the fact that the electricity consumption among residents with the lowest incomes is small, which often fail to reach the upper bound of the first tier. The implementation of TEP, as a result, has little impact on their normal lives. High-income residents are rich enough to afford their high electricity consumption; the premium resulting from TEP would be trivial for them, while for middle-income residents, their electricity expenditure would increase significantly since a large amount of their daily consumption falls within the second tier. They would not care to accept the inevitable increase in expenditure. For another thing, middle-income residents may feel unfairly treated by the proposed TEP considerations. The rated tiers are too few, and the range of the second tier is deemed too large. The disparity in tariff between the second and third tiers is also deemed too small: consequently, middle-income residents may feel that their increased burden is akin to that on high-income residents, and is much more than that of low-income residents. They would, therefore, be reluctant to accept the TEP.

Raising public awareness in the energy crisis and the face of environmental degradation is also important. According to our survey, those respondents concerning themselves with global warming and environmental deterioration, or preferring to use energy-saving products, are more willing to accept TEP. It indicates that a sound social environment needs to be constructed to inspire residents' willingness to engage in energy saving and environmental protection. Typically, educational campaigns around issues of energy scarcity and environmental degradation could be conducted to raise public consciousness of the necessity of TEP reform. Relevant knowledge of the benefits and mechanisms of TEP could be disseminated via social media. Besides, other people's attitudes toward TEP play a positive role and it is necessary to provide convenient conditions for spontaneous activities leading to energy saving among residents. Residents, as a result, can exchange electricity-saving experiences and influence each other's positive attitudes toward TEP reform.

References

Abrahamse, W., and L. Steg. 2009. How do socio-demographic and psychological factors relate to households' direct and indirect energy use and savings? *Journal of Economic Psychology* 30 (5): 711–720.

Ajzen, I. 1991. The theory of planned behavior. *Organizational Behavior and Human Decision Processes* 50 (2): 179–211.

Banfi, S., M. Farsi, M. Filippini, et al. 2008. Willingness to pay for energy-saving measures in residential buildings. *Energy Economics* 30 (2): 503–516.

Bentler, P.M., and G. Speckart. 1979. Models of attitude-behavior relations. *Psychological Review* 86 (5): 452–464.

Borenstein, S. 2013. Effective and equitable adoption of opt-in residential dynamic electricity pricing. *Review of Industrial Organization* 42 (2): 127–160.

Borsch-Supan, A. 1990. On the compatibility of nested logit-models with utility maximization. *Journal of Econometrics* 43: 373–388.

Brandon, G., and A. Lewis. 1999. Reducing household energy consumption: a qualitative and quantitative field study. *Journal of Environmental Psychology* 19: 75–85.

Bruneau, J.F. 2005. Inefficient environmental instruments and the gains from trade. *Journal of Environmental Economics and Management* 49: 536–546.

Busse, M. 2004. Trade, environmental regulations and the world trade organization: new empirical evidence. *Social Science Electronic Publishing* 38 (2): 285–306.

Cameron, T.A. 1985. A nested logit model of energy conservation activity by owners of existing single family dwellings. *Review of Economics and Statistics* 67 (2): 205–211.

Chang, J. 2002. WTO and adaptable innovation to China's environmental legal systems. *Journal of Hunan Public Security College* 14 (1): 23–29. (In Chinese).

Chen H. 2009. Exploration for the implementation of tiered electricity price in household sector. *Journal of Petroleum Educational Institute of Xinjiang* 10: 274–276.

Copeland, B.R., and M.S. Taylor. 2005. Trade, growth and the environment. University of British Columbia. Working Paper No.1222.

Ek, K., and P. Söderholm. 2010. The devil is in the details: Household electricity saving behavior and the role of information. *Energy Policy* 38 (3): 1578–1587.

Feng, D.S., B.K. Sovacool, and M.K. Vu. 2010. The barriers to energy efficiency in China: assessing household electricity savings and consumer behavior in Liaoning Province. *Energy Policy* 38 (2): 1202–1209.

Fredricks, A.J., and D.L. Dossett. 1983. Attitude-behavior relations: a comparison of the Fishbein-Ajzen and the Bentler-Speckart models. *Journal of Personality and Social Psychology* 45 (3): 501–512.

Lai, Y.B., and C.H. Hu. 2008. Trade agreements, domestic environmental regulation, and transboundary pollution. *Resource and Energy Economics* 30 (2): 209–228.

Levinson, A., and M.S. Taylor. 2008. Unmasking the pollution haven effect. *International Economic Review* 49 (1): 223–254.

Lindenberg, S., and L. Steg. 2007. Normative, gain and hedonic goal frames guiding environmental behavior. *Journal of Social Issues* 63: 117–137.

Liu, S.J., and Yang, J. 2010. Research on tiered electricity price. *Price: Theory and Practice* 3: 12–14. (In Chinese).

Long, X.L., K. Oh, and G. Cheng. 2013. Are stronger environmental regulations effective in practice? The case of China's accession to the WTO. *Journal of Cleaner Production* 39: 161–167.

Macey, S.M., amd M.A. Brown. 1983. Residential energy conservation: the role of past experience in repetitive household behaviour. *Environment and Behavior* 15 (2): 123–141.

Mason, C., and W.D. Perreault. 1991. Collinearity, power, and interpretation of multiple regression analysis. *Journal of Marketing Research* 28 (3): 268–280.

Nnorom, I.C., J. Ohakwe, and O. Osibanjo. 2009. Survey of willingness of residents to participate in electronic waste recycling in Nigeria—a case study of mobile phone recycling. *Journal of Cleaner Production* 17 (18): 1629–1637.

Samuelson, C.D. 1990. Energy conservation: a social dilemma approach. *Social Behavior* 5: 207–230.

Sardianou, E. 2007. Estimating energy conservation patterns of Greek households. *Energy Policy* 35: 3778–3791.

Scarpa, R., and K. Willis. 2010. Willingness-to-pay for renewable energy: primary and discretionary choice of British households' for micro-generation technologies. *Energy Economics* 32: 129–136.

Tsen, W.H. 2011. The real exchange rate determination: an empirical investigation. *International Review of Economics & Finance* 20 (4): 800–811.

Thøgersen, J., and A. Grønhøj. 2010. Electricity saving in households—a social cognitive approach. *Energy Policy* 38: 7732–7743.

Viklund, M. 2004. Energy policy options—from the perspective of public attitudes and risk perceptions. *Energy Policy* 32: 1159–1171.

Wang, D.N. 2010. The problems of the proposed tiered electricity price in China and corresponding measures. *Price: Theory and Practice* 4: 42–43. (In Chinese).

Wang, Z., B. Zhang, and Y. Zhang. 2012. Determinants of public acceptance of tiered electricity price reform in China: evidence from four urban cities. *Applied Energy* 91 (1): 235–244.

Xiong, Y., and X. Xu. 2007. The effects of environmental regulation on FDI in China: an empirical analysis of panel data. *Economic Review* 2: 122–124. (In Chinese).

Xu, X.P. 2000. International trade and environmental policy: how effective is "eco-dumping"? *Economic Modelling* 17 (1): 71–90.

Zellner, A. 1962. An efficient method of estimating seemingly unrelated regression equations and tests for aggregation bias. *Journal of the American Statistical Association* 57: 348–368.

Zhang, Y., Y. Fang, K.K. Wei, et al. 2010. Exploring the role of psychological safety in promoting the intention to continue sharing knowledge in virtual communities. *International Journal of Information Management* 30 (5): 425–436.

Zhang, K., J.L. Schnoor, and E.Y. Zeng. 2012. E-waste recycling: where does it go from here? *Environmental Science and Technology* 46: 10861–10867.

Zhang, L.Z., H.T. Huang, and S.R. Gui. 2010. Research on methodology of design of household tiered electricity price in China. *Price: Theory and Practice* 4: 44–45. (In Chinese).

Zhu, Q., and Y. Geng. 2013. Drivers and barriers of extended supply chain practices for energy saving and emission reduction among Chinese manufacturers. *Journal of Cleaner Production* 40: 6–12.

Zhu, Q., and J. Sarkis. 2006. An inter-sectoral comparison of green supply chain management in China: drivers and practices. *Journal of Cleaner Production* 14 (5): 472–486.

Zhu, Q., J. Sarkis, and K.H. Lai. 2007. Green supply chain management: pressures, practices and performance within the Chinese automobile industry. *Journal of Cleaner Production* 15 (11–12): 1041–1052.

Zografakis, N., E. Sifaki, M. Pagalou, et al. 2010. Assessment of public acceptance and willingness to pay for renewable energy sources in Crete. *Renewable and Sustainable Energy Reviews* 14 (3): 1088–1095.

Appendix

6-A. Major Questionnaire Items

- **CO_2 reduction practices**
  - **Innovation**

    R&D investment on energy saving or carbon abatement technology
    Retrofit the production process for energy saving and carbon abatement
    Replacing equipment with energy efficient ones

  - **Residue recycling**

    Recycling residual energy (e.g., residual heat)
    Recycling scrap materials (e.g., blast furnace slag)
    Recycling by-product (e.g., coke oven gas, furnace gas)

  - **Carbon trading**

    Application of CDM projects
    Selling the credits of reduced carbon from CDM
    Construction of carbon reduction accounting system

  - **CO_2 abatement strategy**

    Short-term objective for energy saving or carbon abatement
    Long-term vision for energy saving or carbon abatement
    Clear plans for energy saving or carbon abatement

- **The drivers and barriers**
  - **Regulations**

    Central governmental regulations on energy conservation and CO_2 reduction

Z. Wang and B. Zhang, *Low-Carbon Consumption in China: Residential Behavior, Corporate Practices and Policy Implication*,
https://doi.org/10.1007/978-981-15-2792-0

Regional governmental regulations on energy conservation and CO_2 reduction
Export countries' regulations on energy conservation and CO_2 reduction

- **Pressures from supply chains**

Requirements from customers on energy-saving and CO_2 abatement
Requirements from suppliers on energy-saving and CO_2 abatement
Requirements from material recyclers on energy-saving and CO_2 abatement

- **Imitating effect**

Competitors have taken some practices on energy-saving and CO_2 abatement
Substitution producers have taken some practices on energy-saving and CO_2 abatement
Industrial leaders have taken some practices on energy-saving and CO_2 abatement

- **Financial cost**

High investment of R&D on energy-saving and CO_2 abatement
High cost for equipment replacement and process retrofit
Insignificant benefits from energy-saving and CO_2 abatement in short-term

- **Lack of capacities**

Lack of advanced technology on energy-saving and CO_2 abatemen
Lack of qualified people to solve the issues of energy-saving and CO_2 abatement
Lack of information channels on energy-saving and CO_2 abatement

- **Performance**
  - **Environmental performance**

Reduction of greenhouse gas emission
Reduction of waste discharge
Decrease of frequency for environmental accidents
Improving company's social image

  - **Economic performance**

Decreasing the cost of energy consumption
Decreasing the fee of waste treatment
Improvement on reducing the investment
Improvement on reducing the operation cost

- **Basic information**
  - Your company is (a) state-owned; (b) a joint venture; (c) an FDI enterprise; (d) a private sector
  - The number of employees in your company is (a) below 1000; (b) 1000–5000; (c) 5000–10000; (d) 10000–20000; (e) above 20000

8-A. Major Questionnaire Items

- **Inter-firm cooperation on CO_2 reduction**
  - **Carbon reduction cooperation with suppliers and customers (ZC)**

    Cooperation on energy-saving design of product with suppliers or customers (ZC1)
    Agreement on energy-saving transportation with suppliers or customers (ZC2)
    Cooperation with suppliers or customers on waste energy recycling (ZC3)

  - **Carbon reduction cooperation with competitors and producers (HC)**

    R&D cooperation on carbon reduction with competitors or producers (HC1)

  - **Carbon reduction cooperation through industrial symbiosis (IS)**

    Waste energy and resources exchange among industrial firms in different industrial chain (IS1)

- **Determinants**
  - **Regulation (REG)**

    Central governmental regulations on energy conservation and CO_2 reduction (L1)
    Regional governmental regulations on energy conservation and CO_2 reduction (L2)
    Export countries' regulations on energy conservation and CO_2 reduction (L3)

  - **CO_2 reduction demand from stakeholders within industrial chains (CDI)**

    Energy-saving and CO_2 abatement demand from suppliers (I1)
    Energy-saving and CO_2 abatement demand from customers (I2)
    Waste energy or resource demand from firms in other industrial chain (I3)

  - **Imitating effect (IE)**

    There are some successful cases on carbon reduction cooperation by competitors (M1)

There are some successful cases on carbon reduction cooperation by the leaders in the industry (M2)
There are some successful cases on carbon reduction cooperation by surrogates (M3)

- **Financial pressure (FP)**

  High operation cost on carbon reduction cooperation (E1)
  Unclear benefits from carbon reduction cooperation (E2)
  High financial investment in CO_2 reduction (E3)

- **Defective infrastructure and mechanism (DIM)**

  Defective of risk-cost share mechanism for carbon reduction cooperation (J1)
  Information security problems caused by carbon reduction cooperation (J2)
  "Free-riders" in carbon reduction cooperation (J3)

- **Performance**
  - **Environmental performance**

    Reduction of green house gas emission
    Decrease of frequency for environmental accidents
    Improving company's social image
    Reduction of waste discharge

  - **Economic performance**

    Increasing the rate of investment return
    Increasing profits of the industrial firms
    Decreasing the fee of waste treatment
    Decreasing the expenditure of energy

- **Basic information**
  - Your company is (a) state-owned; (b) a joint venture; (c) an FDI enterprise; (d) a private sector
  - The number of employees in your company is (a) below 100; (b) 100–500; (c) 500–1000; (d) 1000–3000; (e) above 3000
  - Your company belongs to (a) PPC; (b) PSEP; (c) NMP; (d) SPFM; (e) SPNM; (f) RCMCP

References

Abrahamse, W., L. Steg, R. Gifford, et al. 2009. Factors influencing car use for commuting and the intention to reduce it: A question of self-interest or morality? *Transportation Research Part F: Traffic Psychology and Behaviour* 12 (4): 317–324.

Abrahamse, W., L. Steg, C. Vlek, et al. 2007. The effect of tailored information, goal setting, and tailored feedback on household energy use, energy-related behaviors, and behavioral antecedents. *Journal of Environmental Psychology* 27 (4): 265–276.

Anderson, J.C., and D.W. Gerbing. 1988. Structural equation modeling in practice: A review and recommended two-step approach. *Psychological Bulletin* 103 (3): 411–423.

Ang, B.W., F.Q. Zhang, and K.H. Choi. 1998. Factorizing changes in energy and environmental indicators through decomposition. *Energy* 23 (6): 489–495.

Bai, Y., and Y. Liu. 2013. An exploration of residents' low-carbon awareness and behavior in Tianjin, China. *Energy Policy* 61: 1261–1270.

Bamberg, S., M. Hunecke, and A. Blöbaum. 2007. Social context, personal norms and the use of public transportation: Two field studies. *Journal of Environmental Psychology* 27 (3): 190–203.

Baron, R.M., and D.A. Kenny. 1986. The moderator-mediator variable distinction in social psychological research: Conceptual, strategic, and statistical considerations. *Journal of Personality and Social Psychology* 51 (6): 1173–1182.

Barr, S., A.W. Gilg, and N. Ford. 2005. The household energy gap: Examining the divide between habitual- and purchase-related conservation behaviours. *Energy Policy* 33 (11): 1425–1444.

Beck, L., and I. Ajzen. 1991. Predicting dishonest actions using the theory of planned behavior. *Journal of Research in Personality* 25 (3): 285–301.

Bock, G.W., R.W. Zmud, Y.G. Kim, et al. 2005. Behavioral intention formation in knowledge sharing: Examining the roles of extrinsic motivators, social-psychological forces, and organizational climate. *MIS Quarterly* 29 (1): 87–111.

Chaabane, A., A. Ramudhin, and M. Paquet. 2012. Design of sustainable supply chains under the emission trading scheme. *International Journal of Production Economics* 135 (1): 37–49.

Chen, C.J., and J.W. Huang. 2007. How organizational climate and structure affect knowledge management—The social interaction perspective. *International Journal of Information Management* 27 (2): 104–118.

Chen, M. 2016. Extending the theory of planned behavior model to explain people's energy savings and carbon reduction behavioral intentions to mitigate climate change in Taiwan–moral obligation matters. *Journal of Cleaner Production* 112: 1746–1753.

Chin, W.W. 1998. Issues and opinion on structural equation modeling. *MIS Quarterly* 22 (1): vii–xvi.

Z. Wang and B. Zhang, *Low-Carbon Consumption in China: Residential Behavior, Corporate Practices and Policy Implication*,
https://doi.org/10.1007/978-981-15-2792-0

Chin, W.W., B.L. Marcolin, and P.R. Newsted. 2003. A partial least squares latent variable modeling approach for measuring interaction effects: Results from a monte carlo simulation study and an electronic-mail emotion/adoption study. *Information Systems Research* 14 (2): 189–217.

Cong, R.G., and Y.M. Wei. 2010a. Potential impact of (CET) carbon emissions trading on China's power sector: A perspective from different allowance allocation options. *Energy* 35 (9): 3921–3931.

Cong, R.G., and Y.M. Wei. 2010b. Auction design for the allocation of carbon emission allowances: Uniform or discriminatory price. *International Journal of Energy and Environment* 1 (3): 533–546.

Cong, R.G., and Y.M. Wei. 2012. Experimental comparison of impact of auction format on carbon allowance market. *Renewable and Sustainable Energy Reviews* 16: 4148–4156.

Dallmann, T.R., T.W. Kirchstetter, S.J. DeMartini, et al. 2013. Quantifying on-road emissions from gasoline-powered motor vehicles: Accounting for the presence of medium-and heavy-duty diesel trucks. *Environmental Science & Technology* 47 (23): 13873–13881.

David, R., and T. Kenneth. 1998. Mixed logit with repeated choices: Households' choices of appliance efficiency level. *The Review of Economics and Statistics* 80 (4): 647–657.

De Clercq, D., and I.B. Rius. 2007. Organizational commitment in Mexican small and medium-sized firms: The role of work status, organizational climate, and entrepreneurial orientation. *Journal of Small Business Management* 45 (4): 467–490.

De Groot, J.I.M., and L. Steg. 2009. Morality and prosocial behavior: The role of awareness, responsibility, and norms in the norm activation model. *The Journal of Social Psychology* 149 (4): 425–449.

de la Rue, Du, Can Stephane, G. Leventis, A.A. Phadke, et al. 2014. Design of incentive programs for accelerating penetration of energy-efficient appliances. *Energy Policy* 72: 56–66.

Ek, K., and P. Söderholm. 2010. The devil is in the details: Household electricity saving behavior and the role of information. *Energy Policy* 38 (3): 1578–1587.

Färe, R., and S. Grossopf. 2010. Directional distance functions and slacks-based measurement of efficiency. *European Journal of Operational Research* 200: 320–322.

Galarraga, I., M. González-Eguino, and A. Markandya. 2011. Willingness to pay and price elasticities of demand for energy-efficient appliances: Combining the hedonic approach and demand systems. *Energy Economics* 33: S66–S74.

Godin, G., M. Conner, and P. Sheeran. 2005. Bridging the intention–behaviour 'gap': The role of moral norm. *British Journal of Social Psychology* 44 (4): 497–512.

Greene, D. 2007. Rebound: Analysis of U.S. light-duty vehicle travel statistics. *Energy Policy* 41: 14–28.

Greening, L.A., and D.L. Greene. 1998. Energy use, technical efficiency, and the rebound effect: A review of the literature. Oak Ridge, TN: Center for Transportation Analysis.

Guagnano, G.A. 2001. Altruism and market-like behavior: An analysis of willingness to pay for recycled paper products. *Population and Environment: A Journal of Interdisciplinary Studies* 22 (4): 425–438.

Guo, H., and Z. Yin. 2007. Range of recovery of exhaust heat in metallurgical enterprises. *Iron & Steel* 2: 77–81 (in Chinese).

Guo, Z.C., and Z.X. Fu. 2009. Current situation of energy consumption and measures taken for energy saving in the iron and steel industry in China. *Energy* 35 (11): 4356–4360.

Gyberg, P., and J. Palm. 2009. Influencing households' energy behavior—How is it done and on what premises. *Energy Policy* 37 (7): 2807–2813.

Ha, H.Y., and S. Janda. 2012. Predicting consumer intentions to purchase energy-efficient products. *Journal of Consumer Marketing* 29 (7): 461–469.

Harland, P., H. Staats, and H.A.M. Wilke. 2007. Situational and personality factors as direct or personal norm mediated predictors of pro-environmental behavior: Questions derived from norm-activation theory. *Basic and Applied Social Psychology* 29 (4): 323–334.

Honma, S., and J.L. Hu. 2008. Total-factor energy efficiency of regions in Japan. *Energy Policy* 36: 821–833.

Hsu, C., C. Chang, and C. Yansritakul. 2017. Exploring purchase intention of green skincare products using the theory of planned behavior: Testing the moderating effects of country of origin and price sensitivity. *Journal of Retailing and Consumer Services* 34: 145–152.

Hsu, M.L.A., and H.L. Fan. 2010. Organizational innovation climate and creative outcomes: Exploring the moderating effect of time pressure. *Creativity Research Journal* 22 (4): 378–386.

Hu, X., and K. Jiang. 2001. Evaluation of technology and countermeasure for greenhouse gas mitigation in China. Beijing: China Environmental Science Press (in Chinese).

IEA. 2010. *CO_2emissions from fuel combustion highlights*. Paris: International Energy Agency.

Jansson, J. 2011. Consumer eco-innovation adoption: Assessing attitudinal factors and perceived product characteristics. *Business Strategy and the Environment* 20 (3): 192–210.

Jiang, C., W. Zhao, X. Sun, et al. 2016. The effects of the self and social identity on the intention to microblog: An extension of the theory of planned behavior. *Computers in Human Behavior* 64: 754–759.

Jin, S.H. 2007. The effectiveness of energy efficiency improvement in a developing country: Rebound effect of residential electricity use in South Korea. *Energy Policy* 35: 5622–5629.

Kaya, N., E. Koc, and D. Topcu. 2010. An exploratory analysis of the influence of human resource management activities and organizational climate on job satisfaction in Turkish banks. *The International Journal of Human Resource Management* 21 (11): 2031–2051.

Kelly, G. 2012. Sustainability at home: Policy measures for energy-efficient appliances. *Renewable and Sustainable Energy Reviews* 16 (9): 6851–6860.

Kerlinger, F.N. 1973. *Foundations of behavioral research*, 2nd ed. New York: Holt, Rinehart and Winston.

Klöckner, C.A., and E. Matthies. 2004. How habits interfere with norm-directed behaviour: A normative decision-making model for travel mode choice. *Journal of Environmental Psychology* 24 (3): 319–327.

Klöckner, C.A., and S. Ohms. 2009. The importance of personal norms for purchasing organic milk. *British Food Journal* 111 (11): 1173–1187.

Kwak, S.Y., S.H. Yoo, and S.J. Kwak. 2010. Valuing energy-saving measures in residential buildings: A choice experiment study. *Energy Policy* 38 (1): 673–677.

Lee, Y.J., R. De Young, and R.W. Marans. 1995. Factors influencing individual recycling behavior in office settings: A study of office workers in Taiwan. *Environment and Behavior* 27 (3): 380–403.

Liddle, B. 2012. The systemic, long-run relation among gasoline demand, gasoline price, income, and vehicle ownership in OECD countries: Evidence from panel cointegration and causality modeling. *Transportation Research Part D: Transport and Environment* 17 (4): 327–331.

Liu, W.B., W. Meng, X.X. Li, et al. 2010. DEA models with undesirable inputs and outputs. *Annals of Operations Research* 173: 177–194.

Longo, A., A. Markandya, and M. Petrucci. 2008. The internationalization of externalities in the production of electricity: Willingness to pay for the attributes of a policy for renewable energy. *Ecological Economics* 67 (1): 140–152.

Martinsson, J., L.J. Lundqvist, and A. Sundström. 2011. Energy saving in Swedish households. The (relative) importance of environmental attitudes. *Energy Policy* 39: 5182–5191.

Mekaroonreung, M., and A.L. Johnson. 2012. Estimating the shadow price of SO_2 and NO_x for U. S. coal power plants: A convex nonparametric least squares approach. *Energy Economics* 34: 723–732.

Mills, B., and J. Schleich. 2010. What's driving energy efficient appliance label awareness and purchase propensity? *Energy Policy* 38 (2): 814–825.

Min, G.F., and E. Mills. 1997. Energy-efficient lighting in China: Problems and prospects. *Energy Policy* 25: 77–83.

Ministry of Land and Resources. 2005a. China Land and Resources Bulletins 2005. http://data.mlr.gov.cn/tjgb/201807/t20180704_1997935.html [2006-10-09].

Ministry of Land and Resources. 2005b. China Land and Resources Statistical Yearbook 2005. Beijing: Geological Press.

Ministry of Land and Resources. 2006a. China Land and Resources Bulletins 2006. http://data.mlr.gov.cn/tjgb/201807/t20180704_1997936.html [2007-02-04].
Ministry of Land and Resources. 2006b. China Land and Resources Statistical Yearbook 2006. Beijing: Geological Press.
Ministry of Land and Resources. 2007a. China Land and Resources Bulletins 2007. http://data.mlr.gov.cn/tjgb/201807/t20180704_1997937.html [2008-06-09].
Ministry of Land and Resources. 2007b. China Land and Resources Statistical Yearbook 2007. Beijing: Geological Press.
Ministry of Land and Resources. 2008a. China Land and Resources Bulletins 2008. http://data.mlr.gov.cn/tjgb/201807/t20180704_1997938.html [2009-05-14].
Ministry of Land and Resources. 2008b. China Land and Resources Statistical Yearbook 2008. Beijing: Geological Press.
Ministry of Land and Resources. 2009a. China Land and Resources Bulletins 2009. http://data.mlr.gov.cn/tjgb/201807/t20180704_1997939.html [2010-06-10].
Ministry of Land and Resources. 2009b. China Land and Resources Statistical Yearbook 2009. Beijing: Geological Press.
Ministry of Land and Resources. 2010a. China Land and Resources Bulletins 2010. http://data.mlr.gov.cn/tjgb/201807/t20180704_1997940.html [2011-07-17].
Ministry of Land and Resources. 2010b. China Land and Resources Statistical Yearbook 2010. Beijing: Geological Press.
Ministry of Land and Resources. 2011a. China Land and Resources Bulletins 2011. http://data.mlr.gov.cn/tjgb/201807/t20180704_1997941.html [2012-01-12].
Ministry of Land and Resources. 2011b. China Land and Resources Statistical Yearbook 2011. Beijing: Geological Press.
Mizobuchi, K., and K. Takeuchi. 2016. Replacement or additional purchase: The impact of energy-efficient appliances on household electricity saving under public pressures. *Energy Policy* 93: 137–148.
NBSC. 1993. China Energy Statistical Yearbook 1993. Beijing: China Statistics Press.
NBSC. 1996. China Statistical Yearbook 1996. Beijing: China Statistics Press.
NBSC. 1997a. China Statistical Yearbook 1997. Beijing: China Statistics Press.
NBSC. 1997b. China Price Statistical Yearbook 1997. Beijing: China Statistics Press.
NBSC. 1997c. China Energy Statistical Yearbook 1997. Beijing: China Statistics Press.
NBSC. 1997d. China Population Statistical Yearbook 1997. Beijing: China Statistics Press.
NBSC. 1998a. China Energy Statistical Yearbook 1998. Beijing: China Statistics Press.
NBSC. 1998b. China Statistical Yearbook 1998. Beijing: China Statistics Press.
NBSC. 1998c. China Price Statistical Yearbook 1998. Beijing: China Statistics Press.
NBSC. 1998d. China Population Statistical Yearbook 1998. Beijing: China Statistics Press.
NBSC. 1999a. China Statistical Yearbook 1999. Beijing: China Statistics Press.
NBSC. 1999b. China Price Statistical Yearbook 1999. Beijing: China Statistics Press.
NBSC. 1999c. China Energy Statistical Yearbook 1999. Beijing: China Statistics Press.
NBSC. 1999d. China Population Statistical Yearbook 1999. Beijing: China Statistics Press.
NBSC. 2000a. China Energy Statistical Yearbook 2000. Beijing: China Statistics Press.
NBSC. 2000b. China Industry Economy Statistical Yearbook 2000. Beijing: China Statistics Press.
NBSC. 2000c. China Statistical Yearbook 2000. Beijing: China Statistics Press.
NBSC. 2000d. China Price Statistical Yearbook 2000. Beijing: China Statistics Press.
NBSC. 2000e. China Population Statistical Yearbook 2000. Beijing: China Statistics Press.
NBSC. 2001a. China Energy Statistical Yearbook 2001. Beijing: China Statistics Press.
NBSC. 2001b. China Industry Economy Statistical Yearbook 2001. Beijing: China Statistics Press.
NBSC. 2001c. China Statistical Yearbook 2001. Beijing: China Statistics Press.
NBSC. 2001d. China Price Statistical Yearbook 2001. Beijing: China Statistics Press.
NBSC. 2001e. China Population Statistical Yearbook 2001. Beijing: China Statistics Press.
NBSC. 2002a. China Energy Statistical Yearbook 2002. Beijing: China Statistics Press.
NBSC. 2002b. China Industry Economy Statistical Yearbook 2002. Beijing: China Statistics Press.

NBSC. 2002c. China Statistical Yearbook 2002. Beijing: China Statistics Press.

NBSC. 2002d. China Price Statistical Yearbook 2002. Beijing: China Statistics Press.

NBSC. 2002e. China Population Statistical Yearbook 2002. Beijing: China Statistics Press.

NBSC. 2003a. China Energy Statistical Yearbook 2003. Beijing: China Statistics Press.

NBSC. 2003b. China Industry Economy Statistical Yearbook 2003. Beijing: China Statistics Press.

NBSC. 2003c. China Statistical Yearbook 2003. Beijing: China Statistics Press.

NBSC. 2003d. China Price Statistical Yearbook 2003. Beijing: China Statistics Press.

NBSC. 2003e. China Population Statistical Yearbook 2003. Beijing: China Statistics Press.

NBSC. 2004a. China Energy Statistical Yearbook 2004. Beijing: China Statistics Press.

NBSC. 2004b. China Industry Economy Statistical Yearbook 2004. Beijing: China Statistics Press.

Ogunseitan, O.A., J.M. Schoenung, J.D.M. Saphores, et al. 2009. The electronics revolution: From e-wonderland to e-wasteland. *Science* 326: 670–671.

Paris Climate Change Conference. 2015. http://unfccc.int/meetings/paris_2015/meeting/8926.php.

Park, Y., and W.J. Rothwell. 2009. The effects of organizational learning climate, career-enhancing strategy, and work orientation on the protean career. *Human Resource Development International* 12 (4): 387–405.

Patterson, M.G., M.A. West, V.J. Shackleton, et al. 2005. Validating the organizational climate measure: Links to managerial practices, productivity and innovation. *Journal of Organizational Behavior* 26 (4): 379–408.

Podsakoff, P.M., S.B. MacKenzie, J.Y. Lee, et al. 2003. Common method biases in behavioral research: A critical review of the literature and recommended remedies. *Journal of Applied Psychology* 88 (5): 879–903.

Ravi, V., R. Shankar, and M.K. Tiwari. 2005. Analyzing alternatives in reverse logistics for end-of-life computers: ANP and balanced scorecard approach. *Computers & Industrial Engineering* 48: 327–356.

Resende, M. 2002. Relative efficiency measurement and prospects for yardstick competition in Brazilian electricity distribution. *Energy Policy* 30: 637–647.

Ringle C.M., S. Wende, and A. Will. 2005. SmartPLS 2.0 (beta). Hamburg, Germany: University of Hamburg. http://www.smartpls.de.

Rosas, J., C. Sheinbaum, and D. Morillon. 2010. The structure of household energy consumption and related CO_2 emissions by income group in Mexico. *Energy for Sustainable Development* 14 (2): 127–133.

Saunders, H.D. 2000. Does predicted rebound depend on distinguishing between energy and energy service? *Energy Policy* 28: 497–500.

Saunders, H.D. 2008. Fuel conserving (and using) production function. *Energy Economics* 30: 2184–2235.

Saunders, H.D., and J.Y. Tsao. 2012. Rebound effects for lighting. *Energy Policy* 49: 477–478.

Scheel, H. 2001. Undesirable outputs in efficiency valuations. *European Journal of Operational Research* 132: 400–410.

Scherbaum, C.A., P.M. Popovich, and S. Finlinson. 2008. Exploring individual-level factors related to employee energy-conservation behaviors at work. *Journal of Applied Social Psychology* 38 (3): 818–835.

Schmidt J., and N. Helme. 2005. Operational issues for a sector-based approach: Questions and answers. Working Paper prepared for CCAP's Dialogue on Future International Actions to Address Global Climate Change.

Schwartz S.H. 1977. Normative influences on altruism. In *Advances in experimental social psychology*, ed. by L. Berkowitz, 221–279. New York: Academic Press.

Schwartz S.H., and J.A. Howard. 1981. A normative decision-making model of altruism. In *Altruism and helping behavior*, ed. by J.P. Rushton and R.M. Sorrentino, 89–211. Erlbaum: Hillsdale.

Siero, F.W., A.B. Bakker, G.B. Dekker, et al. 1996. Changing organizational energy consumption behaviour through comparative feedback. *Journal of Environmental Psychology* 16 (3): 235–246.

Su, Q. 2012. A quantile regression analysis of the rebound effect: Evidence from the 2009 national household transportation survey in the United States. *Energy Policy* 45: 368–377.

Sun, C., and X. Ouyang. 2016. Price and expenditure elasticities of residential energy demand during urbanization: An empirical analysis based on the household-level survey data in China. *Energy Policy* 88 (1): 56–63.

Sun, C. 2015. An empirical case study about the reform of tiered pricing for household electricity in China. *Applied Energy* 160: 383–389.

Szendro, G., and A. Török. 2014. Theoretical investigation of environmental development pathways in the road transport sector in the European region. *Transport* 29 (1): 12–17.

Tenenhaus, M., V.E. Vinzi, Y.M. Chatelin, et al. 2005. PLS path modeling. *Computational Statistics & Data Analysis* 48 (1): 159–205.

Thakur, T., S.G. Deshmukh, and S.C. Kaushik. 2006. Efficiency evaluation of the state owned electric utilities in India. *Energy Policy* 34: 2788–2804.

Thomas, D., T. Gerald, J. Gilligan, et al. 2009. Household actions can provide a behavioral wedge to rapidly reduce US carbon emissions. *Proceedings of the National Academy of Sciences of the United States of America (PNAS)* 106 (44): 18452–18456.

Tone, K. 2001. A slacks-based measure of efficiency in data envelopment analysis. *European Journal of Operational Research* 130: 498–509.

van Beukering, P.J.H., C.J.M. Jeroen, and J. van den Bergh. 2006. Modelling and analysis of international recycling between developed and developing countries. *Resources Conservation and Recycling* 46: 1–26.

Wang Z.H., and C. Feng. 2014. The impact and economic cost of environmental regulation on energy utilization in China. 46 (27): 3362–3376.

Wang, Y., and Q. Wang. 2013. Factors affecting Beijing residents' buying behavior of new energy vehicle: An integration of technology acceptance model and theory of planned behavior. *Chinese Journal of Management Science* 21: 691–698.

Wang, H., P. Zhou, and D.Q. Zhou. 2013. Scenario-based energy efficiency and productivity in China: A non-radial directional distance function analysis. *Energy Economics* 40: 795–803.

Wei, C., L. Andreas, and B. Liu. 2013. An empirical analysis of the CO_2 shadow price in Chinese thermal power enterprises. *Energy Economics* 40: 22–31.

Wijaya, M.E., and T. Tezuka. 2013. Measures for improving the adoption of higher efficiency appliances in Indonesian households: An analysis of lifetime use and decision-making in the purchase of electrical appliances. *Applied Energy* 112: 981–987.

Wood, R., and M. Lenzen. 2006. Zero-value problems of the logarithmic mean divisia index decomposition method. *Energy Policy* 34 (12): 1326–1331.

Wu, F., L.W. Fan, P. Zhou, et al. 2012. Industrial energy efficiency with CO_2 emissions in China: A nonparametric analysis. *Energy Policy* 49: 164–172.

Wu, C., X. Zhou, and M. Song. 2016. Sustainable consumer behavior in China: An empirical analysis from the Midwest regions. *Journal of Cleaner Production* 134: 147–165.

Wu, K., W. Wang, H. Zhang, et al. 2013. Indirect carbon emissions of Shanghai's residents' consumption and its influence factors. *East China Economic Management* 27 (1): 1–7. (in Chinese).

Wu, M. 2010. *Structural equation model-use and application of AMOS*. Chongqing: Chongqing University Press.

Xu, Z., and S. Ma. 2004. How to fullfill the value of reverse supply chain tactics. *Logistics Technology* 45–47.

Yadav, R., and G.S. Pathak. 2016. Young consumers' intention towards buying green products in a developing nation: Extending the theory of planned behavior. *Journal of Cleaner Production* 135: 732–739.

Yang, H., and M. Pollitt. 2009. Incorporating both undesirable outputs and uncontrollable variables into DEA: The performance of Chinese coal-fired power plants. *Europeans Journal of Operational Research* 197: 1095–1105.

Young, D. 2008. When do energy-efficient appliances generate energy savings? Some evidence from Canada. *Energy Policy* 36 (1): 34–46.

Zaim, O. 2004. Measuring environmental performance of state manufacturing through changes in pollution intensities: A DEA framework. *Ecological Economics* 48: 37–47.

Zhang, N., and Y. Choi. 2013. Total-factor carbon emission performance of fossil fuel power plants in China: A meta-frontier non-radial Malmquist index analysis. *Energy Economics* 40: 549–559.

Zhang, N., P. Zhou, and C.C. Kung. 2015. Total-factor carbon emission performance of the Chinese transportation industry: A bootstrapped non-radial Malmquist index analysis. *Renewable and Sustainable Energy Reviews* 41: 584–593.

Zhao, M., W. Zhang, and L. Yu. 2009. Carbon emission from energy consumption in Shanghai. *Research of Environmental Sciences* 22 (8): 984–989.

Zheng, X., C. Wei, P. Qin, et al. 2014. Characteristics of residential energy consumption in China: Findings from a household survey. *Energy Policy* 75: 126–135.

Zhou, L., and Z. Xu. 2012. Response to waste electrical and electronic equipments in China: Legislation, recycling system, and advanced integrated process. *Environmental Science & Technology* 46: 4713–4724.

Zhou, P., B.W. Ang, and K.L. Poh. 2006. Slacks-based efficiency measures for modelling environmental performance. *Ecological Economics* 60 (1): 111–118.

Zhu Q., and T.Y. Wei. 2013. Impacts of urbanization on carbon emissions from perspective of residential consumption. *China Population, Resources and Environment* 23 (11): 21–29 (in Chinese).

The manufacturer's authorised representative in the EU is Springer Nature Customer Service Centre GmbH, Europaplatz 3, 69115 Heidelberg, Germany. If you have any concerns regarding our products, please contact ProductSafety@springernature.com

Printed and bound by CPI Group (UK) Ltd, Croydon, CR0 4YY
30/09/2026
02225557-0007